' Manufacturing
' 2

...ok covers the standard unit in Manufacturing Technology at Level 2 of the Technician Certificate in Mechanical and Production Engineering. The learning objective structure has been followed, and the content is also suitable for a wide range of TEC programmes having units in workshop technology. The book is the second of a series in Manufacturing Technology, and will immediately be followed by Book 3, which will complete the objectives in Manufacturing Technology up to Certificate and Diploma level.

C. R. Shotbolt is Principal Lecturer in Engineering Technology at Luton College of Higher Education. He is an established author and has written Technician Workshop Processes and Materials I, Workshop Technology for Mechanical Engineering Technicians 1 and 2, and is co-author of Metrology for Engineers. Mr Shotbolt is Chairman of the A5 Programme Committee set up by Bedford College of Higher Education, Dunstable College, and Luton College of Higher Education to coordinate TEC programmes in Mechanical and Production Engineering.

Technician Manufacturing Technology 2

C. R. SHOTBOLT
C.Eng., FIProdE, MIMechE, MIQA, MASQC
Principal Lecturer in Engineering Technology
Luton College of Higher Education

CASSELL · LONDON

CASSELL LTD.
35 Red Lion Square, London WC1R 4SG
and at Sydney, Auckland, Toronto, Johannesburg,
an affiliate of Macmillan Publishing Co. Inc.,
New York

First published 1978

I.S.B.N. 0 304 30005 5

Printed and bound in Great Britain at
The Camelot Press Ltd, Southampton

Preface

This book is written specifically to cover the objectives set in unit U76/056, Manufacturing Technology 2, which is an essential part of most of the programmes produced by the Technician Education Council (TEC) under the heading Mechanical and Production Engineeering (A5). As these programmes are phased in they will replace the National Certificate and City & Guilds of London Institute courses for Technicians that have been so well known for so long throughout the English speaking world.

A major difference between the TEC courses and those they replace is that the new courses are of the modular or unit based type and a student can advance in, or be given credit for, single units or subjects. A particular case is the Level 1 unit, *Workshop Processes and Materials* for which a student can be given credit if he is engaged upon an approved course of first year training as part of his apprenticeship. Thus a student can enter a course of study for the level 2 unit with no educational background and very little experience. For this reason the author has included in this work those chapters dealing with machine tools from the previous Level 1 book where there is a requirement for the work in this unit. At the same time he feels it wrong to embark on a study of the capstan and turret lathe without some background knowledge of the centre lathe and therefore chapter 7 is in two parts, part 1 dealing with the centre lathe and part 2 with capstan and turret lathes.

To students having completed Level 1 and bought the previous book the author apologises for this repetition but trusts that they will find the work with which they are not familiar of value.

Within this book there is some repetition, work from chapters 7, 8, 9 and 10 being used in chapter 6. To avoid tedious references back and forth the author has resorted to repetition which he hopes the reader will find more convenient.

As in other works in this series, this book follows the educational objectives laid down by TEC. They are stated where appropriate in the text and at the end of each chapter there are questions which will help the student to establish whether or not those objectives have been attained. Some require short answers and some longer and more detailed answers. Both types are used in TEC unit assessments but to successfully complete an assessment is not to be proficient in practical engineering. This requires practice and experience which can be gained only in a workshop using machines and equipment. Some work requires only an appreciation of a process and this can be gained in a classroom and from books. The author hopes that this work will help with the theoretical aspects of the former and cover adequately the latter types of work.

Again as in previous works, safety has been emphasised, either in the form of safety notes in the text or as separate sections where that seemed more appropriate. Where space precluded a detailed treatment of safety, for instance in the chapter on presswork, a reference has been given for further and more detailed study.

The author would like to acknowledge the help given by firms and organisations who have provided information and illustrations so generously. These are acknowledged in the text. Thanks are also due to colleagues at Luton College of Higher Education. It is impossible to individually acknowledge all the help given but, collectively, the staff of that establishment have been most helpful and encouraging. Some of the questions used were taken from past examination papers with the kind permission of the City and Guilds of London Institute. The author acknowledges the help of that organisation but points out that most of them were taken from papers written in Imperial units. The author converted these questions to metric units and the responsibility for any errors arising from such metrication is the author's alone.

Finally the author would like to pay tribute to the ladies who, with great difficulty, deciphered the authors handwritten script and typed it in a form which could be interpreted by the publishers. My particular thanks go to Mrs E. Duffield and Mrs F. Swain for this work. Finally, of course, thanks go to my wife, Mollie, without whose encouragement and help none of these words would have got to the type-writer nor any of the diagrams off the drawing board.

Contents

measurement of straightness. Standards of squareness: testing right angles; the sine bar. Safety note. Practical measurements: measurement of plain gap gauges, measurement of plain ring gauges, measurement of taper plug gauges, measurement of taper ring gauges, measurement of taper bores. Comparative measurement: the micrometer as a comparator, the length-bar comparator, bench-type mechanical comparators. Comparative methods of checking diameter, position, parallelism and alignment. Questions.

1 Heat treatment of steel

General Objective: *The student should understand the basic principles of heat treatment as applied to steels.*

THE HEAT TREATMENT OF PLAIN CARBON STEELS

COOLING CURVES

In order to understand the heat treatment of steel it is necessary first to understand the phenomena of cooling and the changes that takes place in a material when it is cooled. If water is placed in the freezing compartment of a refrigerator it becomes solid and is then known as ice; a physical change has taken place. If a thermometer is placed in the water as it cools and the temperature is noted at regular intervals of time, a graph can be plotted of temperature against time. This graph might be expected to be a smooth curve but, in fact, is not. A pause occurs during the cooling, as shown in fig. 1.1. The drop in temperature is arrested for a short period and the pause is known as a *thermal arrest or arrest point.* It is at this temperature that the water changes from a liquid to a solid, i.e. a *physical change* takes place.

A physical change during cooling (or heating) is indicated by an arrest in the cooling (or heating) curve

THE COOLING OF LIQUID SOLUTIONS

It is well known that salt 'melts' ice. What actually happens is that the salt and water form a solution which has a lower melting point than that of water. If a cooling curve is plotted for a salt/water solution, as shown in fig. 1.2, two differences are apparent from the cooling curve for pure water:

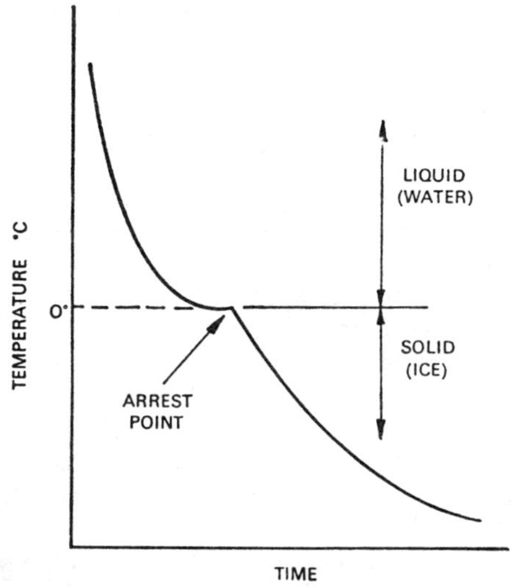

Fig. 1.1. Cooling curve for pure water

1. there are two arrest points;
2. the temperature of the upper arrest point is lower than that of the single arrest point for pure water.

Observation during cooling shows that, at the upper arrest point, solid particles start to appear in the liquid solution and that these particles are solid water, or ice. As the temperature continues to drop, more and more pure water solidifies out until the lower arrest point is reached, when the remaining solution of salt and water freezes. The final solid therefore consists of:

1. crystals of pure solid water (ice) plus
2. crystals of a *solid solution*[1] of salt and water.

[1] It is important to note that many materials can dissolve or remain dissolved in each other in the solid state to form what is known as a *solid solution,* in contrast to the liquid solutions which are more commonly encountered in everyday life.

Fig. 1.2. Cooling curves for pure water and a weak salt/water solution

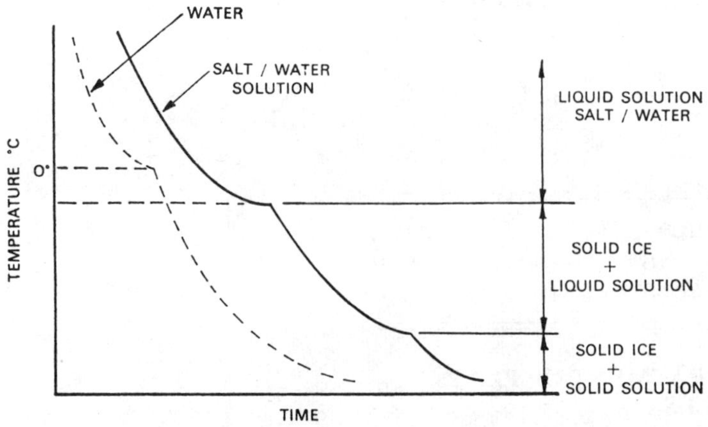

If a second salt/water solution, containing a higher percentage of salt, is made up and the cooling curve is plotted, the curve will again be found to have two arrest points, but the temperature of the upper arrest will have dropped while that of the lower arrest will be the same as before. As the arrests are closer together, there is less time for the pure water to freeze out between them and although the final solid will again consist of a mixture of crystals of pure water and crystals of a solid solution of salt and water, there will be fewer crystals of water and more of the solid solution.

As more salt is added, the two arrests draw closer together, with less pure water in the final structure, until eventually the solution has only one arrest point. This particular solution is known as the *eutectic*, and has the following characteristics:

1. it changes directly from a liquid to a solid, with no partially solid stage;
2. the solid structure consists wholly of crystals of the solid solution;
3. the solution has the lowest freezing point of all solutions of these two materials.

If the percentage of salt is increased beyond the eutectic composition, the upper arrest point reappears. In this case it is solid salt which freezes out between the arrests, and the final structure consists of grains of pure salt and crystals of the solid solution of salt and water.

During the cooling of a solution of *less* than eutectic composition (*hypo-eutectic*), the effect of freezing out pure water is to *increase* the percentage of salt in the remaining solution. This is readily apparent if one considers a bag containing 50 red balls and 50 green balls, i.e. 50% of each. If 25 red balls are removed, the percentage of green balls left in the bag will have increased to 67%.

Similarly, as a solution of *greater* than eutectic composition (*hyper-eutectic*) cools and the salt freezes out, the percentage of salt remaining in the solution decreases. In both cases, the change in the composition of the solution is such that, when the lower arrest point is reached, the remaining solution is of eutectic composition. Thus the solid materials produced by freezing any solution of salt and water will consist of *grains of solid eutectic plus crystals of one of the pure constituents*, depending on whether the original solution was of hypo-eutectic or hyper-eutectic composition.

A family of cooling curves could therefore be produced, as in fig. 1.3, and the condition of the solution during each phase of cooling would be as shown. However, life is too short to draw a cooling curve for every possible solution of two materials, and by producing a few curves it is possible to draw a graph of arrest temperature against composition. Such a graph is called a *thermal equilibrium diagram*, that for salt and

Fig. 1.3. 'Family' of salt/water solution cooling curves

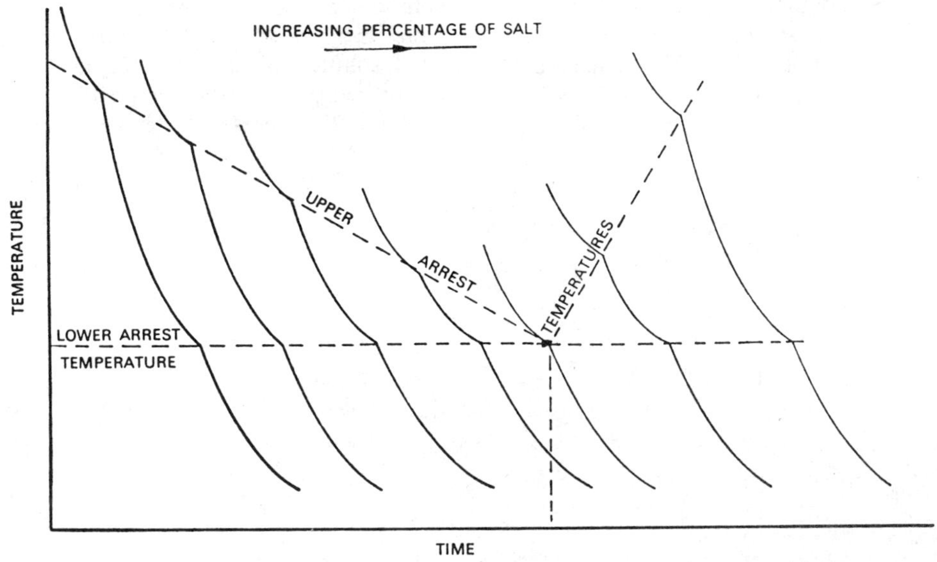

water being shown in fig. 1.4, with the condition of the solution in the various phases of the diagram also indicated.

THERMAL EQUILIBRIUM DIAGRAMS

Bearing in mind all that has been said about the cooling of salt and water solutions, consider now the cooling of an alloy more commonly encountered in the workshop, that of *lead* (Pb)

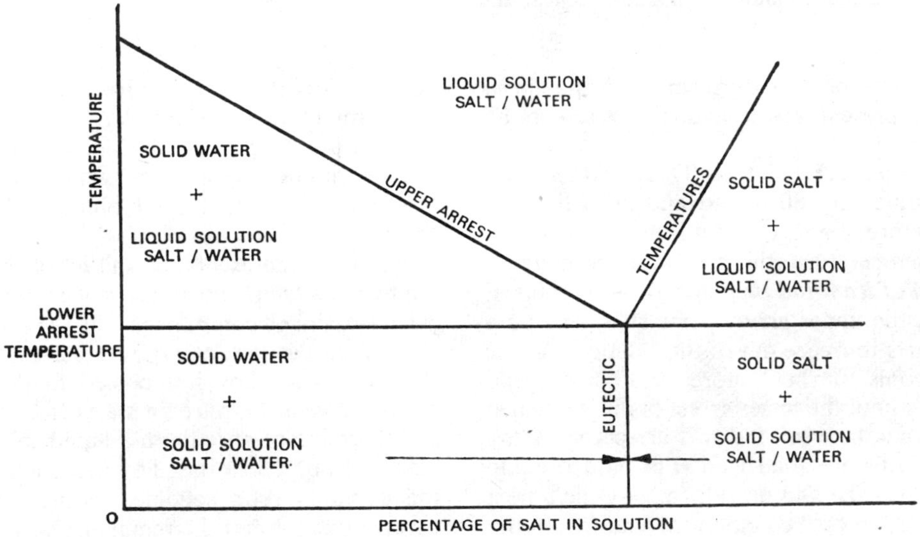

Fig. 1.4. Thermal equilibrium diagram for salt/water solutions

and *tin* (Sn) an alloy more commonly known as *solder*.

The thermal equilibrium diagram for solder is shown in fig. 1.5. The melting point of pure lead is 327°C and that of tin is 232°C. The eutectic alloy, which contains 67% tin and 33% lead, melts at 183°C. This information provides the main points on the diagram, and the lines, which represent arrest temperatures, can be drawn in.

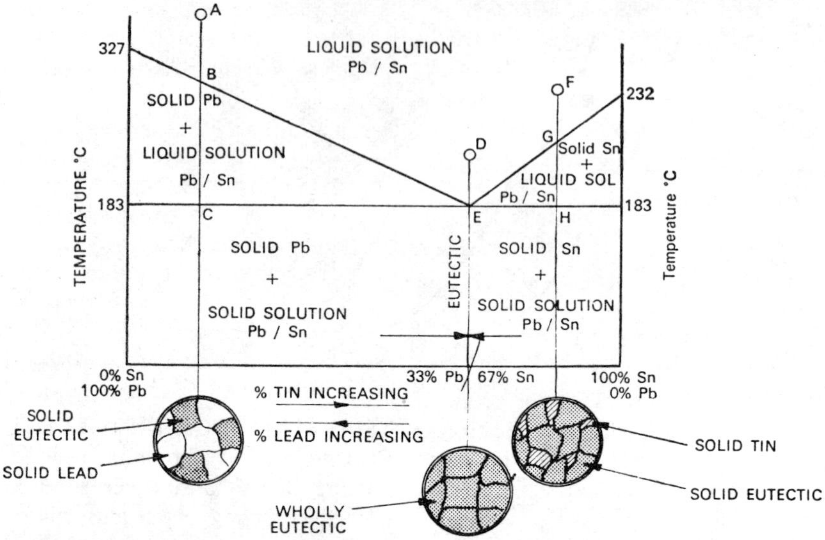

Fig. 1.5. Thermal equilibrium diagram for lead-tin alloys (solders)

Now consider point A. This represents an alloy containing 80% lead and 20% tin at a temperature of 350°C. As it is above the upper arrest temperature, the alloy will be a *liquid solution of lead and tin* (Pb/Sn). As it cools, it reaches the upper arrest point at B, and pure lead starts to freeze out of the solution. As the alloy cools further, more and more lead solidifies until the lower arrest point is reached at C. Thus between the two arrests the alloy consists of pure solid lead (Pb), and a liquid solution of lead and tin (Pb/Sn). By the time it has cooled to C, the amount of lead frozen out is such that the percentage of tin in the remaining solution has increased to the eutectic composition of 67%. Further cooling below this temperature causes the remaining liquid solution to freeze, so that the final solid alloy consists of grains of pure lead and grains of the solid solution of lead and tin, all these latter grains being of eutectic composition.

Point D represents an alloy of eutectic composition at a temperature of 200°C, i.e. in the form of a liquid solution. As it cools it reaches point E, where the single arrest point occurs, immediately after which the alloy freezes and becomes a solid solution of lead and tin, of eutectic composition.

Point F represents an alloy of high tin content at a temperature above the upper arrest point. Again, by comparison with the salt/water cooling curves described previously, it is clear that after the alloy has cooled to the arrest temperature at G, pure tin starts to freeze out, giving grains of pure tin in a liquid solution of lead and tin. When the alloy has cooled to H, the amount of tin solidified is such that the percentage of tin in the remaining solution has dropped to the eutectic composition. Thus, when this alloy cools to below 183°C and becomes solid, it consists of grains of pure tin and grains of the solid solution, again of eutectic composition.

Diagrams of the three structures are shown in the appropriate positions under the equilibrium diagram in fig. 1.5.

As stated earlier, an alloy of lead and tin is a solder, and it is of interest that there are two basic types of solder:

1. Tinman's solder, used for soldered assembly joints;
2. Plumber's solder, used for wiped joints in pipes and cables, and 'puddling', or covering and smoothing over welds in motor car bodies, etc.

A solder used for assembly work, soldered connections in electronic equipment and the like must be one that sets, or solidifies, rapidly. For this reason, tinman's solder is of approximately eutectic composition. Above 183°C it is liquid, below this it is solid, and the joint sets almost immediately it is made, if not overheated.

On the other hand, a solder for smooth, wiped pipejoints must be capable of being manipulated as it sets and must go through a stage when it is neither liquid nor solid, but pasty. A solder having a high lead content goes through just such a pasty stage as it cools between the upper and lower arrest temperatures, and plumber's solder is therefore of approximately 70% lead–30% tin composition.

THE IRON/IRON-CARBIDE EQUILIBRIUM DIAGRAM

Specific Objective: *The students should be able to* (i) *Construct a simplified equilibrium diagram of the solid steel portion of the iron-carbon diagram, and* (ii) *define the terms ferrite, austenite, cementite, martensite and critical cooling rate.*

By now the reader will probably be wondering what salt water and solder have to do with the heat treatment of steel. There is a similar diagram for alloys of iron and carbon, called the *iron/carbon equilibrium diagram*, dealing with these alloys in the molten state at temperatures above 1100°C and with carbon contents up to 5% or more. We are, however, concerned only with the 'steel' part of this full diagram that is, with alloys which have carbon contents up to 1.8% and *have already changed from the liquid to the solid state.* For this we use a second equilibrium diagram, called *the iron/iron-carbide diagram*, which shows what happens as plain carbon steels, in their solid state, cool from about 900°C. The diagram is very similar to that for lead/tin alloys and for convenience the two are shown side by side in fig. 1.6. There is one important difference, however, for the arrest points of the lead/tin diagram denote *changes from liquid to solid* whereas the arrest points of the iron/iron-carbide diagram denote *changes in the internal structure of the solid steel* as it cools.

The *crystals* or *crystal grains* of which iron and steel are composed are built up of exceedingly small cubes and the arrangement or spacing of the atoms of carbon and iron in these cubes, called the *space-lattice*, changes as the steel cools. Above the upper arrest point, the steel is a solid solution of iron and iron carbide, called *austenite*, and it is the relative proportions of iron and iron carbide which determine the temperatures of the upper arrest points, just as the relative proportions of lead and tin determine the temperatures of the upper arrest points in solders. The arrest points are sometimes called *critical points* or *change points.*

With the lead/tin alloys, there was one alloy, the eutectic, which solidified completely without the formation of separate crystals of pure lead or pure tin. In the same way, there is one solid solution of iron and iron carbide which does not produce separate crystals of iron or iron carbide as it cools. This, since we are dealing with a solid solution, is called the *eutectoid*, where the percentage of carbon in the steel is 0.83%. Note that although the carbon in steel is always in the form of iron carbide, we classify the steel by the percentage of *carbon*, not iron carbide, which it contains.

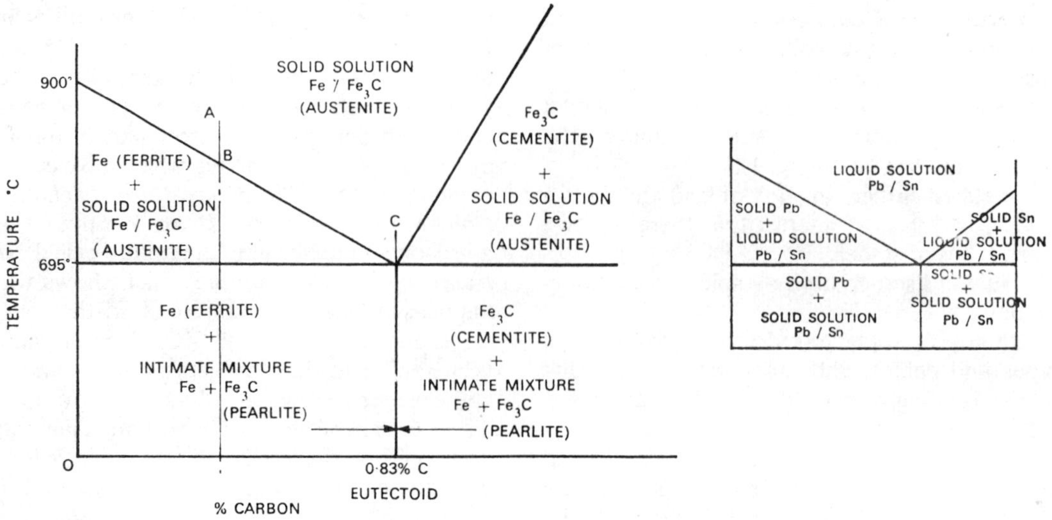

Fig. 1.6. Iron/iron-carbide thermal equilibrium diagram (with lead-tin diagram for comparison)

In steels with a lower percentage of carbon than 0.83%, there is an 'excess' of iron and at the upper critical point crystals of pure iron, called *ferrite*, separate out within the austenite. In steels with a higher percentage of carbon than 0.83% there is an 'excess' of iron carbide and at the upper critical point crystals of iron carbide, called *cementite*, separate out.

The lower critical point, 695°C, is common for all steels and at this point the solid solution of austenite changes into an intimate mixture of iron and iron carbide called *pearlite*. The grains of pearlite consist of alternate and exceedingly thin layers, or laminates, of iron and iron carbide, and this laminated structure has a 'mother-of-pearl' or 'pearly' appearance under the microscope.

The eutectoid, 0.83% carbon, having no excess of ferrite or cementite, is pure pearlite. Steels with less than 0.83% carbon are a mixture of pearlite and ferrite; those with more than 0.83% carbon are a mixture of pearlite and cementite.

For reference purposes:

Austenite is the solid solution of iron and iron carbide, Fe/Fe_3C, above the critical range;

Ferrite is almost pure iron, Fe, and is very soft;

Cementite is iron carbide, Fe_3C, and is very hard;

Pearlite is an intimate mixture of fine alternate layers of iron and iron carbide, $Fe + Fe_3C$, and is the eutectoid of carbon steel, containing 0.83% carbon.

Bearing in mind the previous analysis of the cooling of salt/water solutions, and particularly of lead/tin alloys, consider point A on the iron/iron-carbide diagram (fig. 1.6). This represents a steel containing about 0.4% carbon and at a temperature of about 850°C. It is well above the upper critical temperature and will be in the form of a solid solution of iron and iron carbide (Fe/Fe_3C, austenite). As it cools, no changes occur until the temperature drops to the upper arrest point at B. Further cooling causes iron (Fe, ferrite) to come out of solution and the percentage of carbon left in the remaining solution is thus increased. By the time the temperature has dropped to 695°C, the lower arrest temperature, the amount of ferrite precipitated is such that the carbon content of the *remaining solution* has risen to eutectoid composition of 0.83%C.

Further cooling causes this remaining solution to break down into the finely-divided layers of iron and iron carbide ($Fe + Fe_3C$) known as pearlite.

Thus, for a steel containing less than 0.83% carbon, the structures at the various phases of the diagram are:

1. *Above the upper critical point:* solid solution of iron and iron carbide (austenite);
2. *Between the arrest points:* mixture of iron and the solid solution (ferrite plus austenite);
3. *Below the lower arrest point:* mixture of iron and the finely-divided layers of iron and iron carbide (ferrite plus pearlite).

This final structure is shown, as it would appear under a microscope, in fig. 1.7(a).

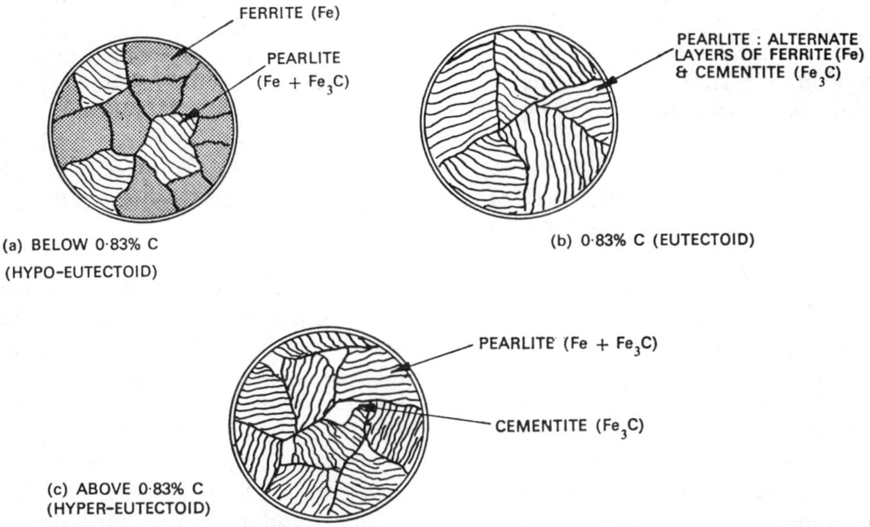

Fig. 1.7. Structures of slowly cooled steels of different carbon contents

Consider now point C on the iron/iron-carbide diagram. This represents a steel containing 0.83% carbon at a temperature of 750°C. It is at a higher temperature than the single arrest point which occurs at this eutectoid composition and its structure will be that of a solid solution of iron and iron carbide (austenite). As it cools, no change occurs until it reaches the arrest temperature, at which point the solid solution breaks down into the mixture of the two constituents in finely-divided layers, which we know as pearlite. In this case there is no precipitation of either iron or iron carbide, and the final structure is 100% pearlite, as shown in fig. 1.7(b).

When the carbon content is greater than 0.83%, iron carbide (Fe_3C, cementite) comes out of solution as the steel cools between the arrest points. As all the carbon in the steel is contained in the form of this constituent it follows that the percentage of carbon in the remaining solution is reduced until, by the time the temperature has fallen to the lower arrest point, the remaining solid solution is of eutectoid composition. Further cooling causes this solution to break down to finely-divided pearlite. Thus the structures occurring at the different tenperature zones will be:

1. *Above the upper arrest point:* solid solution of iron carbide (austenite);
2. *Between the arrest points:* mixture of iron carbide and the solid solution (cementite plus austenite);
3. *Below the lower critical point:* mixture of iron carbide and the finely-divided mixture of iron and iron carbide (cementite plus pearlite).

The final micro-structure is shown in fig. 1.7(c).

It is important that this analysis of the iron/iron-carbide equilibrium diagram should be clearly understood if the heat treatment of steel is to be appreciated, and for this reason the main points of the foregoing explanation are again emphasised as follows:

1. All structures of steel are brought about by a breakdown of the solid solution of iron and iron carbide.
2. All slowly cooled structures of steel contain pearlite.
3. Pearlite is always of eutectoid composition, i.e. it always contains 0.83% carbon.
4. It follows that as the carbon content of the steel approaches 0.83% from either above or below this figure, the amount of pearlite contained in the steel increases until, at this carbon content, the steel is all pearlite. This state of affairs is shown graphically in fig. 1.8.

APPLICATION OF THE IRON/IRON-CARBIDE EQUILIBRIUM DIAGRAM TO THE HEAT TREATMENT OF STEEL

Specific Objective: *The student should be able to state the procedures for hardening, annealing, normalising and tempering plain carbon steels and relate these procedures to the iron/iron-carbide diagram.*

The 'steel' portion of the iron/carbon diagram, or the iron/iron-carbide diagram as we have called it, is the key to the heat treatment of plain carbon steels. Before it can be used, however, we must delve a little deeper into its implications and into the effects of the various combinations of the constituents involved. In order to understand these effects we must clearly appreciate the terms *mixture, solution* and *chemical compound.*

(a) *Mixture*

To reduce the risk of accidents, icy roads are spread with a *mixture* of salt and sand. This is a typical mixture and has the characteristics of all mixtures. If a quantity of the two materials is shaken up in a test tube and examined, however intimately they are mixed it is found that:

(i) the separate grains of each can readily be identified;
(ii) with a pair of tweezers and a magnifying glass we can separate the two by picking out the separate grains;
(iii) when separated completely, the salt and sand are completely unaffected by having been mixed together.

(b) *Solution*

A simpler way to separate the mixed salt and sand would be to add water. The salt will dissolve in the water which can then be poured off and filtered, leaving the sand. The water, however, will have become salty and, of course, the salt solution will have taken the characteristics of a liquid. Furthermore, it takes very little salt to make water salty to the taste.

In a solution each material affects the other

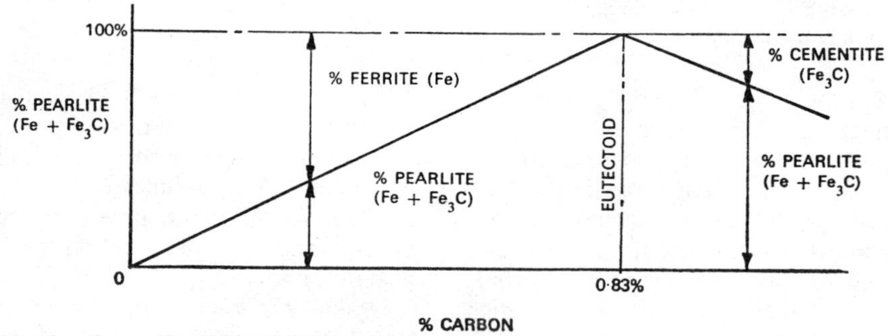

Fig. 1.8. Graph of pearlite (Fe + Fe_3C) content against carbon

and the solution has some of the characteristics of both.

(c) *Chemical compound*
Salt is a typical chemical compound. It consists of the elements sodium (Na) and chlorine (Cl) combined chemically to produce sodium chloride (NaCl), common salt. Salt is an interesting material; we flavour our food with it, our health would suffer if we lacked it and, in submarines, factories and hot, humid climates, salt tablets are issued to combat heat fatigue. In short, we could find it hard to do without. However, if we took the sodium and the chlorine separately, the results would be very different.

If the metal sodium is dropped in water it fizzes about and gives off hydrogen. If swallowed its effect on the gastric juices could only be imagined! Chlorine is a dirty greenish brown gas which is extremely poisonous and was, in fact, used as a poison gas in the 1914–18 war. Thus, here we have two materials which if taken internally, separately, would undoubtedly be fatal but, if taken in the correct chemically-combined form, the resulting compound, salt, is pleasant and beneficial.

Two materials combined chemically produce a 'new' material whose characteristics are different from those of its constituents.

In the iron/iron-carbide diagram all three of these combinations are encountered.

Iron carbide is a chemical compound of iron, which is soft, grey, ductile and malleable, and carbon, which is soft, black, and brittle (in its graphitic form, carbon is a very good lubricant). Iron carbide, the compound, is white, hard and brittle.

Steel is, of course, predominantly iron and contains a maximum of 1.8% carbon, so that even a high-carbon steel cannot contain much iron carbide. Thus, if the iron and iron carbide are in the form of a simple mixture, the physical properties of the material will be similar to those of iron, i.e. soft and malleable. As the amount of carbon increases, and hence the iron-carbide content increases, there is a change in hardness, strength and ductility as shown in fig. 1.9, but

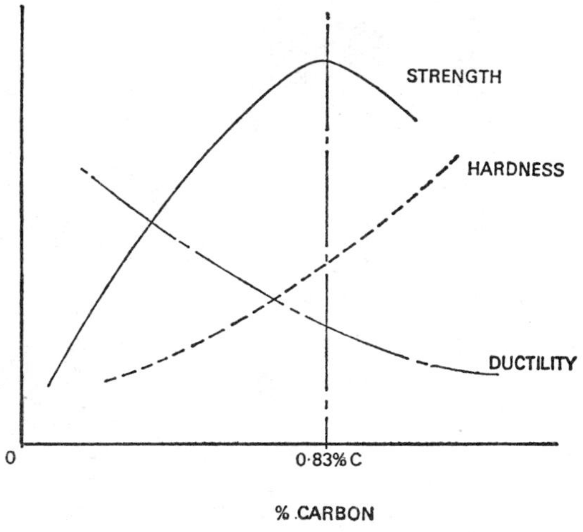

Fig. 1.9. Effect of carbon content on strength, hardness and ductility of normalised plain carbon steel (compare with fig. 1.8.)

the change in hardness is small compared with that obtainable by heat treatment. Note that strength falls off when the carbon content exceeds 0.83%. The strength depends on the amount of pearlite in the steel, and the strongest plain carbon steel is wholly pearlitic with 0.83% carbon content.

In order to get the full effect of the hardness of the iron carbide it is necessary to ensure that all the iron is in solution with the hard constituent. This can be achieved by heating the steel to temperatures above those of the shaded zones of the diagram shown in fig. 1.10(a). The steel must now be cooled so rapidly that the solution does not have time to break down into its former constituents. This, again fortunately, is easily done by a simple quench in water. We can draw a simplified diagram, fig. 1.10(b), showing the temperature zones to which the steel must be heated for the desired effects to take place. The main effects we shall consider are:

1. HARDENING
The steel is heated to the temperature zone shown in fig. 1.10(b) and quenched in water. This does not allow the solid solution, austenite,

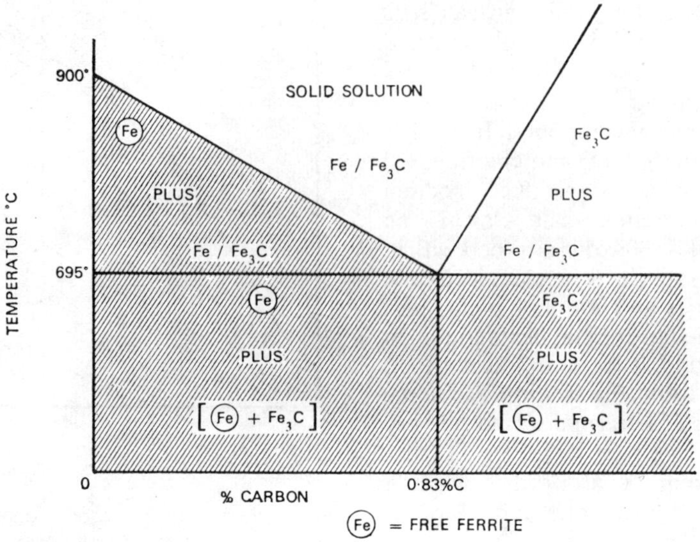

Fig. 1.10(a). Phases of iron/iron-carbide diagram which must be cleared to ensure that iron (Fe) is in solution with iron carbide

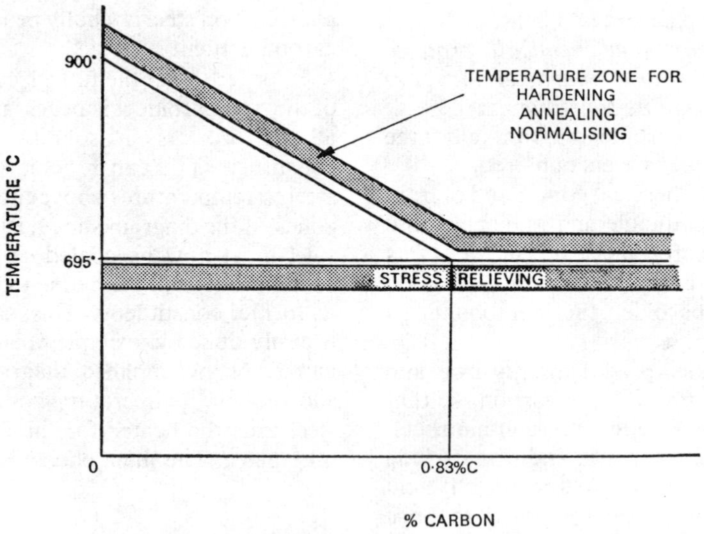

Fig. 1.10(b) Simplified iron/iron-carbide diagram showing temperatures to which plain carbon steel must be heated for various heat treatments

to break down into pearlite and a new, hard structure called *martensite* is formed.

2. TOUGHENING

If steel is quenched rapidly it becomes hard. If it is cooled slowly it stays soft. It follows that there is a speed of cooling above which hardening occurs and below which it does not. This is called the *critical cooling speed.* If the steel is cooled just below the critical cooling speed, more slowly than in water, a partial but incomplete breakdown of the austenite takes place, giving a very fine-grained structure, much softer than martensite but very much tougher. This slightly slower cooling can be brought about by using an oil quench.

To toughen, heat the steel to the temperature zone shown in fig. 1.10(b) and quench it in oil.

3. ANNEALING

This process is used to make a hardened piece of steel as soft as possible. If the steel is heated to the hardening temperature the martensite changes back to austenite and, if the steel is then cooled very slowly, the austenite breaks down completely to a soft structure containing pearlite. The prolonged heating and slow cooling cause the individual grains to grow into each other, giving a large-grained structure which, apart from being soft, is very weak.

To anneal, heat the steel to the temperature zone shown in fig. 1.10(b), 'soak' it at that temperature and cool it very slowly. This may be done by cooling at a controlled rate in the furnace, by switching off the furnace and allowing the furnace insulation to slow down the cooling, or by plunging the steel into dry sand or ashes.

4. NORMALISING

This process is used to correct abnormal grain structures in steel. Such structures may be the enlarged grains of a fully annealed steel, which is too soft to machine well, or the distorted grain structure of a material which has been coldworked, e.g. cold-rolled or pressed, and which might split it further cold work was carried out. To correct these unsatisfactory structures, the distorted grains must be allowed to re-form, which can be achieved by causing a solid solution to form through heating the steel to the temperature zone shown in fig. 1.10(b) and allowing it to cool naturally in air. This allows the austenite to break down into a normal pearlitic structure of the correct grain size, which is ideal for machining or will allow further cold work to be carried out.

To normalise, heat the steel to the temperature zone shown in fig. 1.10(b) and cool it in still air.

5. STRESS-RELIEVING, SUB-CRITICAL ANNEALING OR PROCESS ANNEALING

These are all names given to a process which has certain advantages over normalising for low-carbon steels. The normalising process calls for cooling in air and, at the temperatures used, the oxygen in the atmosphere attacks the surface of the material, causing scaling. However, if the steel is heated to just below the *lower* critical point and cooled slowly, the grains of ferrite re-form, and since a low-carbon steel contains little other than ferrite, this process can be regarded as a normalising type of process for such a steel. Apart from its lower temperature, the process calls for slow cooling, which can be carried out in the furnace. If the furnace atmosphere is controlled and air is excluded, scaling will not occur and the parts will come out with nothing more than a slight tarnish. Subsequent operations can then be performed without the trouble of preliminary descaling or cleaning, which is particularly valuable where stress-relieving is carried out between operations, e.g. in a press shop on large-scale production.

To stress relieve, heat the steel to just below the lower critical temperature and cool it slowly, as indicated in fig. 1.10(b).

6. TEMPERING

If *hardened* plain carbon steel is reheated after hardening, the hard martensite starts to break down and the steel loses some of its hardness and gains in toughness. It is important to note that the loss of hardness and gain in toughness depend upon the temperature to which the steel is reheated. If a graph of hardness against tempering temperature is plotted it will be of the type shown in fig. 1.11 and will also show the two temperature zones used for different types of work:

(a) low-temperature tempering
(b) high-temperature tempering

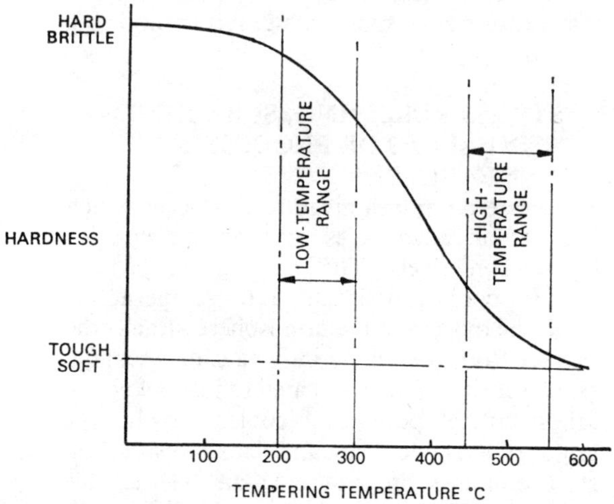

Fig. 1.11. Effect of tempering temperature on hardness

(a) *Low-temperature tempering*
This is normally used for the range of work where the steel must be hard but must have enough brittleness removed to enable the tool to stand up to its task. Thus the temperature chosen must be related to the type of work the tool is required to do.

A scriber must retain its point without dulling for long periods but suffers little in the way of shock loads which might break it. It is therefore tempered at a low temperature of 200°C which, as the diagram shows, barely affects its hardness but just takes the edge off its brittleness.

A centrepunch has a slightly more arduous function, since it is struck lightly with a hammer and the point must not chip. It is thus tempered at a higher temperature of 260°C. Similarly, a cold chisel receives harder blows and is tempered at a still higher temperature.

SAFETY NOTE

The head of any tool which is struck with a hammer must be soft. If the head is hard, a blow might cause it to chip and a splinter might enter the eye. The author knows a man to whom this happened, both of whose eyes became affected; as a result of the accident the man is permanently blind.

Such tools can be tempered by heating the head of the tool with a gas blowlamp and allowing the heat to be conducted up the shank of the tool until the cutting edge attains the correct temperature. This is indicated by the changing colour of the film of iron oxide which forms on the polished surface of the tool during the heating and may be used as a guide to tempering temperature. The different colours, their associated temperatures and the type of work for which each is used are shown in the table opposite.

It is important to note that this method of tempering is not particularly reliable and should only be used for plain carbon steels, since the oxide colours differ with the composition of the steel. For accurate tempering, the steel should be heated in a suitable furnace where the temperature can be closely controlled. The furnace may be an oil-bath, salt-bath, lead-bath or hot-air furnace. If, after tempering, the tool head has to be softened, the head can be rapidly heated and then allowed to cool slowly while the cutting end is kept unchanged by immersion in a quenching medium.

SAFETY NOTE

After prolonged use, the head of any tool which is struck with a hammer will 'mushroom' from the effect of the repeated hammer blows. The head has, in fact, been cold-worked and will therefore have become hard, brittle and liable to splinter. To avoid accidents, the head should be reground to its normal shape and re-annealed. Any punch or chisel with a mushroom head is a dangerous tool and its use can cause serious

Table of tempering temperatures and oxide colours

Temperature deg. C	Tool	Colour
200	Scribers, gauges	Faint straw
230	Lathe tools	Pale straw
240	Drills	Dark straw
250	Taps	Brown
260	Centre punches	Brown/purple
270	Press punches	Purple
280	Cold chisels	Dark purple
300	Springs	Blue

injury to the user, who ought to have more sense than to use it, and also to any innocent passer-by, since the splinters will fly in no predictable direction.

(b) *High-temperature tempering*
If *hardened* plain carbon steel is heated to a temperature of 450°C to 550°C, the breakdown of the hard martensite is more complete and a very fine-grained, tough structure is produced. Thus, to make a steel as tough as possible it should first be hardened, preferably using an oil quench, and then reheated to 550°C by immersion in a lead-bath or a salt-bath of neutral salt, for even heating, and allowed to cool. If the part is to be stabilised it can then be treated as described in part 7 without the toughened structure being affected.

STRUCTURES FORMED DURING HEAT TREATMENT

The above has covered the five basic processes used in heat treatment, viz., hardening, toughening, normalising, annealing and tempering. It will be noted that the structures produced are all formed by the breakdown of the solid solution of iron and iron carbide (Fe/Fe_3C) called austenite, and the type and properties of the structure depend upon the rate of cooling from the austenitic stage. These different structures are all given different names by the metallurgist, many of which have deliberately not been used in the text for the sake of simplicity. Fig. 1.12 shows in diagrammatic form the effects of the various treatments and also names the structures formed.

It now remains to discuss certain more specialised types of heat treatment but, as will be shown, these are really variants of the basic processes already mentioned.

7. STABILISING
Heating followed by sudden cooling of the steel sets up internal stresses which, over a period of time, may cause the component to distort. If this happens to a gauge, the gauge may become inaccurate and start to reject good work, or worse still, accept work which should be rejected. Distortion must be prevented by relieving the internal stresses but the stress-relieving process described on page 21 cannot be used since it would soften the gauge, which must be hard to resist wear. Thus the temperature adopted must be that of a low-temperature tempering process. If the gauge is heated to 150°C for a period of five hours and cooled slowly it will still be of the required hardness but will also be extremely stable, and will retain its shape and size throughout its normal working life.
To stabilise, heat the steel to 150°C for five hours and cool it slowly.

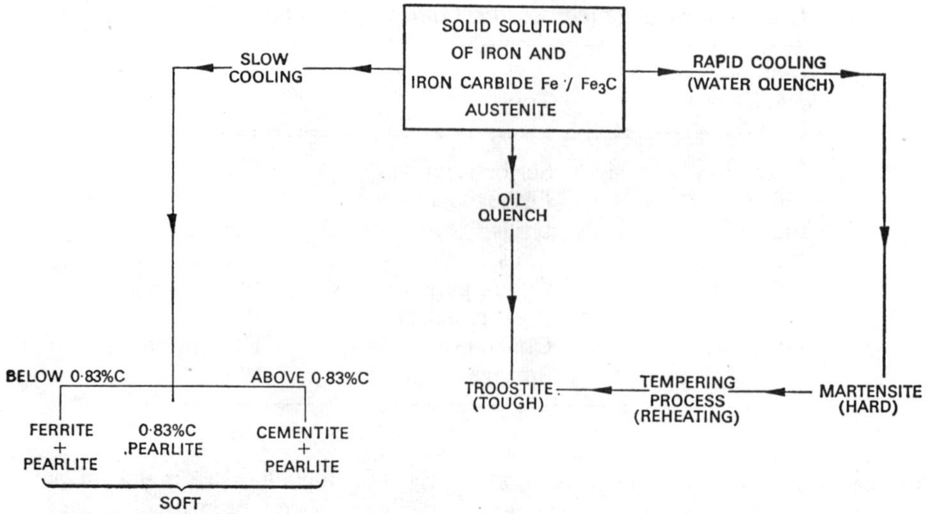

Fig. 1.12. Constituents formed in plain carbon steel through different speeds of cooling

8. SURFACE HARDENING

Specific Objective: *The student should be able to list the procedures for surface hardening plain carbon steel.*

The hardness of a material is a measure of its ability to resist wear but it is an unfortunate fact that if a plain carbon steel is hardened enough to resist wear it is then usually too brittle to resist shock loads. Therefore, if we require a steel to stand up to both wear and shock loads in service we must arrange for it to have a hard surface to resist wear and a tough core to withstand blows and shock loads.

There are two ways of bringing about this condition in a piece of plain carbon steel:

(a) To use a high-carbon steel and harden only the skin to the depth required, leaving the core tough. This is usually done by means of a sophisticated electrical process called *high-frequency induction heating* in which the heating is confined to the surface of the steel and is so rapid that the core does not have time to get hot enough to become hardened when the steel is quenched. The process is outside the scope of this book and is generally limited to large-production fully-automated heat-treatment shops with expensive electronic equipment, the complete treatment taking only a few seconds per component.

(b) To use a low-carbon steel and raise the carbon content of the case, or outer skin, to the required depth so that it can be hardened. This is the more common process, usually known as case-hardening, which we shall consider in some detail. The process requires three stages, which are usually separate:

(i) carburising, to raise the carbon content of the case;
(ii) core-refining, to toughen the core;
(iii) case-hardening, to harden the case.

If a low-carbon steel of, say, 0.1% carbon content is heated to above the upper critical temperature in intimate contact with a carbon-rich material such as charcoal, a chemical reaction takes place and iron carbide is formed in the surface of the steel, producing a skin of high-carbon steel. During the period of high temperature, some of this iron carbide may spread into the core, raising its carbon content slightly so that the component may now have a skin containing 1.1% carbon and a core containing 0.3%. In most cases, the steel will have been held at above the upper critical temperature for a considerable time and its condition will be similar to that of fully annealed steel, viz., soft, weak and with an enlarged grain structure. It must now be treated to produce the tough core and hard case required, the temperatures for these treatments being shown on the heat-treatment diagram which is reproduced for convenience in fig. 1.13.

critical temperature and reheated to below this critical temperature when the case was hardened.

Various methods of surface hardening low-carbon steel are used, the stages described above being common to all of them. The main difference in the processes is the method of carburising, which can affect the subsequent treatment, but in all methods the general idea is the same, to raise the carbon content of the case

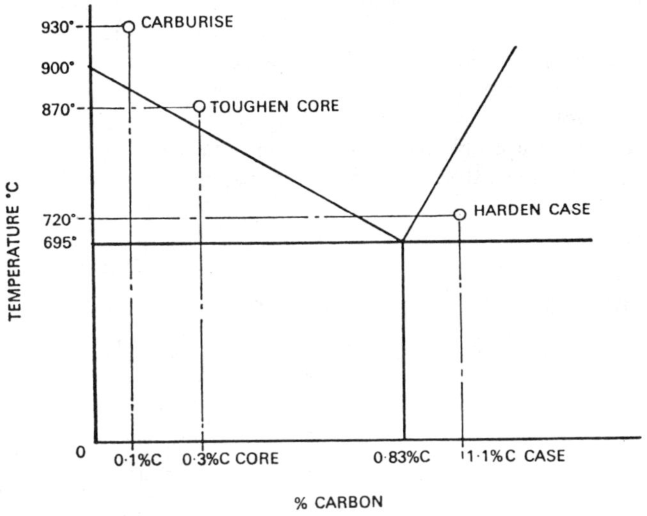

Fig. 1.13. Application of iron/iron-carbide diagram to case-hardening

and then harden it.

Methods of carburising differ in the material used to provide the carbon, which can be solid, liquid or gaseous.

Thus, the temperatures and quenches used in the process are:

(i) carburising: 950°C
(ii) core-refining: 850°C with an oil quench
(iii) case-hardening: 750°C with a water quench

Finally, a low-temperature tempering at 200°C may be used with prolonged heating and slow cooling, for the purposes of stress-relieving and stabilising.

The student should now be able to decide for himself the nature of the structures of the case and core, the case being a high-carbon steel, water-quenched from above the lower critical temperature, and the core a medium-carbon steel, oil-quenched from above the upper

9. CARBURISING

Specific Objective: *The student should know the procedures for carburising by pack, salt-bath and gaseous methods.*

(a) *Pack carburising (solid media)*

The workpieces are packed in charcoal, to which barium carbonate has been added to act as a catalyst, in a cast iron box which is sealed with fireclay, through which test pieces are left protruding. The box is then placed in the furnace at the required temperature of 950°C and left until the required case-depth is estimated to have been achieved, when a test piece is removed, quenched and fractured to check the depth of case. If the depth is correct the box is allowed to cool, the carburising material acting as insulation and enforcing slow cooling, which produces the weak annealed structure previously mentioned. The parts are now removed from the box, core-refined and case-hardened.

The rate of penetration varies with the depth of case, being about 0.3 mm per hour up to 0.5 mm depth and then reducing as the depth increases, as shown in the graph of fig. 1.14. Examination of the graph shows that it is not really practicable to exceed a case depth of 1.5 mm because of the time required. Nor in fact is it desirable, since greater case depths are prone to produce cracking and peeling, the case separating from the core. A common depth of case is 0.75 mm to 1 mm, to allow up to about 0.75 mm to be removed from the diameter by grinding after hardening. This allowance should be ample to cover any distortion and allow the workpiece to be cleaned up and brought to size.

The pack-carburising process is most commonly found in toolrooms and small heat-treatment shops since, although it is a lengthy process, it can be carried out in standard furnaces, the only extra equipment needed being a cast iron box, a supply of carburising charcoal and some fireclay.

(b) *Cyanide carburising (liquid media)*

Sodium cyanide is a salt extremely rich in carbon which melts at about 500°C, thus providing an extremely active carburising agent. It is heated to 950°C, the normal temperature for carburising, and the workpiece is suspended on a piece of wire and immersed in the bath. The liquid provides very even heating and gives minimum distortion with rapid penetration, ten minutes' immersion producing 0.25 mm depth of case. These values are for small parts, large components taking longer since they cool the salt on immersion and an initial heating period is then necessary to bring the bath up to the temperature required.

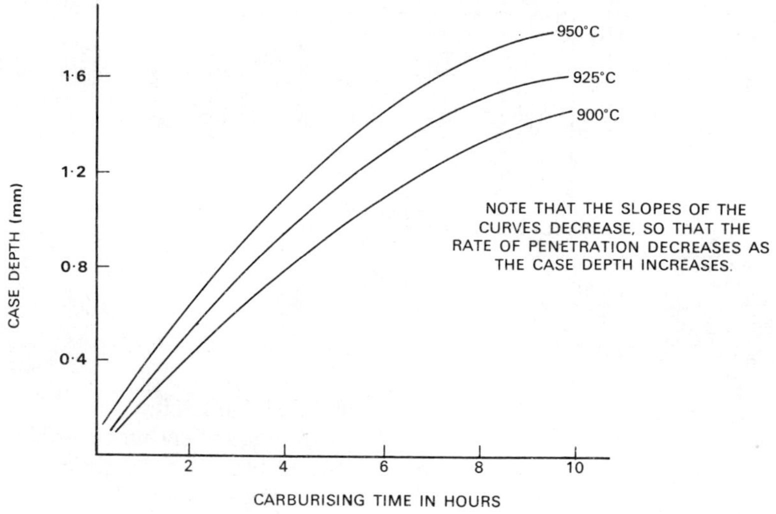

Fig. 1.14. Effects of carburising time and temperature on depth of case produced

The speed of the process does not allow the usual grain growth to take place and in many cases the core-refining treatment can be omitted. This reduces still further the time required and also reduces the cost of case hardening where the type and quantity of work are suitable.

The cyanide process is widely used where large numbers of small parts are to be treated but requires specialised equipment of a comparatively expensive nature and is thus normally found only where the quantity of work warrants its installation.

SAFETY NOTE

Sodium cyanide is a deadly poison. An antidote must be kept in a readily accessible position in the shop where it is used, and instructions in the safe handling of the cyanide and its storage, and in the use of the antidote, must be displayed and obeyed. Common sense and personal hygiene can save a lot of grief; washing the hands after handling the material, and never allowing familiarity to breed contempt are both important. DON'T TAKE RISKS.

(c) *Gas carburising (gaseous process)*
In this process the workpiece is heated in an electric furnace which is filled with enriched town gas or a bottled carbon-rich gas such as propane. This is essentially a production process, usually involving a large and expensive installation. The parts travel through the furnace on a conveyor-type hearth and fall into a quenching bath outside.

THE HEAT TREATMENT OF HIGH-SPEED STEEL

In this book it is not intended to consider alloy steels as such, but the book would be incomplete without mention of high-speed steel, which is so widely used for cutting tools that a young engineer becomes acquainted with it as soon as he enters a machine shop.

In our consideration of plain carbon steels we noted that a hardened plain carbon steel loses some of its hardness if it is reheated to a relatively low temperature, a process we know as tempering. This is a disadvantage if the steel is to be used as a cutting tool in machining, because the heat generated during cutting will produce the same effect and soften the tool, unless the cutting speed is kept so low that the small amount of heat produced can be easily dissipated.

If a steel is to be used for cutting at higher speeds it must be capable of retaining its hardness at higher temperatures, the reasoning that led to the development of high-speed steel by F. W. Taylor and his colleagues.

The commonest form of high-speed steel is known as 18–4–1 steel, the figures referring to the percentages of tungsten (W). chromium (Cr) and vanadium (V) in the steel. Each of these elements has its purposes but the combination produces a steel which retains its hardness when hot. The heat treatment for hardening the steel is more complex than that for plain carbon steel and is carried out in three stages:

1. heating slowly and carefully up to 850°C;
2. heating rapidly up to 1280°C;
3. quenching in oil.

Even then the steel is not fully hardened but may be hardened further by reheating to 600°C, a process known as *secondary hardening*. If this secondary hardening is not deliberately carried out it will probably occur to some extent early on in the life of the tool through the high temperature attained at its normal cutting speed, which is roughly three times that possible for a similar job with a high-carbon steel tool.

HEAT-TREATMENT EQUIPMENT

Steel can be hardened, tempered, annealed and normalised with no more equipment than a blow-torch and a bucket of water, but the variety of work would be limited and the results inconsistent. To carry out properly the full range of work for more than the simplest of tools it is necessary to use up-to-date equipment, consisting of:

1. *Furnaces,* for correct heating of the work;
2. *Temperature-measuring equipment,* to ensure that correct temperatures are employed;
3. *Quenching equipment,* to enable the hot steel

to be quenched in the correct way under the best conditions;

4. *Equipment for testing the properties of steel,* to check that the required results have been achieved.

1. HEAT-TREATMENT FURNACES

Specific Objective: *The student should recognise and describe the function of the main features of muffle, non-muffle and salt-bath furnaces.*

Three basic types of furnace are required in a general heat-treatment shop: muffle furnaces for general work, salt-bath furnaces for more specialised work, and tempering furnaces. A wide range of work can be carried out by only a muffle furnace, but the range is still limited for present-day requirements and it is better if all three types of furnace are available.

Muffle furnaces

A muffle furnace is one in which the source of heat does not directly contact the work but circulates around the outside of the chamber, or muffle, in which the work is placed. This is true for either gas or electric furnaces as shown in figs. 1.15(a) and 1.15(b). Semi-muffle furnaces of the gas-fired type are available, in which combustion takes place beneath the furnace hearth and the products of combustion pass through the work chamber, as shown in fig. 1.15(c). In this type of furnace the atmosphere is controlled by varying the proportions of gas and air which are burned together to produce the required temperature. Excess air, above that required for complete combustion of the gas, produces an oxidising atmosphere, while excess gas gives a reducing atmosphere. A neutral atmosphere, which is neither oxidising nor reducing, is obtained by using the correct mixture of air and gas for complete combustion. Thus, control of the combustion process gives some degree of atmospheric control. In the 'straight' muffle furnace the atmosphere is always oxidising unless the air is excluded by introducing an inert gas at a pressure slightly higher than atmospheric to keep out air.

The high-speed steel muffle furnace

As stated on page 27, high-speed steel must be heated slowly up to 850°C and then rapidly up to 1280°C. This is usually done by using a special type of muffle furnace, illustrated in fig. 1.15(d), which consists of two chambers, one above the other, but with a single source of heat. The lower, high-temperature chamber is heated directly but the upper chamber is heated by the waste gases from the high-temperature chamber and is therefore at a lower temperature. The work is usually placed on the sill of the lower-temperature, or pre-heating, chamber to warm up gradually and is then edged into the chamber. When the work has attained the required temperature both doors are opened simultaneously and it is transferred quickly to the hot chamber, for rapid heating to hardening temperature.

The salt-bath furnace (fig. 1.16)

This type of furnace consists basically of a pot to contain the molten salt, a means of heating and an extraction system to remove the dangerous fumes. The furnace may be heated by gas, when combustion takes place beneath and around the pot in an enclosed combustion chamber, or by electricity. The electric type usually has a non-metallic pot into which a pair of electrodes project. The current passes between the electrodes through the salt and the resistance of the salt produces the required heating effect. In its solid state, the salt is non-conductive and the furnace is started by wedging a carbon rod between the electrodes. The rod is heated by the passage of the current and a pool of molten salt forms around it between the electrodes, in which the carbon rod dissolves. This conductive pool continues the heating process until the whole bath is molten.

The tempering furnace (fig. 1.17)

This furnace runs at a lower temperature than either a muffle or a salt-bath furnace. It usually has a chamber in which the air is heated either

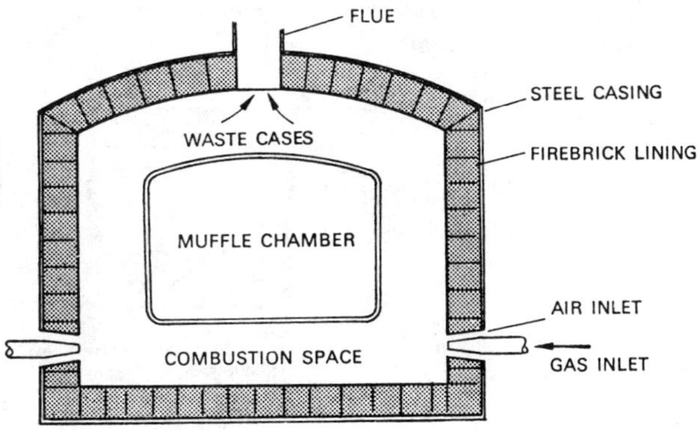

Fig. 1.15(a). Natural-draught gas-fired muffle furnace

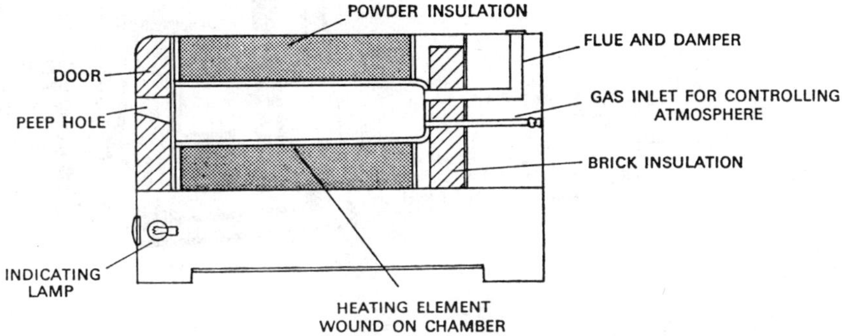

Fig. 1.15(b). Electric muffle furnace with provision for atmospheric control (by courtesy of Wild-Barfield Ltd.)

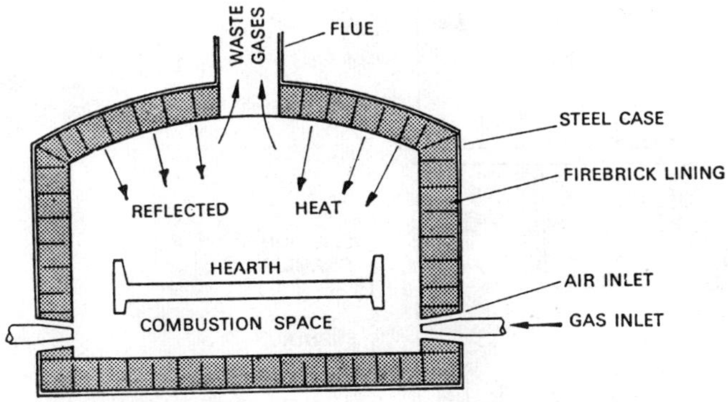

Fig. 1.15(c). Natural-draught gas-fired semi-muffle furnace

Fig. 1.15(d). High-speed steels hardening furnace (by courtesy of the Wellman Incandescent Furnace Co. Ltd.)

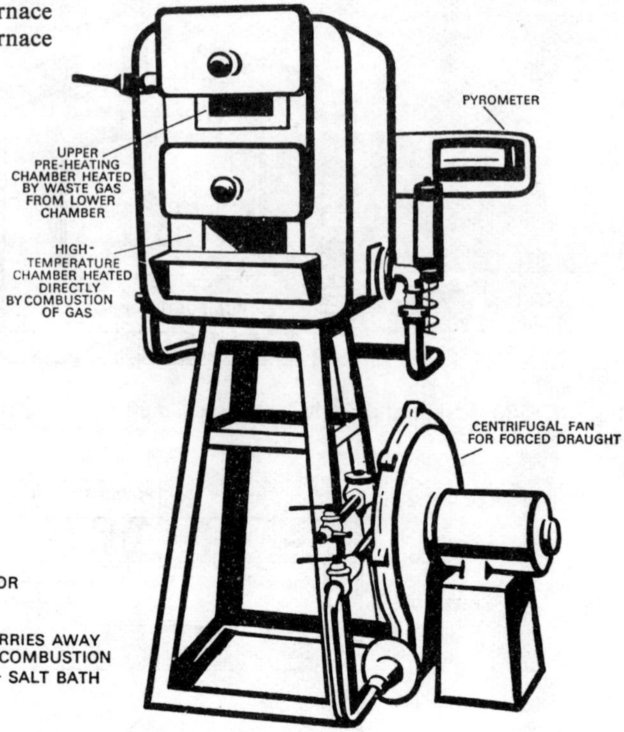

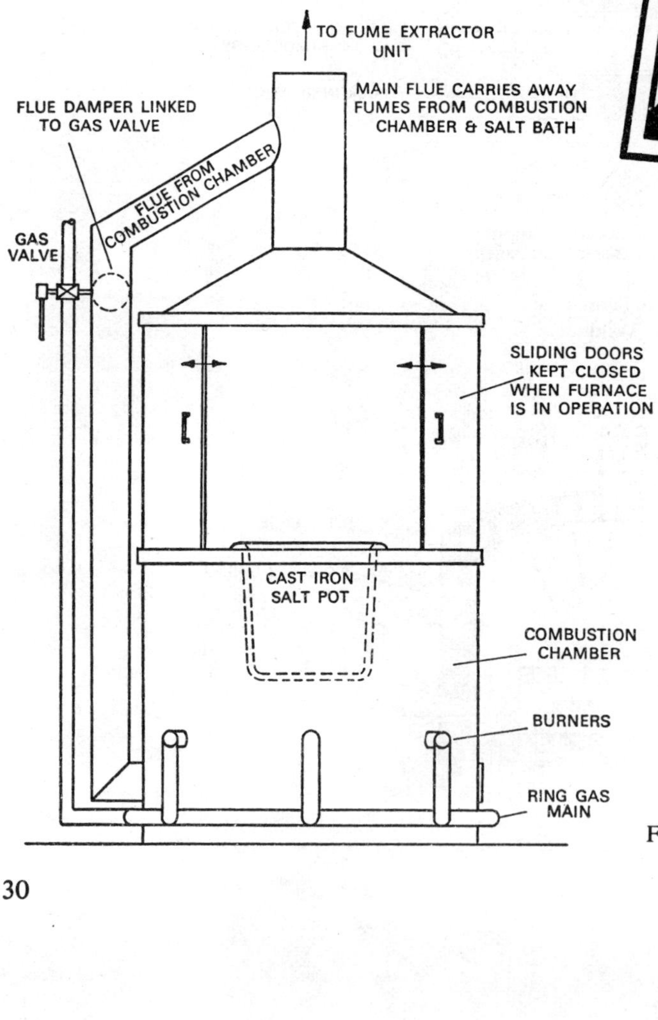

Fig. 1.16. Natural-draught salt-bath furnace

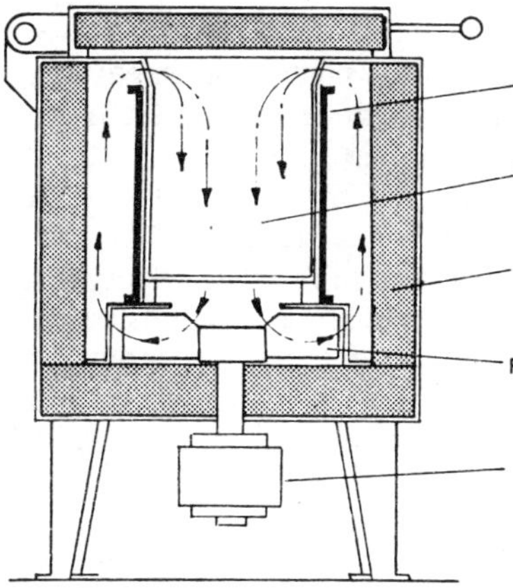

Fig. 1.17. Tempering furnace with forced-air circulation

electrically or by gas, and a built-in fan to give good circulation of air and even heating.

In some cases, as in the secondary hardening of high-speed steel, where it is important that the atmosphere should be neutral and the temperature higher than normal, the tempering is carried out in a salt-bath of neutral salt or in a bath of molten lead.

Furnace fuels

Specific Objective: *The student should be able to compare the advantages of oil, gas and electricity as heat sources.*

The three sources of heat available for heating are, as stated in the objective, oil, gas and electricity. Each has its advantages and limitations, mainly concerned with the temperature attainable, cost, control, cleanliness, storage, and convenience. In general it is cheaper to burn a fuel and use if for direct heating rather than to pay somebody else to burn the fuel, convert it into electricity, and transmit it to the consumer who then converts it back into heat, as long as the fuel is burned efficiently in the furnace. Thus in terms of cost, oil and gas have the advantage. At the same time higher temperatures are available from efficient direct combustion. It follows that furnaces operating at very high temperatures, such as those for hardening high-speed steels, are usually gas or oil fired. A problem with oil as a fuel is that it must be stored by the consumer and regular supplies must be available. These problems do not occur with gas, so most heat treatment furnaces are gas or electrically heated, those operating at very high temperatures being gas fired.

Electrically heated furnaces are cleaner, lend themselves to atmosphere control and, as most temperature control devices are electrically operated, are easier to control.

Much of this is a matter of opinion and doubtless the manufacturers of different types of furnaces would each produce advantages for the type he manufactures. However, summarising the above it would appear that:

(a) for low to medium temperatures, electricity has the advantages of convenience, control and cleanliness; this applies to the heat treatment of plain carbon steels;

(b) for high-temperature operation, gas has the advantages of high temperatures attainable, cost and convenience of supply.

2. TEMPERATURE-MEASURING EQUIPMENT

Specific Objective: *The student should be able to describe the construction, application, and principle of operation of the thermocouple pyrometer.*

'Heat it to a cherry red and dip it in water' is the instruction sometimes given to a young man who wishes to harden a piece of steel. Such an instruction is not good enough because, as we have seen, the temperature required depends upon the carbon content of the steel and, in any case, the colour radiated by the hot steel will vary with the background lighting. In a dull room a bright red might be too cool and in a bright room a dull red might be too hot. Temperature can be judged by colour, with experience and knowledge of the conditions, but gaining the experience can be an expensive process. It is far better to start properly with accurate equipment for the measurement of temperature.

There are many forms of equipment available for temperature measurement and indeed for low-temperature work such as tempering there is much to be said for the use of the ordinary mercury thermometer, except for the high breakage rate which could be expected. The commonest instrument in use today is the thermocouple pyrometer.

Thermocouple pyrometers
These are based on a phenomenon known as the '*Seebeck effect*'. If two dissimilar metal wires are joined at their ends to form a closed loop and the temperature of one joint, or junction, is raised above that of the other, an electric current will flow round the loop. The magnitude of the current will depend upon the *difference in temperature* between the hot and cold junctions. If an electrical measuring instrument is inserted into the loop so that its terminals form the cold junction, it can be calibrated to read in units of temperature, thereby producing a thermocouple pyrometer.

In practice it is not quite so simple, because the reading depends on the temperature *difference* between the junctions, and if the temperature of the cold junction varies, the reading correspondingly varies. It is also usually necessary to place the measuring instrument some distance away from the thermocouple to obtain a more convenient position for reading. This enables a simple means of cold-junction correction to be used, by making the leads from the thermocouple to the instrument also of dissimilar metals so that they form, in effect, a second thermocouple to compensate for changes in temperature at the cold junction. These leads are thus called compensating leads.

The wires of the thermocouple must be insulated from each other by small ceramic insulators, or quills, and in the furnace the thermocouple is protected from damage by a silica sheath; the installation is shown in diagrammatic form in fig. 1.18.

Thermocouple materials The materials used for the wires of a thermocouple must be chosen so that:

(a) the current produced varies uniformly with the temperature throughout the working range;
(b) the melting points of the materials are above the working range;
(c) the materials have good corrosion resistance at high temperatures.

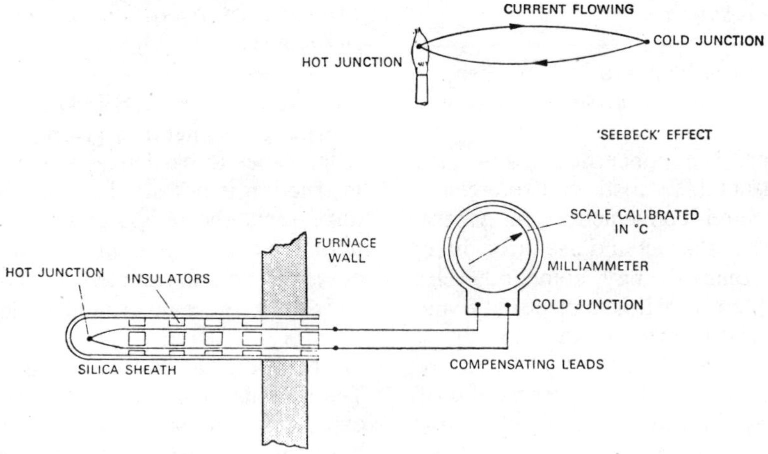

Fig. 1.18. Thermocouple pyrometer

Materials in common use and their working temperature ranges are:

(i) iron/constantan: up to 700°C
(ii) chromel/alumel: up to 1100°C
(iii) platinum/platinum-rhodium: up to 1200°C

Constantan is an alloy containing 45% Ni and 55% Cu.
Chromel is an alloy containing 90% Ni and 10% Cr.
Alumel is an alloy containing 98% Ni and 2% Al.
Platinum/rhodium is an alloy containing 90% Pt and 10% Rh.

Calibration of thermocouples It must be remembered that a thermocouple is a measuring instrument, that is to say, it compares an unknown standard with a known one. Over a period of time the accuracy of the instrument may fall off and it must therefore be checked at intervals. If a standard thermocouple of known accuracy is available, the thermocouple in question can be checked throughout its range by immersing the two side by side in a bath of molten metal and comparing the two readings as the metal cools.

If a standard thermocouple is not available, another method is to immerse the thermocouple in a bath of molten material of known freezing point; for example, pure sodium chloride, which freezes at 801°C. If a cooling curve for the salt is plotted, the difference between this temperature and that of the arrest point recorded by the thermocouple will be the error in the instrument *at this temperature*. The check can be repeated, using other materials, at other suitable fixed points over the range of the thermocouple.

A word of warning on the use of thermocouples is needed at this point. The instrument does not necessarily give the temperature of the metal in the furnace, but that of the end of the thermocouple in its protective sheath. This temperature depends upon the position of the thermocouple in the furnace and upon whether the furnace temperature is uniform over the furnace volume. It will also depend upon how rapidly the furnace has been heated or cooled, since there will always be some lag between actual temperature and instrument reading unless the furnace has been held at constant temperature for a period. Thus, the skill and experience of the heat-treatment technician are important; even when he has good measuring equipment he must still be familiar with the performance of his furnaces to get the best results.

3. QUENCHING EQUIPMENT

Specific Objective: *The student should be able to compare the advantages and limitations of oil, water, and brine as quenching media and justify vertical quenching and agitation procedures.*

As shown previously, different conditions are produced in a piece of steel by different methods of cooling, and these involve different cooling media. The cooling media are commonly:

(a) an air blast, usually from the compressed-air line of the workshop;
(b) a water quench in either fresh water or brine;
(c) an oil quench.

The latter two involve quenching tanks which should be designed so that the oil and water stay as nearly as possible at a constant temperature. This can be achieved by containing the oil tanks within the water tanks as shown in fig. 1.19 and having an overflow pipe, or sill, so that a steady flow of water carries the heat away.

A bath of fresh water is not an ideal quenching medium. Steam is formed at the surface of the steel, which forms an insulating barrier and prevents the heat being carried away by the water. If common salt is dissolved in the water, it reduces the tendency to form steam and this allows better water contact and more even cooling. Brine does not give a more drastic quench in that it produces greater hardness, but it gives more uniform hardness and eliminates soft spots. This uniform cooling has another effect in that, contrary to popular

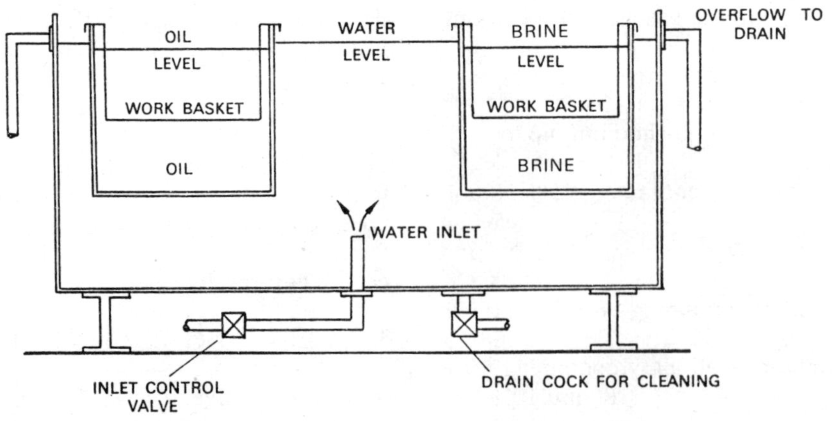

Fig. 1.19. Water-cooled oil quenching tank

opinion, plain carbon steel is *less* likely to crack if quenched in brine rather than fresh water.

The manner in which the workpiece is presented to the quenching medium is also important. However rapidly it is submerged in the bath, the first area to make contact with the liquid is cooled first. Thus if a long, flat piece of hot metal is dipped into a tank horizontally, the bottom surface cools first, contracts and distortion or cracking occurs. Such pieces should be plunged in end first and kept completely submerged until cool. Any agitation or movement of the part should be a gentle up and down motion rather than a rapid stirring motion which can only increase the chance of bubbles at the surface and hot spots.

4. EQUIPMENT FOR TESTING THE PROPERTIES OF STEEL

Specific Objective: *The student should be able to recall common tests for hardness, toughness, ductility and strength.*

Hardness testing

The easiest way to check whether a piece of steel is hard is to try to file it. Unfortunately this does not tell us how hard it is, and methods of measuring hardness based on numerical values are required so that comparisons can be made. Hardness is defined as the ability to resist indentation ot scratching, and most machines designed to measure hardness do so by measuring the ability of a material to resist indentation by either a diamond or a hard steel ball under a standard load. Another method is based on a common phenomenon observed by anybody who has played a ball game, that a ball bounces higher from a hard surface than from a soft one. Yet another method, used at the National Physical Laboratory, is based on the fact that a different 'feel' is observed when a standard stylus is drawn over materials of different hardness.

Indentation methods

(a) *The Brinell test* The Brinell hardness tester uses a hard steel ball as an indenter, which is impressed under a standard mass into the material to be tested. The hardness number is given by dividing the *applied mass* by the *surface area* of the indentation produced, i.e.:

$$\text{Brinell hardness} = \frac{W}{1.571\,D(D-\sqrt{D^2-d^2})}$$

where D = diameter of ball, mm
d = diameter of impression, mm
W = applied mass, kg

Normally, a mass of 3000 kg is used with a 10 mm ball but different balls and loads are used for soft materials and thin sheets. The test is not really suitable for very hard materials since the deformation of the ball affects the results, but a tungsten-carbide ball is sometimes used for such work.

(b) *The Vickers pyramid harness test* A diameter indenter is used, in the form of a pyramid with a square base and an included angle of 136°. The hardness number is again given by the surface area of the indentation divided into the load, i.e.:

$$\text{Vickers pyramid no. (V.P.N.)} = \frac{2W \sin \theta/2}{d^2}$$

where W = applied mass, kg
θ = included angle of pyramid, degree
d = width of indentation *across the diagonals*, mm

The testing machine is fitted with a measuring microscope which is automatically aligned with the indentation, the operator's view through the eyepiece being as shown in fig. 1.20. The two knife edges are adjusted in the micrometer until they span the diagonal of the impression, and the micrometer reading is noted and converted into a hardness number, either by the use of the above formula or by reference to tables.

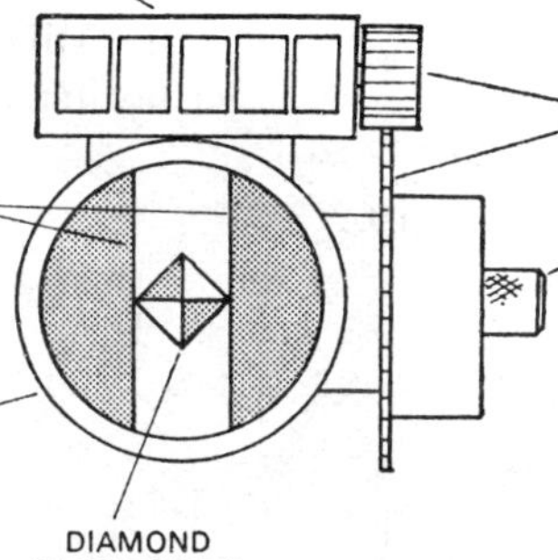

Fig. 1.20. Measuring microscope on Vickers hardness-testing machine

(c) *The Rockwell hardness test* Both the Brinell and the Vickers tests give a hardness number which is based on the load per unit surface area of indentation, and a calculation or reference to a table of values is required after the test has been made. The Rockwell machine is different in that it is direct-reading, one reason for its popularity in industry, although in laboratories the Vickers or Brinell machines are usually preferred.

The Rockwell tester uses a conical diamond penetrator, with a 120° included angle and a 150 kg mass for harder materials on scale 'C', or a steel ball of 1.6 mm diameter with a 100 kg mass on scale 'B'. An initial mass of 10 kg is applied, to reduce the effects of slackness in the machines and any surface defects. The appropriate main mass is then applied and the *depth of penetration of the indenter* is measured. This measurement is shown on a dial gauge, built into the machine and calibrated in the 'B' and 'C' scales for the appropriate penetrator and load.

(d) *'Bounce' tests* The *Shore Scleroscope* consists of a glass tube graduated into 140 equal parts, down which a diamond-tipped hammer of mass 7.09 g is allowed to fall freely from a height of 250 mm on to the work. The height of the first rebound as shown on the scale is used as the Shore hardness number. If the material is hard, little energy will be expended in causing deformation and the hammer will bounce back to almost its original height. On the other hand, the deformation produced in a soft material will absorb a greater amount of the energy of the moving hammer and the hammer will not bounce back as far.

(e) *Scratch tests* The standard test for the hardness of minerals is the *Moh scratch test,* in which the mineral under test is scratched with stylii of graduated hardness, and the number allocated is one less than that of the stylus which just scratches the mineral. The scratch leaves a mark on the surface and the test is often regarded as not suitable for engineering purposes. However, a test which leaves no surface defect is described by the National Physical Laboratory.[1] A piece of silver steel is hardened and ground at one end to a smooth radius of about 0.50 mm, and mounted in a handle. The 'scriber' is drawn across the work surface and then across the surface of a number of graduated hardness standards of similar surface finish, the resulting marks being compared. This is particularly useful for testing the working surface of a gauge, for although a mark is produced no surface damage results.

[1] N.P.L. Notes on Applied Science, No. 1 – Gauge making and measuring.

Toughness testing
Toughness is the ability of a material to withstand shock loads or impact. The easiest way to compare the toughness of different pieces of metal is to put them in a vice and hit them with a hammer. Those that fracture or bend most for a blow of given strength are the least tough, and those on which the hammer has least effect are the most tough. The hammer blows must be kept as consistent as possible and this, for an accurate test, is the difficulty. It is, however, quite simple to arrange for the blow always to have the same force by fixing the hammer to a pendulum and allowing it to swing from the same height for each test. The problem then is to judge the effect of the blow on the specimen. It is easier, in fact, to judge the effect on the hammer since, at the moment of impact, it contains energy equal to its effective mass multiplied by its original height. At impact some of this energy is used to break the specimen and therefore the hammer, as it continues its swing, will not reach its original height. The distance it swings short of its original height is a measure of the energy lost at impact.

Both the *Izod* and the *Hounsfield* impact testing machines operate on this principle, as

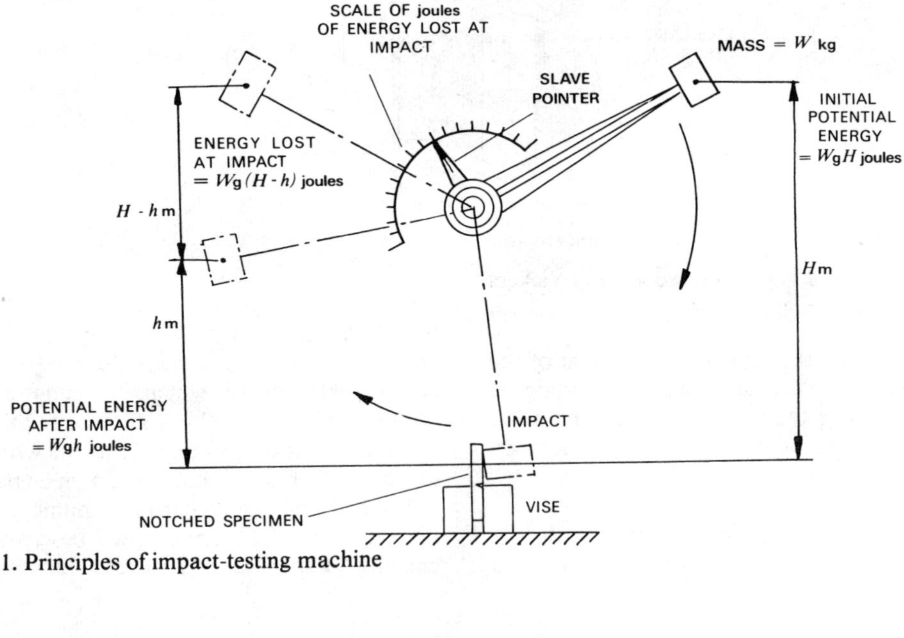

Fig. 1.21. Principles of impact-testing machine

shown in fig. 1.21. A slave pointer records the height to which the pendulum rises, reading against a scale of joules of energy lost in fracturing the specimen. The Izod machine is a large floor-mounted one with a single heavy pendulum delivering 160 joules of energy to a notched specimen which is held in a vice. The Hounsfield machine is smaller and has two pendulums, one swinging through the other, which carries the specimen. The energy lost at impact is again recorded by a slave pointer and is read off the scale in joule units.

SAFETY NOTE

These toughness tests are both dynamic tests and if the energy contained in the moving parts is expended on any part of the human anatomy the results are likely to be painful, if not disastrous, and may easily be fractures. The machines should be properly guarded and used with respect at all times.

Strength testing

The most common test for evaluating the strength of a piece of material is the tensile test. A specimen of the material is gripped in jaws of the type shown in fig. 1.22, a force F being exerted by the machine to draw the jaws apart. Due to the inclined faces of the jaw blocks a gripping force is exerted which increases as the force F increases. The gripping faces of the

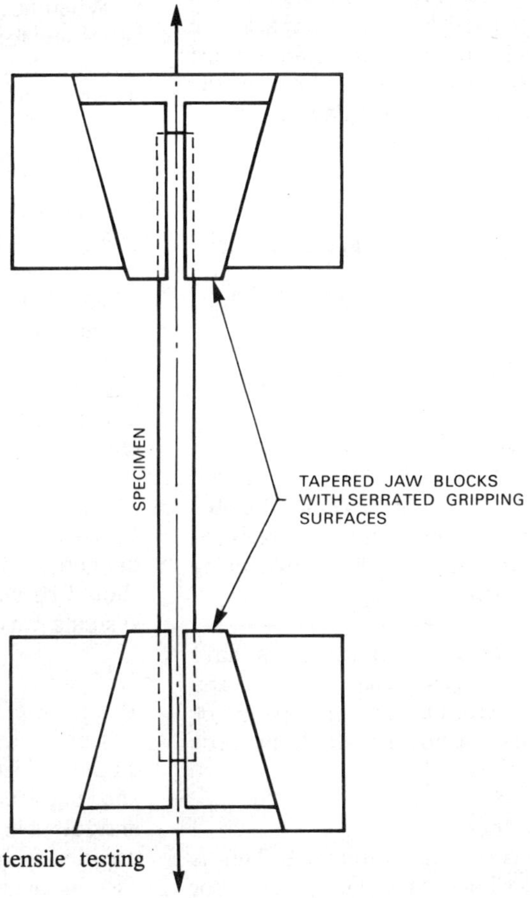

Fig. 1.22. Jaws and specimen in tensile testing machine

jaws are serrated to prevent the specimen from slipping.

The testing machine may be equipped to plot a graph of load against extension or simply with a load gauge with a slave pointer which enables the maximum load (not necessarily the same as the breaking load) to be determined. The tensile strength then calculated is:

$$\text{Ultimate tensile strength} = \frac{\text{Maximum load carried (MN)}}{\text{Original cross section area (m}^2\text{)}}$$

Ductility testing
Ductility is the property of a material which allows it to be permanently deformed under a tensile load without failing. It is measured by determining the percentage elongation which takes place during a tensile test. Two marks are made on the specimen a known distance apart so that when the specimen is broken the extension can be determined by placing the two pieces together and measuring the distance between the marks. Then:

$$\text{Elongation} = \text{Final test length} - \text{original test length}$$

and

$$\%\ \text{Elongation} = \frac{\text{Elongation} \times 100\%}{\text{Original test length}}$$

The greater the percentage elongation the more ductile is the material.

SAFETY NOTES

Specific Objective: The student should be able to name major sources of danger and relevant precautions which must be taken when using heat treatment equipment.

Within this chapter some safety notes have been given where appropriate. There are, however, some general hazards which must be recognised so that accidents which may be caused can be avoided.

1. *Hot metal is dangerous*
It will cause severe burns if touched. This is stating the obvious but the fact that metal is hot is not always obvious. It only appears hot when it starts to glow at about 600°C. A specific and clearly marked area should be set aside for leaving metal to cool but in any case never assume that any metal in a heat treatment area is cool. Approach it carefully if is is not known to be cold.

2. *Fire hazards*
Wherever fuel is burned or temperatures raised there is a fire hazard. Extinguishers should be provided and they must be of the correct type. Some extinguishers should never be used on electrical equipment and water certainly should not. Make sure that you know the location of the appropriate type of fire extinguisher for a given purpose.

When hot metal is quenched in oil there is a tendency for some of the oil to vaporise and flare up. If a fire in a quenching bath looks like getting out of hand it can be snuffed out by cutting off the air supply – a well fitting lid should be provided for oil quenching baths.

3. *Explosions*
If a gas fired furnace becomes charged with a gas/air mixture *before* lighting it will explode when the lighter is thrust in. Always have the lighter alight and by the jets *before* the gas is turned on. If the lighter goes out or the furnace jets do not ignite at once turn off the gas immediately. Never let the combustion chamber fill with gas.

4. *Fumes*
Gas fired furnaces and quenching baths give off dangerous fumes. The heat treatment area should be well ventilated, and fume extraction systems provided and used.

5. *Poisons*
The toxic nature of some of the materials used in heat treatment has been dealt with separately on p. 27. Briefly, such materials must be stored and used according to the regulations, the antidote kept available and instructions for its use clearly displayed – preferably never to be used because nobody gets poisoned.

QUESTIONS

1. Sketch the family of cooling curves for solutions of salt and water from which the equilibrium diagram for these two materials can be constructed and describe the cooling of:
   (a) a hypo-eutectic solution
   (b) a eutectic solution
   (c) a hyper-eutectic solution.

In each case *name* the composition of the solution which finally solidifies.

2. Explain why a solder containing 33% lead and 67% tin is preferred for electrical work and small assembly operations. To answer this question reference should be made to the thermal equilibrium diagram for lead and tin.

3. Sketch a simplified version of the steel portion of the iron/carbon equilibrium diagram and state the main differences (apart from the constituents) between this and the tin/lead diagram.

4. Describe the changes which take place during the slow cooling of steels containing:
   (a) 0.4% carbon
   (b) 0.83% carbon
   (c) 1.2% carbon.

Sketch the structures formed, naming the constituents and state the mechanical properties of each constituent.

5. Sketch a graph showing the relationship between the percentage carbon and the percentage pearlite in slowly cooled steel. Show, with the aid of a simple graph how the constituents in the micro-structures effect the mechanical properties of the material.

6. Draw a simplified version of the iron/iron carbide diagram to show the temperatures at which the commonly used heat treatment processes should be carried out and use the diagram to help to explain how:
   (a) a 0.4%C steel should be made as tough as possible
   (b) a 0.85%C steel should be made as hard as possible
   (c) a hardened piece of 1.2%C steel should be made soft as possible
   (d) a hardened piece of 0.9%C steel should be put in the best condition for machining.

7. Explain what is meant by the term tempering and what the process is used for. What process must be carried out on all high carbon steel parts before they can be tempered? State what tempering temperature would be used for tempering the following:
   (a) scriber;
   (b) locating pin to be as tough as possible;
   (c) plain plug gauge to be stabilised;
   (d) letter punch.

8. Surface hardening is a complex process compared with the through hardening of high carbon steel. This being so explain why surface hardening processes are so often used.

9. List the procedures in the order in which they are carried out for case-hardening by the pack carburising process. By reference to the iron/iron carbide diagram specify the temperatures at which they are carried out.

10. Sodium cyanide is a very dangerous material but it is frequently used in case hardening. In view of the hazards associated with the material explain why it is used so much. Describe the process of cyanide hardening and sketch the type of furnace in which it may be carried out. List the safety precautions which must be observed when handling sodium cyanide.

11. State the properties required of each of the following components. Assuming them to be made from plain carbon steel, select a steel of suitable carbon content and describe the heat treatment necessary:
   (a) screw plug gauge;
   (b) crane hook;
   (c) pivot pin for car brake pedal;
   (d) machine lead screw.

12. Make neat diagrams to show the difference between muffle and non-muffle gas fired furnaces. List the advantages and limitations of each.

13. Compare the advantages and limitations of electricity, gas, and oil as fuels for heat treatment furnaces.

14. Name the principle upon which the operation of a thermo-couple pyrometer is based. Sketch a thermo-couple installation and

explain what is its source of power. Why are compensating leads necessary?

15. Oil and water are commonly used quenching media, each being used for a specific purpose in the heat treatment of plain carbon steel. State the application of each. What are the advantages of brine as a quenching medium compared with water?

16. When work is being quenched during heat treatment it seems natural to agitate it in the quenching medium. Such agitation may be a rapid horizontal motion or a gentle vertical motion. State which is preferred giving reasons for your answer.

17. List three different methods for testing the hardness of a part stating which tend to be preferred in laboratories and which in workshops. Describe briefly the principle of each of the three tests.

18. Strength and ductility can be assessed from different aspects of the same test. Describe the equipment used in such a test and how the results are evaluated.

19. Toughness can be evaluated by putting a piece of steel in a vice and hitting it with a hammer. Describe a better method and state the units in which toughness would be measured.

20. List five sources of danger in a heat treatment shop and for each hazard explain one safety precaution which should be taken.

2 Primary forming processes

General Objective: *The student should appreciate the basic primary forming processes.*

INTRODUCTION

If the young engineer looks around he sees many components which are cast, forged, rolled, drawn, extruded or pressed.[1] In most cases, further observation shows that had the part been machined from a solid piece or billet of metal, an enormous heap of swarf would have been produced of comparatively small value as scrap, with a relatively small amount of metal remaining in the component. A good example of this is the cylinder barrel of an air-cooled motor cycle engine. From a solid billet the bore would have to be machined out and the cooling fins machined leaving a small volume of metal around a large volume of air and a great pile of swarf. It is worth noting that swarf is expensive to produce and of little value when made. It therefore pays us to consider ways in which the part can be made with as little swarf as possible and to do this manufacture of the component proceeds in two stages.

1. Primary process: casting, forging, rolling, drawing or extruding the part to the approximate shape and size required.
2. Secondary process: machining important sizes and working surfaces to the accuracy and surface finish necessary for the efficient functioning of the component.

In each of the processes to be considered the material must possess particular properties to make it suitable for production by the chosen method and, at the same time, be suitable for the functioning of the component when made and in service. These properties of materials were considered with regard to functional requirements during Level 1 but they will be briefly reconsidered in the light of manufacturing requirements.

[1] Presswork will not be considered in this chapter – see Chapter 3.

PROPERTIES OF MATERIALS

Specific Objective: *The student should be able to state the properties of a raw material used in the primary manufacturing processes, e.g. fluidity, plasticity, ductility, malleability, and toughness.*

The major properties required of materials for them to be suitable for manufacturing by these methods are:

1. FLUIDITY

If a metal is to be cast it is poured into a mould in the molten state. In order to completely fill the mould and produce a sound casting it must flow easily, i.e. have the property of fluidity. Water has good fluidity, thick oil has less fluidity and treacle has still less fluidity. Similarly, when compared with other ferrous metals, cast iron has good fluidity while most steel does not and steel castings are therefore more difficult and expensive to produce than those made of cast iron.

2. PLASTICITY

For a material to be forged, hot rolled or hot-stamped it must be plastic when hot (not molten). It must be capable of considerable deformation under compressive loads. The opposite to plasticity is 'hot-shortness', i.e. the material is unable to undergo deformation under these conditions and cracks.

3. DUCTILITY

Ductile materials are capable of great elongation without failure, usually under tensile loads. This property is necessary for a material to be drawn into wire or tube, or extruded.

4. MALLEABILITY

A malleable material is capable of deformation without failure under compressive loads. It follows that parts made by cold forming processes such as cold stamping or coining must be malleable. Non-malleable materials are said to be 'cold short'.

5. TOUGHNESS

A material which can withstand shock loads is said to be tough. This is more usually a requirement of the functional needs of the part, than of the process by which they are to be made.

It is found for some parts that their shape dictates the manufacturing method to be used but their functional requirements are not normally found in parts made by such processes. It has therefore been necessary to develop materials and processes to meet both requirements. Consider the bell-housing containing the clutch of a motor car. The gearbox is attached to the engine by the bell-housing and the drive transmitted through the clutch. If the clutch is released suddenly, shock loads are transmitted through the bell-housing which must be capable of withstanding them and therefore be tough; but the most economic material from which to make the part is in cast iron. Here then, is a requirement for a tough cast iron. Such material is called, wrongly, malleable cast iron. It is not malleable by definition but has been subjected to a heat treatment process to make it tough.

SUPPLY OF RAW MATERIALS

Specific Objective: *The student should be able to describe the form of supply of raw materials for the primary forming processes.*

The most common materials used in engineering are still the ferrous materials, steel and cast iron. This section will therefore consider the raw materials and manufacturing methods in iron and steel production in a little detail but it must be remembered that other metallic materials are also produced from ores which are smelted and refined so that they can be delivered in a form to be conveniently used in the primary forming processes. These forms of supply are:

1. INGOTS

When the material is first produced it is cast into large tapering slabs called ingots. Steel ingots may be up to 2 metres in length and have a mass of up to 10 000 kg. They usually have a defect called 'pipe' at the top due to uneven cooling rates and shrinkage during solidification and before subsequent operations this top section is removed and returned as scrap for remelting.

2. BILLETS

The ingots are reheated or held in soaking pits to be passed to the first rolling mills for reduction in size to billets. These are up to 300 mm square and are of course correspondingly longer. They are cut to lengths suitable for the subsequent process to be used and then hot rolled into rolled steel sections, further reduced by rolling and drawn into rods and bars or rolled into sheet and strip ready for presswork. Alternatively the billets may be delivered in billet form of suitable size for large forgings.

3. RODS

As mentioned above, the billets can be reduced by rolling into rod form for drawing into the rods and bars of various sections to be machined. Alternatively the rod, which up to 6 mm diameter can be supplied in coils, can itself be drawn down further, usually cold, into wire.

4. PIGS

The raw material for iron castings is produced in the form of small slabs called pigs which are further melted and refined before casting in moulds into the shape and size required for the finished product.

The above notes refer generally to the forms of supply of ferrous materials but it must be remembered that other materials are delivered in similar ways for similar purposes. Aluminium alloys, copper, brass and the other metals used in industry are cast into suitable ingots for rolling into sheet and strip, drawing into rods, wires and tubes, billets for hot or cold forging and pieces similar to pigs which can be conveniently handled and remelted for casting.

THE MANUFACTURE OF IRON AND STEEL

Ferrous materials are still the most used materials in industry. Fig 2.1 is a simple flow chart of the production of steel and cast iron, on which the blocks represent the types of furnace used while the materials entering the furnace and those obtained from it are indicated by arrows.

THE BLAST FURNACE (fig. 2.2)

The source of all ferrous materials is *iron ore,* a chemical compound of iron and oxygen. The separation of the iron from its oxide, known chemically as a *reducing process,* is carried out in a *blast furnace.* The oxide is heated and brought into contact with the gas carbon monoxide, which combines with the oxygen in the ore to form carbon dioxide, leaving the molten iron to sink to the furnace hearth. The heat for the reaction is obtained by burning *coke,* an air blast being required to give the necessary temperature, and the resulting carbon monoxide reduces the iron ore. *Limestone* is included in the charge to combine with some of the unwanted impurities in the form of a slag, which floats on top of the molten iron and is tapped off separately.

The molten metal from the blast furnace is either tapped into ladles and used for immediate conversion into steel, or cast into conveniently sized slabs known as *pigs* for future conversion

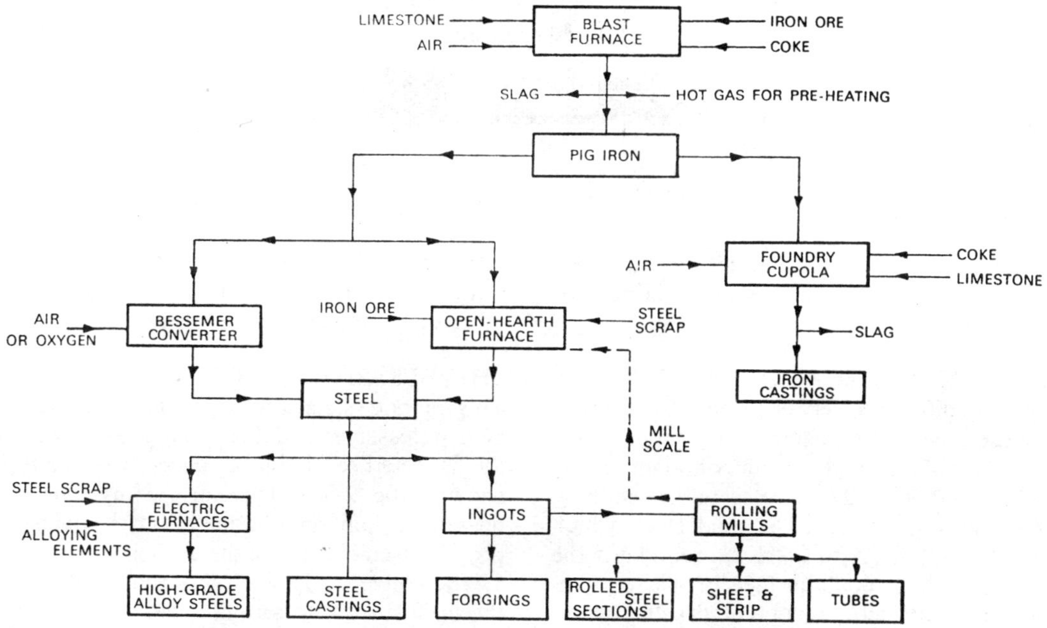

Fig. 2.1. Flow chart of the production of steel and cast iron

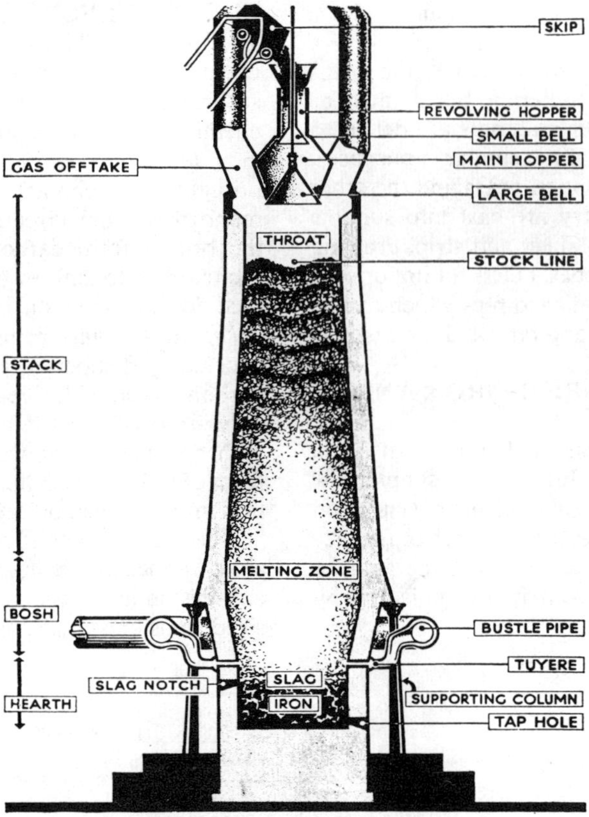

Fig. 2.2. The blast furnace. (Reproduced by permission of the British Iron and Steel Federation)

into cast iron or steel. In either case the iron contains too many impurities to be suitable for immediate use without further refining.

THE MANUFACTURE OF STEEL

The main difference between steel and cast iron is in the amount of carbon they contain. Steel is an alloy of iron and carbon containing up to 1.8% of carbon, and the greatest proportion by far of all steel produced is mild steel, which contains only 0.1% to 0.2% carbon. Thus the pig iron, which contains nearly 4% carbon, must have its carbon content drastically reduced to be converted into steel. There are two major steelmaking processes in use today, as shown in fig. 2.1, and although they differ widely in method of operation they are identical in principle. In both cases, the excess carbon and other unwanted impurities are oxidised out.

THE BESSEMER PROCESS

This method of steelmaking was invented by Sir Henry Bessemer and does not use a furnace, in the accepted sense, but a *converter* of the type shown in fig. 2.3. It does not use any source of heat other than that obtained from the oxidising process used to remove the carbon.

The converter is filled with molten pig iron, the blast turned on and the whole converter is tilted to the upright position. The air blast, in passing through the molten steel, oxidises away the carbon and other impurities, leaving almost

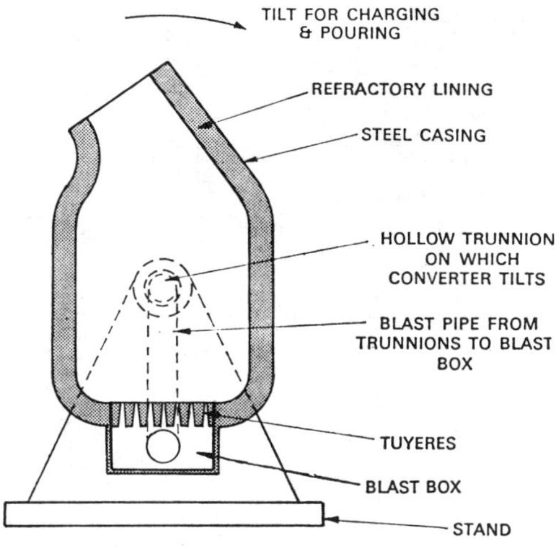

Fig. 2.3. The Bessemer converter

pure iron. It now remains to deoxidise the metal and add the precise amount of carbon necessary to produce the steel required.

For many years Bessemer steel was considered inferior to other types of steel for two reasons:

1. the rapidity of the process was such that the only control available was the furnace master's judgement of the flame;
2. the use of air for the blast caused metallic compounds of nitrogen to be formed, which reduced the quality of the steel.

Rapid methods of automatic analysis have largely overcome the first problem, while modern Bessemer converters are blown with a mixture of oxygen and steam to prevent the formation of harmful nitrides.

The great advantages of the process are its speed and flexibility. The converter can pour about 4000 kg of steel each hour and although it must be preheated it does not have to be fired in the accepted sense.

THE OPEN-HEARTH PROCESS

In this process the carbon is again oxidised out of the pig iron but the oxidising agent is iron oxide in the form of iron ore and of mill scale, the surface scale which flakes off the ingots in steel rolling mills.

The charge consists of pig iron, iron ore and mill scale, the heat necessary for the process being obtained in the conventional way by burning fuel. The furnaces are commonly gas fired though many oil-fired furnaces are now in operation. Normally a gas/air combustion process will not generate enough heat to melt the iron and to overcome this difficulty the furnace is made *regenerative.* As the hot gases leave the furnace they pass through large chambers which contain chequered stacks of firebrick. The firebrick becomes hot and the flow of gas is then reversed, the new unburnt gases entering through the hot chambers and thus becoming preheated. The furnace valves shown in fig. 2.4 are controlled to give optimum conditions and the gases enter the furnace at up to 1200°C. The energy of combustion, added to the internal energy which the gas and air already possess through preheating, is ample to melt the iron.

As the charge is melted, the carbon and other impurities are oxidised out by reacting with the iron ore and mill scale.

There are four major differences between this process and the Bessemer process:

1. *Size:* furnace capacities of up to 200 tonnes are not uncommon;
2. *Speed:* the process may take anything up to fourteen hours, which allows time for analysis and control during the process;
3. *Method of oxidisation of impurities:* see above;
4. *Source of heat:* see above.

In many steelworks there are large batteries of open-hearth furnaces, backed up by Bessemer converters which may be brought into action at short notice to increase the manufacturing capacity.

THE 'BASIC' STEELMAKING PROCESS

Phosphorus is the steelmaker's enemy, for a content of more than 0.02% of it has disastrous effects upon the mechanical properties of the steel. The processes already described do not remove the phosphorus, and for many years large sources of iron ore could not be exploited because of their high phosphorus content. The *basic process,* which overcame this defect, was

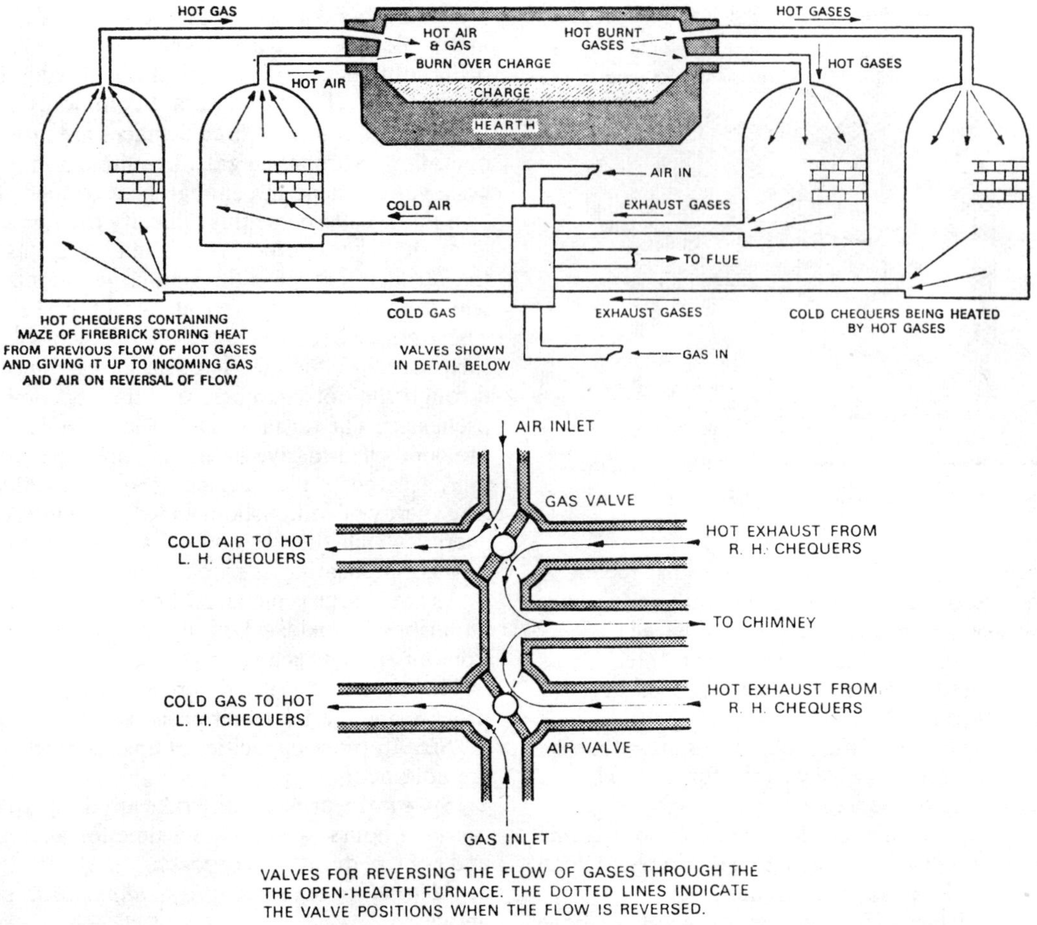

Fig. 2.4. Diagram of open-hearth furnace

developed specifically to use these hitherto uptapped sources of ore. In both the Bessemer and open-hearth *basic* processes, a special furnace lining is used and, before the deoxidation of the steel, large amounts of limestone are added to the charge, combining with the phosphorous to form a basic slag, which is later used as an agricultural fertiliser.

THE MANUFACTURE OF CAST IRON

THE FOUNDRY CUPOLA

Cast iron is pig iron which has been refined by remelting in a foundry cupola and cast in a mould to the approximate shape and size required for the finished article.

The cupola is simply a miniature blast furnace, as shown in fig. 2.5, with certain modifications:

1. The charge consists of pig iron
   coke
   limestone
   scrap cast iron

accompanied by the air blast;

2. The process is not continuous, and only sufficient iron is produced to meet the immediate need;
3. To allow for complete emptying, the cupola

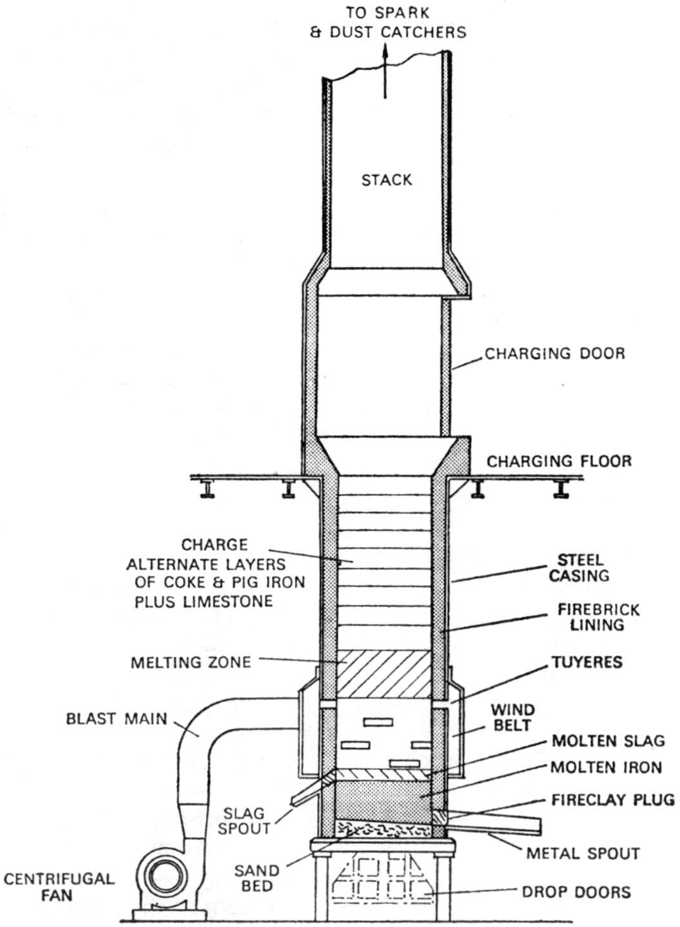

Fig. 2.5. Diagram of foundry cupola

is normally mounted on a stand and has a 'drop bottom'.

The cupola is first tapped to remove the slag and the cast iron is then tapped and stored in a ladle until poured into the mould. The process of moulding will be dealt with from p. 48.

HIGH QUALITY STEELS

High-grade steels and alloy steels are now usually made in electric furnaces, heated by either an electric-arc or an induction-heating process. The charge used is not pig iron, as in the other processes, but carefully selected scrap. This is further refined by the remelting and the addition of the required alloying elements. Both types of furnace are sealed and, burning no fuel, require no air. Thus there is no contamination of the charge through atmospheric attack.

Each of the furnaces has its own advantages; electric-arc furnaces can be made 'basic' and will then reduce the phosphorus content in the original charge. This is not the case in the high-frequency induction furnace, but the method of heating has the effect of stirring the charge, giving good mixing, which is beneficial for alloy steels.

From these steels are made the press tools,

cutting tools and precise measuring equipment which are used in the manufacturing processes we shall discuss.

PRIMARY PROCESSES: CASTING, FORGING and ROLLING

Specific Objective: *The student should be able to describe with the aid of diagrams the principles of sand casting, forging, rolling, drawing and extrusion.*

THE MANUFACTURE OF CASTINGS

One of several methods may be used to make a casting, depending upon the metal to be cast, the type and size of job, the number of parts required, and the speed at which they are to be produced. All the methods used, however, are common in that a cavity or *mould* is made, approximately of the size and shape of the required part, and is filled with molten metal which is then allowed to solidify. The mould is then opened and the casting removed. Methods of casting commonly used are:

1. *Sand casting* (including plate moulding)
2. *Shell moulding*
3. *Investment casting*
4. *Die casting*
5. *Centrifugal casting.*

In this work only the *sand casting* process will be considered.

Before manufacture starts, the designer of the casting must bear in mind certain points of importance in the production of a good sound casting, and incorporate them in his design. Amongst these are:

1. *Ease of manufacture*
In most cases, the pattern for the casting must be made so that it can be removed from the mould *without damaging the mould.* As shown in fig. 2.6, the designer can assist by:
(a) avoiding undercuts
(b) providing adequate 'draft' angles to allow the pattern or casting to free itself readily from the mould or die, in the same way that a pudding frees itself from a basin. The shape of a pudding basin is quite deliberate, to provide 'draft'.

2. *Designing to avoid cracks*
Cracks in casting usually start at sharp corners, due to stress concentrations on cooling, or at changes in thickness, due to the uneven cooling rates of thick and thin sections. The part shown in fig. 2.7(a) is badly designed and its faults are corrected in fig. 2.7(b).

Sand casting
A sand casting is made by producing a cavity in moulding sand by ramming damp sand around a pattern of the shape required. The pattern is removed and the cavity filled with molten metal.

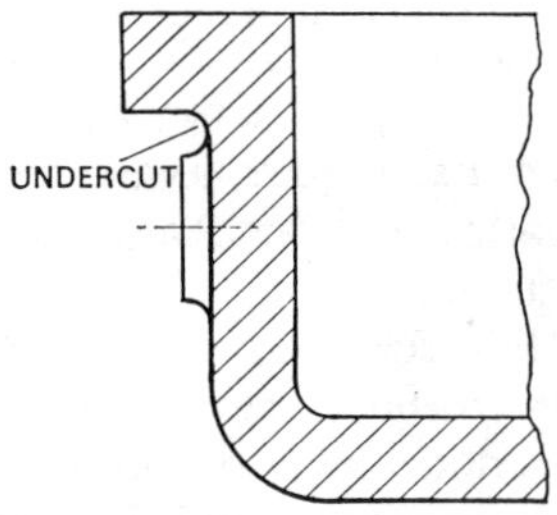

(a) CASTING WITHOUT DRAFT AND WITH AN UNDERCUT

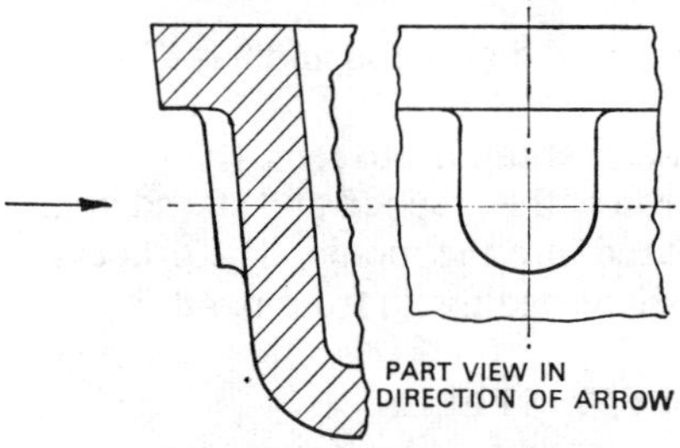

(b) DRAFT ADDED AND UNDERCUT ELIMINATED BY 'WASHING BOSS INTO FLANGE

Fig. 2.6. Casting modifications to ease withdrawal of pattern from mould

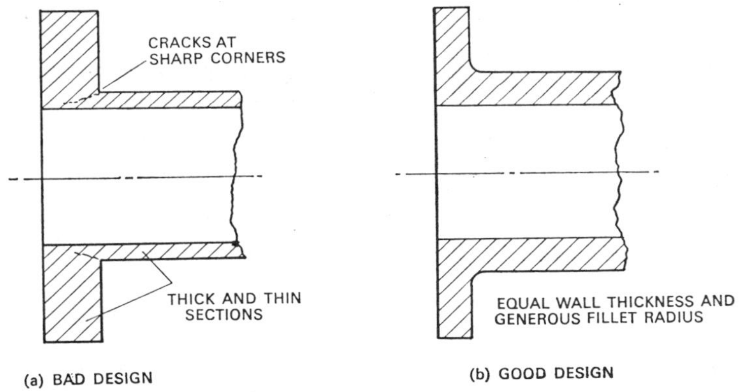

Fig. 2.7. Modifications in casting to reduce likelihood of cracks.

Any holes in the casting are made by inserting *cores* in the mould, and provision must be made for feeding the metal into the mould and allowing the air in it to escape. The stages in manufacturing a simple flanged pipe, as shown in fig. 2.8, by sand casting are as follows.

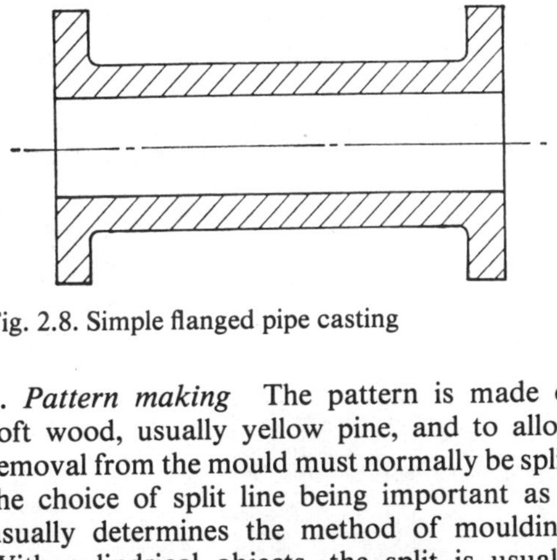

Fig. 2.8. Simple flanged pipe casting

1. *Pattern making* The pattern is made of soft wood, usually yellow pine, and to allow removal from the mould must normally be split, the choice of split line being important as it usually determines the method of moulding. With cylindrical objects, the split is usually axial along a diameter and the two halves of the pattern are dowelled so that they locate accurately together.

The patternmaker must also provide the draft and fillet radii, and add to the pattern any necessary *core prints*. These make cavities in the mould to locate and fix the *cores* which produce the holes required in the casting. Cores are made separately by ramming a special sand in a *core box*, which is usually in two halves and has a cavity the shape of the core required. On removal, the core is baked in an oven and becomes a stiff rod of sand, ready for insertion in the mould. When making the pattern for the casting, the patternmaker also makes any necessary core boxes.

As the casting cools it contracts, and the pattern must be made oversize to allow for the contraction. To save calculations the patternmaker uses a *contraction rule* which has four scales, each nominally of 300 mm but whose length and graduations are increased by an amount depending on the contraction allowance for the metal being cast. These allowances are:

cast iron, 10.67 mm/m.
brass, 16 mm/m.
aluminium, 21.3 mm/m.
steel 16 mm/m.

Thus a 300 mm scale allowing a contraction of 21.3 mm/m for aluminium would actually be 306.39 mm long, divided into 30 equal parts for centimetres with the cm subdivided equally into millimetres. Each subdivision would therefore allow for the contraction of the casting during cooling.

The pattern for the part shown in fig. 2.8 would therefore appear as shown in fig. 2.9(a),

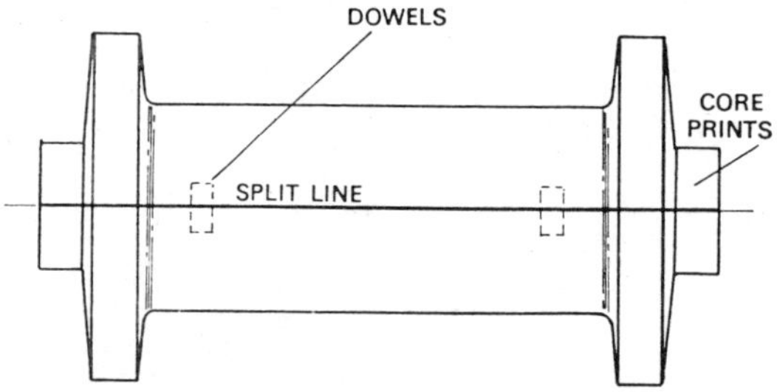

Fig. 2.9(a). Pattern for flanged pipe

with the core box as shown in fig. 2.9(b). When completed, patterns are usually painted and varnished to preserve a smooth surface.

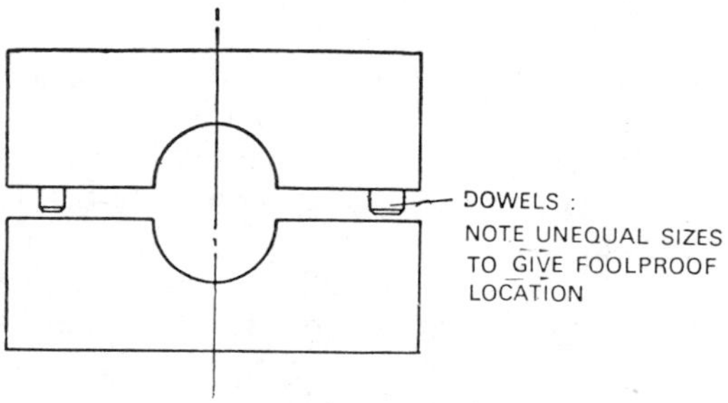

Fig. 2.9(b). End view of simple core box for flanged pipe casting. The halves are shown separated

2. *Moulding* With the pattern and core box made, the mould and core can now be produced and assembled ready for pouring. The mould is made in a two-part box, split along the same line as the pattern, for simple castings. The stages in the moulding process are clearly defined and the names of the materials and equipment used are given as the stages are described, below.

(a) *Lower half-mould* Half of the pattern is placed face downwards on a *turnover board* and the lower half of the box, or *drag,* is placed around it. The pattern is packed around with *facing sand,* a fine sand to give a good finish, and the remainder of the cavity is rammed up with *green sand.* This is not green in colour but is so called because it is damp and contains clay to bond the grains together. At this stage the half-mould is as shown in fig. 2.10.

(b) *Top half-mould* The lower half-mould is inverted and the turnover board removed. The upper part of the pattern is dowelled into place on its lower part and the top half of the moulding box, or *cope,* is fitted to the drag. The cope and drag are 'registered' by dowels, as shown in fig. 2.11. The sand face is sprinkled with *parting sand* to prevent the two halves of the mould from sticking to each other. Facing sand and green sand are again filled in and rammed up, but this time provision must be made for pouring the molten metal into the mould and allowing the hot air and gases to come out. *A runner* must be made as an inlet and *risers* formed wherever air will be

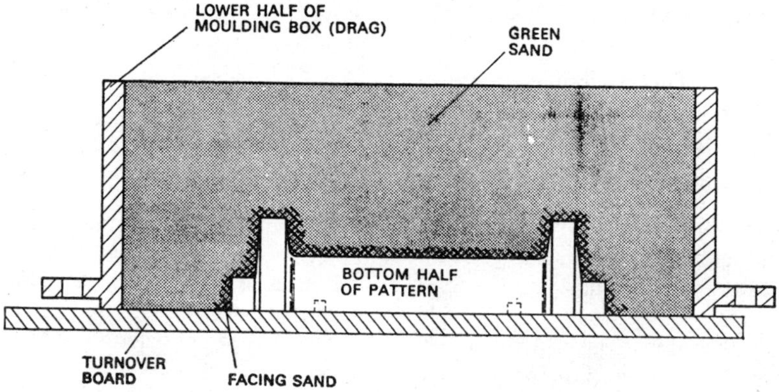

Fig. 2.10. First stage in making the mould

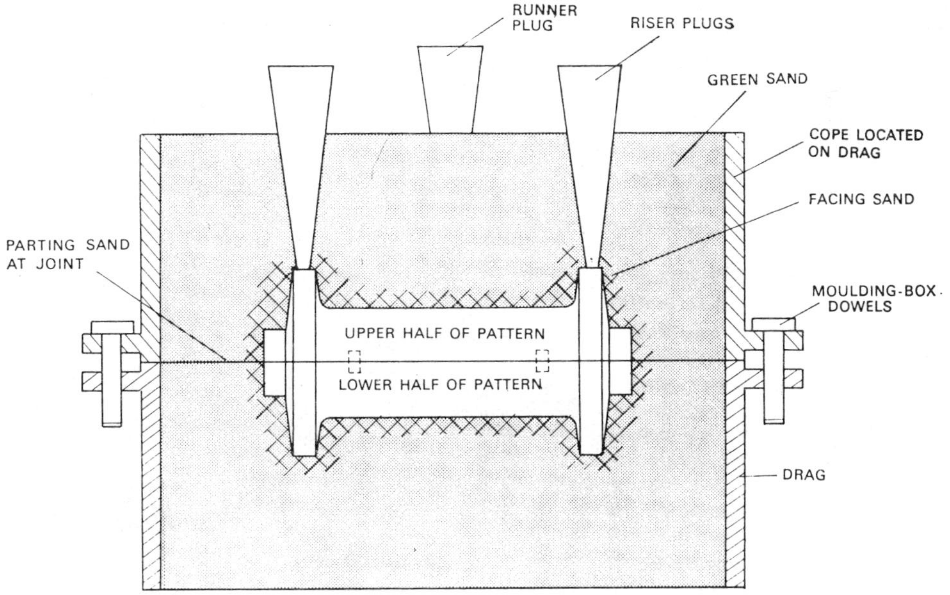

Fig. 2.11. Second stage in making the mould

entrapped, in this case on top of the two flanges. The runner and risers are formed by placing tapered wooden plugs in position and ramming the sand around them. To provide an easy run-in, the metal will be poured in the side of the mould cavity about halfway along, so that it does not drop directly on to and break the core yet spreads easily through the mould. Sections through the mould at this stage are shown in fig. 2.11.

The core will have already been prepared in the core box from a special bonded sand and baked hard to give it the necessary strength.

(c) *Finishing the mould* The cope is gently lifted off and turned over to reveal the pattern halves enbedded in the sand. A large wood screw or spike is driven into each half of the pattern and tapped gently in all directions. This *rapping* slightly enlarges the mould and allows the pattern halves to be easily and cleanly removed. An *ingate* is cut at the bottom of the runner in the lower half to provide a clean run-in for the metal to the mould, and a *pouring cup* is formed about the runner plug in the top half. Similar cups known as *headers* are formed about the risers and the plugs are removed. Wires are poked through the mould to provide small vents so that gases generated by the action of the hot metal on the damp sand can escape. The mould is then finished off, loose sand is removed, any damage made good and the cavity is painted or dusted with plumbago or black lead to give a good finish to the casting. The core is placed carefully in position and the two halves of the mould are assembled, as shown in fig. 2.12, the cope and drag being held together by cotters through the moulding-box dowels. To prevent the top half of the mould 'floating' when the molten metal is poured in, weights are usually placed on it before pouring begins.

(d) *Pouring* The molten metal is poured into the runner as rapidly as possible, without overflowing the pouring cup, until it appears in the riser headers, indicating that the mould is full. Apart from this, the headers maintain a

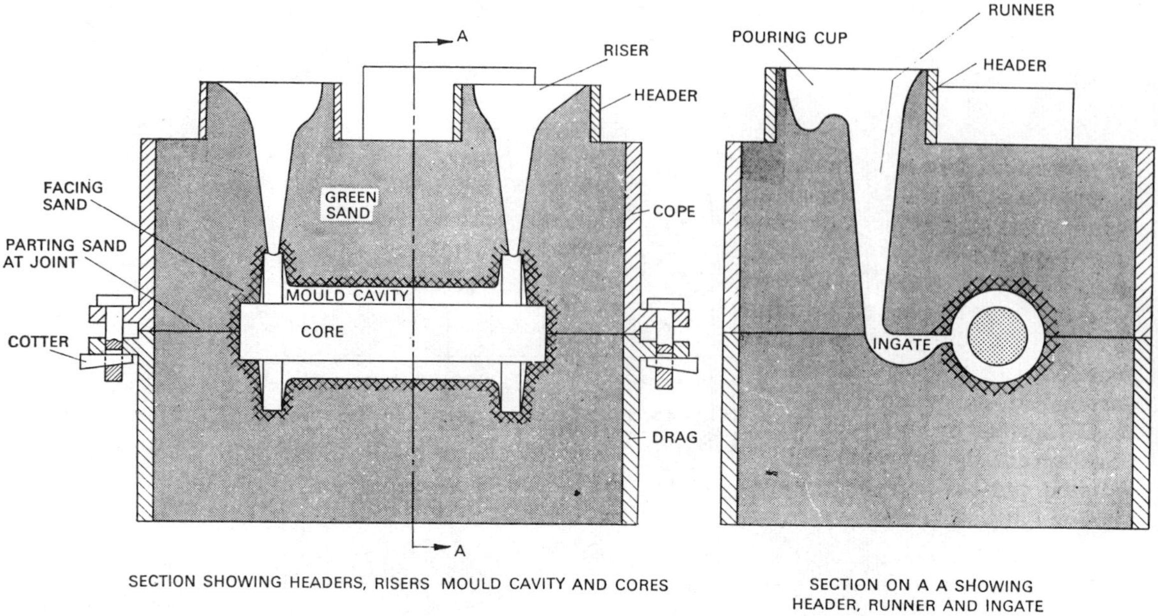

Fig. 2.12. Finished mould ready for pouring

pressure, or head, of molten metal ensuring that the mould remains full. Since any bad metal, slag, bubbles of gas and displaced sand tend to float to the surface, they float into the risers, ensuring good metal in the mould and a sound casting.

(e) *Finishing the casting* When the casting has cooled, the boxes are vibrated to break up the sand and the rough casting is removed, looking rather different from the finished product. The *sprues* or sticks of metal formed by the runner and risers are broken off. The casting is pickled in a sulphuric acid solution to remove the scale, and then washed to remove the acid. Further dressing, or *fettling,* may be carried out by grinding, chipping and sand blasting until the casting is fully cleaned up. It is not yet ready for final machining, however, since it may warp and distort over a period of time as the internal stresses set up during cooling are dissipated. Castings are often stored in the open for this period and the treatment is known as *weathering*. When it is necessary for a casting to remain stable it is often first roughly machined and weathered in the open for a few months, to remove and release internal stresses, and is then finally machined to size. Weathering can be accelerated by heat treatment and by vibration, and small castings may be *tumbled* or *barrelled,* or shot blasted. In the barrelling process, the castings are placed in barrels or drums with small pieces of scrap, and the drums are rotated for about half an hour. The vibrations and light blows to which the castings are then subjected relieve the internal stresses and prevent subsequent distortion.

The above briefly outlines the process of sand casting. Points to note are:

(i) the mould is used only once and if more than one part is required a new mould is necessary for each;

(ii) sand casting is essentially a process for small numbers of parts but can be used for all sizes of castings;

(iii) it can be used for all materials that readily cast;

(iv) it does not give castings accurate to

close limits, and even such functional details as hexagonal flats to fit spanners cannot be cast accurately enough to obviate machining.

Plate moulding

This is a modification of the sand casting process used for the mass-production of small parts. The two halves of a metal pattern are set opposite each other on the two sides of a metal plate which is placed between the boxes which both are rammed together. Thus when the plate is removed its thickness, separating the pattern halves, is 'lost' and the two mould halves come together as shown in fig. 2.13.

It is not generally worthwhile making a plate pattern for a single casting and the plate usually carries patterns for a number of castings. These patterns are connected by raised sections which form channels to allow the molten metal to flow to all the moulds. Thus a simple plate pattern for moulding our flanged pipes four at a time would appear as in fig. 2.14. The plate-moulding process is therefore a sand-casting process about which the following points should be noted:

(i) the mould is used only once;
(ii) the process lends itself to mechanisation, i.e. to machine ramming, and is essentially a production process;

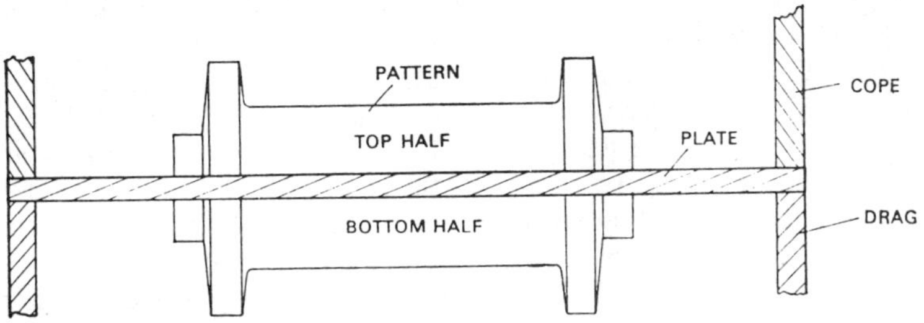

Fig. 2.13. Plate pattern for flanged pipe. When the plate pattern is removed and the cope and drag are brought together again, the thickness of the plate is 'lost'

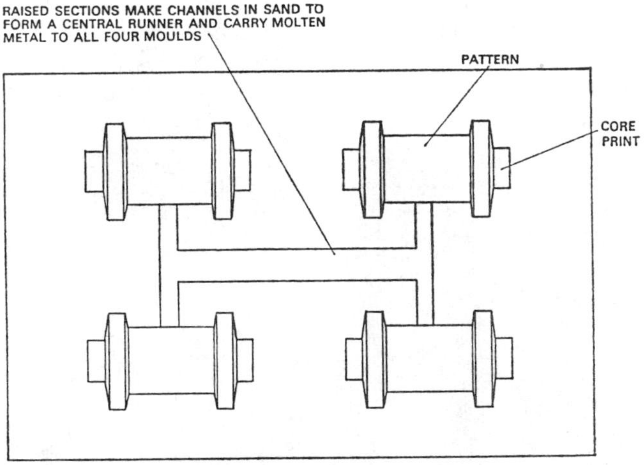

Fig. 2.14. Plan view of plate pattern for four flanged pipes

(iii) the pattern plates are more expensive than single patterns, but their cost is more than offset by the quantity production of castings.

THE FORGING PROCESS

Forging and casting are complementary to each other as primary processes, but whereas castings are made by pouring molten metal into a mould, a forging is made by heating the metal until it is plastic and hammering, pressing, or bending it into shape.

Forgings may be made by hand by a blacksmith, often assisted by his mate, or by various types of power-operated machinery. Quantity production of small and medium-sized forgings is usually performed by *drop hammers* or *forging machines,* in both of which the hot metal is hammered or squeezed between metal dies in which impressions have been cut to produce the shape of forging required. Larger forgings are generally shaped by *power hammers* or *hydraulic presses,* some of which can handle forgings of very large size, e.g. big gun barrels and ships' mainshafts. Smaller power hammers fitted with dies are sometimes used in place of drop hammers. A recent development is that of high-energy-rate forming, where the tup, or hammer, strikes a billet in the die at very high speed, and the energy thus generated is used to produce a good quality forging in one stroke.

Hand forging

Some people think that a blacksmith heats a piece of metal and hammers it into intricate shapes simply by using a hammer and his anvil. This is a misconception, for the blacksmith uses a variety of tools, each having its own particular purpose, as does any other skilled man. The most-used tools are undoubtedly the hammer and anvil, but frequently the anvil is used as a mount for pairs of tools between which the work is forged by hammer blows. Many forging operations require two men, the smith and his mate or *striker.* The smith is the skilled man, holding the work in tongs and positioning the upper tool on it. By a system of signals, and an understanding built up during years of working together, he instructs the striker where, when, and with what force the blows are to be struck.

The smith and the striker use a range of hammers, of different weights and shapes to suit the operation in hand. The smith's own hand hammer is usually about 4 kg in mass of head, while the striker generally uses heavier hammers, called *sledge* hammers, from 8 kg to 14 kg, or more, mass of head.

Before considering the various operations performed by the smith, let us examine the anvil and the use of its various parts (see fig. 2.15). The main body of the anvil is made of mild steel, with a hardened *top face* welded on. The *beak* is soft and, with an increasing diameter of cross-section, is useful for producing bends of

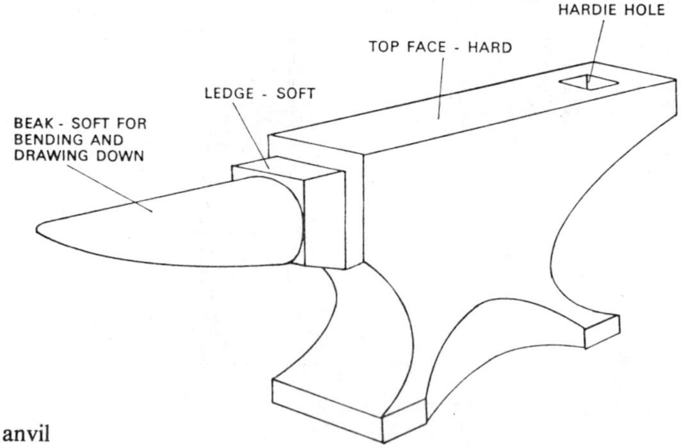

Fig. 2.15. A smith's anvil

different radii. The *ledge* between the beak and the anvil face is soft and can be used as a base for cutting operations with *hot chisels*, i.e. chisels shaped to cut hot metal. Cutting should not be done on the anvil face since this might spoil the face whose hardness might also damage the edge of the chisel. Near the end of the anvil face is a square hole, or *hardie hole,* used for holding the shank of the lower member of a pair of tools.

A large anvil may have a mass up to 250 kg and its face is positioned at a comfortable working height, usually about 0.7 m. This is achieved by mounting it on a large solid block of wood or a heavy metal stand whose shock-absorbent qualities help to spread the shock loads and reduce damage to the floor.

Hand forging operations

Most of the forgings produced by a blacksmith are the result of a sequence of fairly standard operations. These and the tools used are considered in turn below.

1. *Cutting* Cutting is normally done by chisels. These may be *hot chisels* for cutting hot metal or *cold chisels* for cutting cold metal. Both are normally fitted with a long handle, the chisel being held on the work and struck with a hammer.

The cold chisel is similar to a fitter's chisel, except that it is longer and has a handle. Its edge is hardened and tempered, and its point angle is 60°.

The hot chisel has a point angle of 30° and is soft. To harden and temper it would be pointless, since its contact with hot metal would soften it.

Chisels are often used in pairs, the lower tool then being known as a *hardie,* which has a square shank to fit the hardie hole. The work is positioned between the hardie and the top chisel by the smith, as shown in fig. 2.16, and the top chisel is struck by the striker.

2. *Upsetting* This process consists of increasing the cross-section of the material

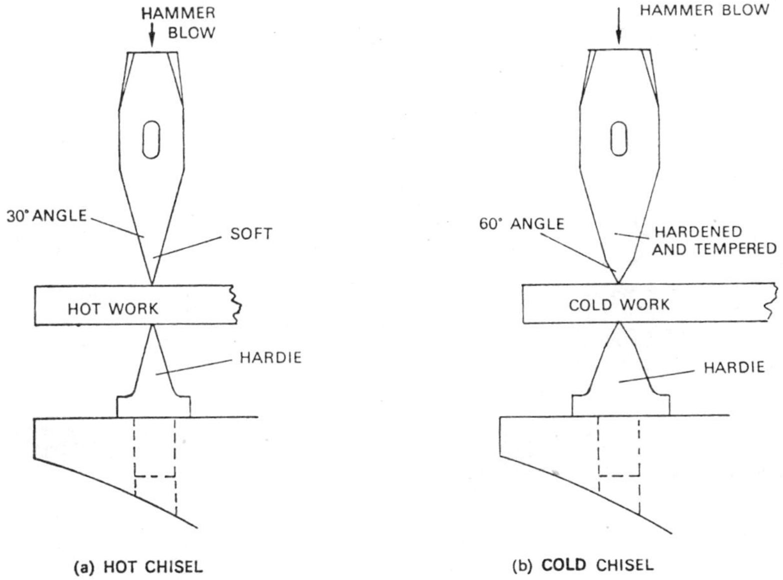

Fig. 2.16. Use of chisels in forging

locally at the expense of its length. Only the part to be upset is heated to forging temperature, and the bar, or work, is then struck at the end, usually between the hammer and the anvil, as shown in fig. 2.17.

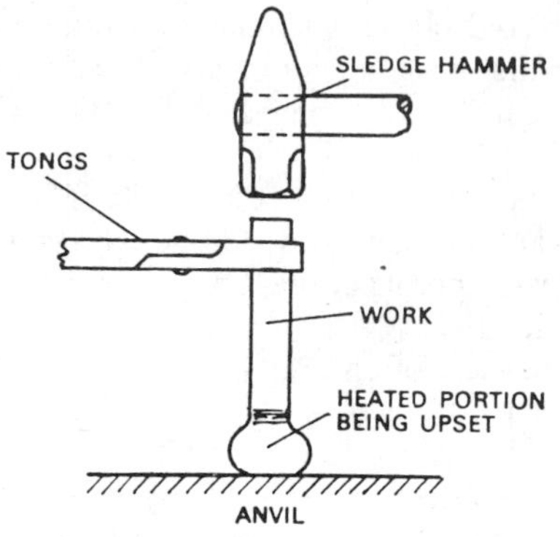

Fig. 2.17. Upsetting

3. *Drawing-down* This is the opposite of upsetting, i.e. the section of the bar is reduced locally, with an increase in length. The point where drawing-down is to start is usually first notched, or necked, by *fullers,* and drawing-down is continued from that point either between a pair of fullers or against the corner edge of the anvil. This leaves an undulating surface which is usually finished with a *flatter* or *sethammer,* as shown in fig. 2.18.

4. *Swaging* Swaging is the process of finishing a round or hexagonal section of bar between a pair of swages of the appropriate size. These may be separate tools, top and bottom, or held together by a long spring handle, as shown in fig. 2.19. In place of the bottom swage, the smith may use a *swage block.* This is a large square block of cast iron mounted on a heavy stand; around its four sides are semicircular, half-hexagon and vee-shaped grooves, of increasing sizes, which correspond to bottom swages and other bottom tools. In the face of the swage block is a series of holes of various shapes which are convenient for holding bars during bending operations.

5. *Bending* Bends may be made to a sharp angle or to a radius. Sharp bends can be made in a vice, or by hammering the work over the anvil edge, but in either case the metal should be thickened by upsetting on the *outside* of the bend before the bend is started. This compensates for the distortion which takes place at the corner, the metal being compressed

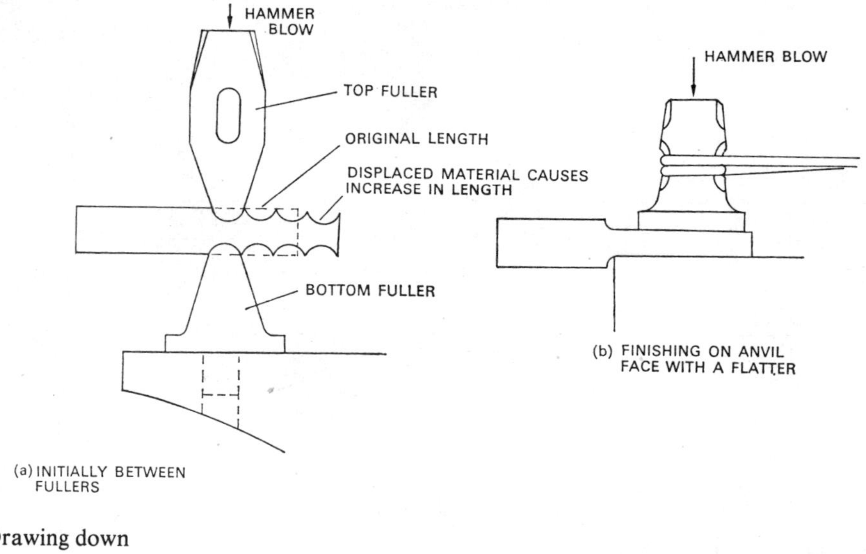

Fig. 2.18. Drawing down

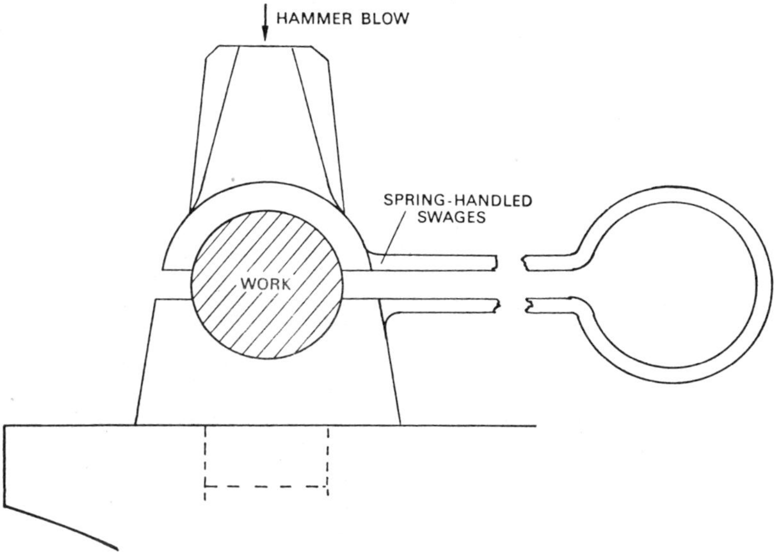

Fig. 2.19. Swaging

on the inside and stretched on the outside. The effect can be shown by bending a piece of indiarubber and observing the bend produced. The metal bulges wider at the inside and is stretched further at the outside, with local thinning.

Radiused bends can be made over the beak of the anvil, around a bar in a vice or round a mandrel in a bending fixture. There is again a change in cross-section but the tendency is less severe, and the change in width is corrected after bending.

6. *Piercing* The smith often produces a hole in the work by piercing the metal with a *punch,* over the hardie hole or over a *die,* which is a block of metal with a hole through it corresponding in size to the punch. Use of a punch and die gives a more accurate hole than using the punch alone. Holes other than round can be produced by first making a round hole and then driving through a tool called a *drift,* which tapers from round at the point to the size and shape required at the shank. Alternatively, a round hole can be filed or machined to a different shape when the workpiece has cooled.

Machine forging operations

Hand forging processes are suitable for relatively small work but when larger pieces of metal have to be forged the energy available from a hand-operated sledge hammer is not enough. Consequently, power hammers are used, driven by water power, compressed air, steam or friction. In most of these the sledge hammer is replaced by a heavy power-operated upper anvil called the *tup*, which is lifted up and then falls, or is driven, down on to the lower anvil which is built into the base of the hammer. When the tup falls under its own weight, its potential energy at the top of the stroke is changed to kinetic energy as it gains speed. When impact occurs, this kinetic energy is released and the work done on the hot metal is equal to it.

Similar pairs of tools to those used in hand forging can be used and a similar range of operations carried out.

Drop forging This is a machine forging process used to produce large numbers of similar parts. A pair of dies with appropriate impressions are fitted to the tup and anvil and accurately aligned by using a dummy component to set them in position. The hot blank to be forged is often pre-formed to a suitable size and shape before it is placed in the lower die. The top die, attached to the tup, then drops on to the blank and the energy of its fall squeezes the metal until it fills the impressions of the dies, any excess forming a flash, or fin, which is clipped off later. This process is used for a wide diversity of parts whose sizes vary from rocker arms for internal-combustion engines to wheels for railway trucks and carriages. The dies for a single-sided crank are shown in fig. 2.20.

It should be noted that in this process, as in casting, allowance must be made for shrinkage, and for extracting the forging from the dies. Draft and a suitable split line are necessary, since there must be no undercuts. Examination of a drop forging will usually reveal the split line of the dies, where the fin has been trimmed off, and the draft angle will show in which direction the draft has been made.

A modern development of this process is *high-energy-rate forming* in which the two halves of the die are impacted together by the explosive release of compressed air, or, in some cases, by an actual explosion. The parts are often produced by one blow, a typical machine giving 17 000 J per blow.

A high degree of accuracy and very good quality of forging are achieved, and a collet

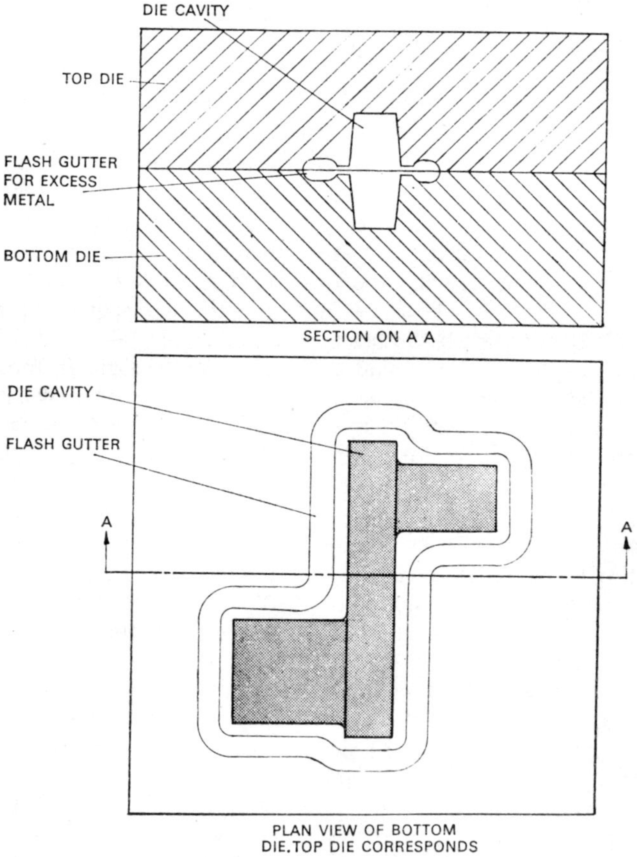

Fig. 2.20. Dies for drop-forging a single-sided crank

blank produced by this method is shown in fig. 2.21.

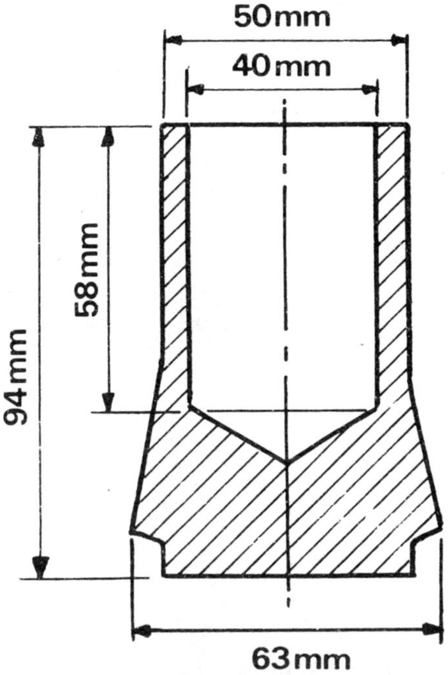

Fig. 2.21. Collet blank produced from 0.60% carbon steel in one blow by a high-energy process. Billet size: 49.3 mm dia. × 50.8 mm long. Concentricity held within 0.08 mm

THE ROLLING PROCESS

Hot rolling

Molten steel from the Bessemer or open-hearth process is normally cast into *ingots* of roughly square section and weighing up to 10 tonnes, while most of the material required by the engineer, other than castings or forgings, is of much smaller section. The ingots are usually reduced to the size and section required by passing between pairs of cylindrical rolls under high pressure in a *rolling mill* to reduce the thickness and width, with increase in length. This process takes place in stages, the ingot first being heated to 1200°C in a soaking pit, and then reduced to *blooms* or *billets* which are cut to length before reduction to the section finally required.

The rolling process may require up to twelve sets of rolls of different sizes and shapes, each set approaching nearer to the section wanted, and the billet is passed through each set in turn. If the rolling is not continuous the various reductions are made by different-sized grooves along the length of roll. The bar, or stock, moves forward on the first pass, back on the second, forward on the third and so on. The final rolls for a channel section are shown in fig. 2.22.

If steel plate or strip is required the process is similar, except that the blooms first pass through a *slabbing mill* to produce slabs, which are then reduced between successive parallel rolls to the thickness required. For steel strip, used by the mill is often employed, the slab passing through a succession of rolls each of which produces a reduction in section and an *increase in length*. The speed of each set of rolls must be carefully controlled in relation to the one preceding it to maintain the correct tension in the strip. At the final stage the strip, moving at speeds of up to 100 km/h, is wound into coils ready for despatch to the user.

Cold rolling

This process is basically similar to the hot rolling process except that the material is rolled cold, i.e. at room temperature. As in hot rolling, the section of the material is reduced and the length correspondingly increased. The material is initially hot rolled near to size and pickled in acid to remove the scale. It is then cold rolled between rolls whose surface has a high finish, to produce a corresponding finish on the rolled metal. The pressures are, of course, very much higher since the cold metal cannot be 'moved' so easily, and the reduction per pass is less, but a much greater degree of control can be exercised and the thickness can be kept accurate to less than 0.03 mm.

COLD DRAWING OR WIRE DRAWING

Much of the bar and rod used in engineering is known as *bright-drawn mild steel,* or B.D.M.S. The metal is first hot rolled to produce *black* bar which is pickled to remove the scale, and then

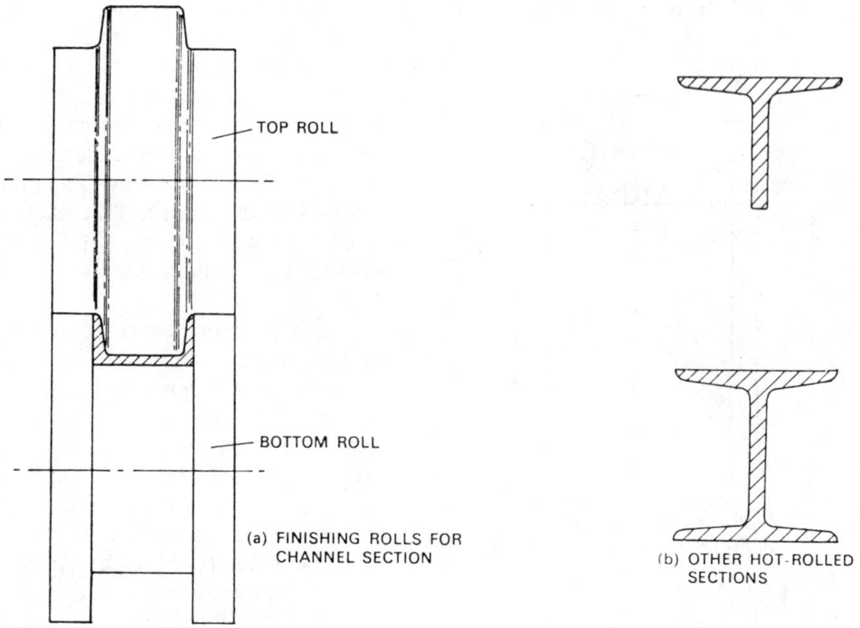

Fig. 2.22. Hot rolling

pulled cold through dies so that the section is reduced, and the length increased.

The die holes are conical in shape and must, of course, be very hard to maintain their size and finish against wear through the high pressures involved. Small dies for thin wires are made from diamond or tungsten carbide, larger sizes from hardened tool steel and the very large sizes from chilled cast iron.

With cold drawing, as with the rolling processes, all the materials produced have a 'grain', i.e. the grains of metal are elongated in the direction of drawing or rolling and tend to increase the strength of the metal in this direction. Thus if a piece of bright mild steel is bent across the grain it will normally withstand the bending stresses, but if the bend is along the bar, in the same direction as the grain, the bar may crack unless it has first been normalised.

THE EXTRUSION PROCESS

Just as toothpaste is squeezed from a tube and comes out with the same shape as the hole, metal may be forced through a die under pressure and will emerge with the same shape as the die. A much higher pressure is required than for toothpaste, of course, and the process is normally limited to brass, copper, and the aluminium alloys, all of which have a fairly low tensile strength. Extremely complex sections, solid or hollow, may be extruded, a typical section being shown with a diagram of the process in fig. 2.23. These sections can be extremely thin and of insufficient strength to withstand wire drawing, yet the same high degree of accuracy can be achieved.

Impact extrusion

If a metal which is extremely plastic when cold is struck by a punch moving at very high speed, the material 'splashes' back up the punch and produces a tube. This process is called impact extrusion and is used for the manufacture of toothpaste tubes, electrical-condenser cans, ignition-coil cans for motor cars, cigar tubes, etc. All of these are made from aluminium

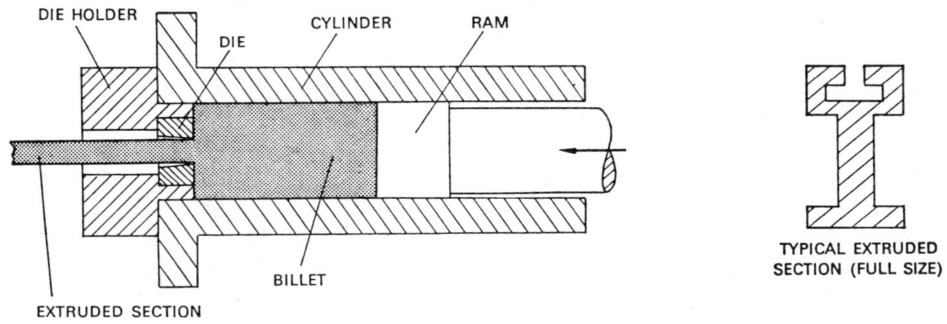

Fig. 2.23. Simple forward extrusion and a typical extruded section

supplied in the form of slugs, the tube shown in fig. 2.24 being produced from the slug by a single blow.

THE PROPERTIES OF CASTINGS AND FORGINGS

Specific Objective: *The student should be able to describe some special characteristics, advantages and limitations of parts produced by the primary forming processes.*

For ease of comparison the characteristics of these processes will be discussed under the same headings for each. These will be: cost of equipment used; speed of process; accuracy and finish produced; materials capable of undergoing the process; special characteristics, advantages and limitations of the process.

Sand casting

Small castings of low melting point material can be produced with low cost equipment. If large castings in quantity are required the equipment

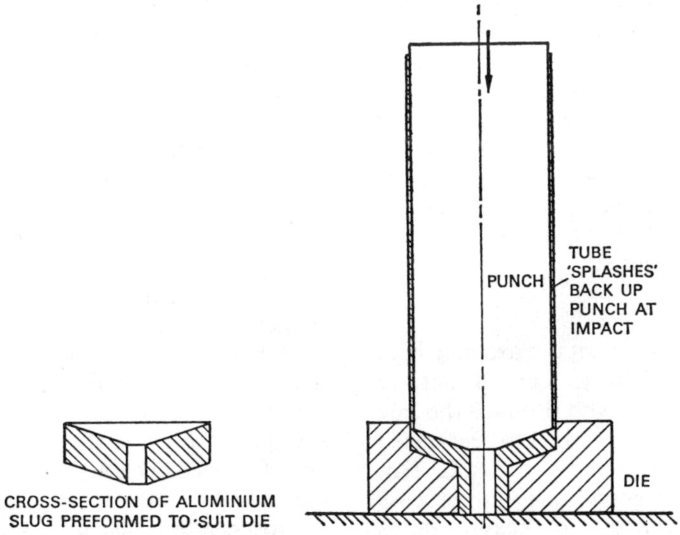

Fig. 2.24. Impact extrusion. A toothpaste tube is produced by a single blow on a slug of aluminium

cost is high and a foundry is usually only set up where high utilisation of the equipment can justify it. For this reason even large quantities of castings are bought from firms specialising in this type of work. Even with machine moulding equipment the process is slow and finish and accuracy is poor (±0.4 mm). The materials used should be of high fluidity such as cast iron and brasses, bronzes and aluminium alloys suitable for casting. Steel castings can be made, but are more difficult and expensive. Castings are used for complex shapes and parts where the cost of the casting is more than offset by saving in machining costs. Iron castings are brittle and should not be subject to shock loads or tensile and bending loads.

Forging
The cost of hand forging equipment is low but large forgings and small ones made in large quantities require specialist equipment, and are usually purchased from specialist manufacturers. Hand forging is a slow process but a single hand forging can be produced much more rapidly than a casting which would require the making of a pattern and mould. Machine forgings or hot stampings are produced very rapidly. The accuracy is better than casting but still of the order of ±0.2 mm. The materials required must be plastic when hot and steel and certain brasses, bronzes and aluminium alloys can be forged. In general the shapes produced by forging cannot be as complex as castings but the process gives a fine grain structure product having greater strength and an ability to withstand shock loads not usually found in castings.

Rolling
The cost of rolling equipment is extremely high and this process is always carried out by specialists, usually those who produce the raw material. Rolling is a high speed process used for large quantities of similar material. Hot rolled stock is not of good finish, the process normally being confined to heavy gauge materials such as girders and rolled steel joists. Cold rolling is used for sheet production in ductile materials and the order of accuracy is high, the thickness of thin sheet being held to ±0.01 mm or better. The raw material must be ductile but, depending on the amount of cold work carried out and the material being used it can finish with a spring temper. Materials produced by this process include sheet steel, brass, copper, phosphor bronze and aluminium alloys. Cold rolled sheet has directional qualities – see Chapter 3.

Drawing
Most of the foregoing equally applies to drawing. This process is used to manufacture rod of various cross sections and seamless tube, a typical example of the latter being the copper pipe used in domestic heating installations. The accuracy of bright drawn rod is of the order of ±0.02 mm, the materials used being ductile such as low carbon steel, copper, brass and aluminium. Very fine wire is drawn through a series of dies to a high degree of accuracy. Again this process involves the use of expensive equipment and is invariably carried out by specialists in the field.

Extrusion
Of all the primary forming processes extrusion is the one that could be carried out by a non-specialist manufacture. It is a single stage process which, unlike rolling and wire drawing, does not involve a series of operations all interconnected and requiring close control between the stages. Impact extrusion particularly could be carried out in general engineering companies if the demand for the product was high enough. However, to the authors knowledge, it is so specialised a process that extrusion, either as a continuous process or impact extrusion, is carried out by specialist manufacturers having the equipment and expertise. The process is carried out at high speed in both cases and the material used must be soft, malleable and ductile. Aluminium and the ductile brasses are used for extrusion purposes as are plastics, the plastic insulation used on electric cable often being extruded over the wire. The main advantage of the process

compared with rolling is that it can be used to produce sections with undercuts as shown in fig. 2.23. However, as the billet in the die gets shorter and the length of the extruded part increases, the force required increases and complete extrusion is not possible; there will always be a short length of billet wasted which must be cut off and scrapped before reloading the cylinder.

SAFETY NOTE

Specific Objective: The student should be able to name the main hazards in Primary Process work, and the necessary relevant precautions.

Most of the work described in this chapter refers to hot metal; it may be molten or at forging temperature. In either case, it can maim, or even kill unless proper precautions are taken.

A drop of molten metal can burn through normal clothing. ALWAYS wear proper safety boots and overalls – leather aprons are usually provided. A large spill of molten metal can maim either the operator or a passer-by whatever clothes are worn. Always keep out from under ladles, and handle them with the proper equipment – never improvise.

The glare from molten iron and steel can damage the eyes. If necessary wear *darkened* safety glasses. In general, primary process areas are eye hazard areas. Always wear safety glasses to protect the eyes from sparks and flying scale.

Molten metal coming into contact with moisture can generate steam at an explosive rate. Foundry sand used in moulding should be of the correct moisture content and protected to ensure that it remains correct.

Much of the equipment used in extrusion and forging is pneumatic or hydraulic in operation, working at high pressure. It must be carefully and regularly maintained and checked. An oil line bursting at the pressures involved can be deadly.

In rolling mills, the metal is not only hot, but is moving at high speed. It must always be handled correctly, using the right equipment. Again, improvisation is dangerous.

The work carried out in these areas can be dangerous if it is not treated with respect and common-sense. Horseplay, distractions from work, and especially, treating the material with contempt can be fatal. Above all, if a hazard is seen, take action or see that action is taken. This is now the legal duty of all, employers and employees alike.

QUESTIONS

1. List the essential properties of a material for it to be suitable for manufacturing by each of the following processes.
   (a) Casting
   (b) Hot forging
   (c) Cold stamping
   (d) Extrusion

For each process name one material which is suitable and one which is not suitable for working by that process.

2. Name the furnace used for the manufacture of pig iron and list the materials entering the furnace and those leaving it. The pig iron produced may be used for two purposes – what are they?

3. Name the furnace used for the manufacture of cast iron. List the materials entering the furnace and those leaving it. Explain how the furnace, the materials entering the furnace, and those obtained from it differ from those considered in question 2.

4. Explain two ways in which a casting designer can make it easy for a pattern to be removed from a mould and two details in casting design which tend to reduce the likelihood of cracks.

5. The manifold shown in fig. 2.25 is to be made by sand casting. Sketch:
   (a) the pattern and
   (b) the core.

On the sketch of the pattern shown any modifications which may be made to ease the moulding process.

6. Explain with the aid of diagrams the process of producing a mould for the part shown in fig. 2.25. Your sketch should show clearly the

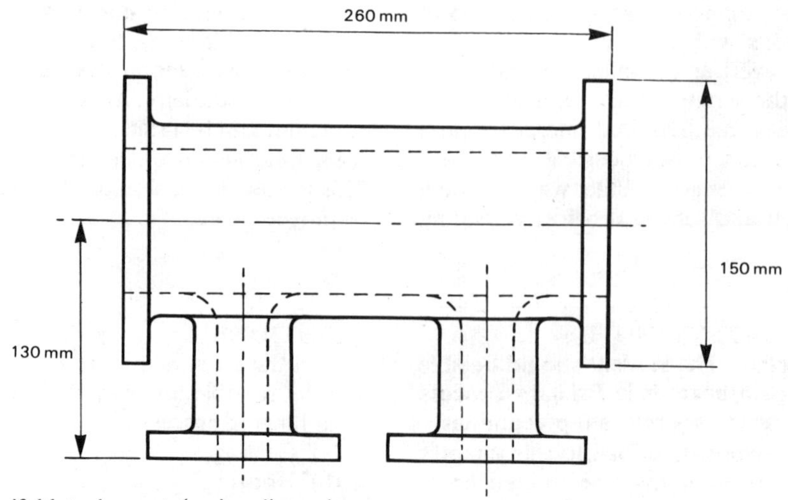

Fig. 2.25. Manifold to be cast (major dimensions only are given to indicate size)

position of the runner(s), riser(s) and the core(s) in position.

7. If large numbers of the part shown in fig. 2.25 are required how would the process be changed? Name the process likely to be used and sketch the pattern.

8. State the difference between the casting and forging processes and the differences in the mechanical properties of parts produced by the two processes. Assuming that the parts in the following list could be produced either by casting in iron or forging in steel state which method of manufacture you would chose for each giving reasons:

(a) carpenter's claw-hammer
(b) motor-cycle cylinder block
(c) motor-cycle brake pedal lever
(d) ring spanner
(e) gear type coolant pump body
(f) lathe handwheel.

9. State the purpose of the following tools used in hand forging:

(a) cold chisels
(b) hot chisels
(c) fullers
(d) flatters
(e) swages
(f) hardie.

10. Name the processes used to manufacture the following items stating in each case whether the process is carried out hot or cold:

(a) steel joists used in the construction industry;
(b) toothpaste tubes;
(c) reinforcing rods used in reinforced concrete;
(d) aluminium curtain rails;
(e) sheet steel for motor bodies.

3 Press work

General Objective: *The student recognises the principles of simple presswork.*

INTRODUCTION

The types of press considered in this chapter will be the simple types used in the manipulation of sheet metal. The student may be familiar with the mandrel press which is used, as its name implies, for the pressing of mandrels in and out of components and for some assembly work, but this type of press is not under consideration here. In this chapter will be considered the *flypress* and the simple *powerpress* which lend themselves to blanking, piercing and simple bending operations.

In both presses note that the energy is supplied fairly slowly to the press mechanism which stores it and delivers it rapidly to the metal being worked, through a suitable tool. The tool usually consists of a punch, attached to the press ram, and a die mounted in a bolster, which is bolted or clamped to the press bed. In this way the load produced by the force necessary to perform the operation is distrubuted over the press bed, which is not then locally loaded under shock conditions.

TYPES OF PRESS

Specific Objective: *The student should be able to explain with the aid of sketches the construction, advantages and limitations of the fly press and power press.*

THE FLY PRESS

The fly press is a hand operated press whose configuration is shown in fig. 3.1. It consists of an open fronted cast iron frame. Through a bush in the top of the frame passes the ram screw, a large diameter, multi-start, square-threaded screw on which is mounted the operating handle. This has a spigot at each end on which can be mounted a large cast iron ball; as the operator swings the handle to bring the ram down his energy is stored in the handle and ball, and is then given up to the work metal when the punch contacts the work in the die. The punch cannot simply be mounted in a rotating screw, and the actual ram is actuated by the screw but is constrained to move in a straight line by guides machined and attached to the press frame.

The die is mounted in a cast iron bolster which is clamped to the press bed, being positioned suitably relative to the tool when the press is set up.

For most operations the distance the punch moves into the die is limited and a stop is provided by having a fine thread cut over the square thread of the ram screw. The stop, with an internal thread to suit the fine thread, can then be positioned on the ram screw and, being in effect, a split nut, is locked in place with a pinch bolt.

The operator raises the ram, positions the work correctly on the die and swings the handle which flies round, brings the punch into contact with the work and is stopped at the appropriate position by the stop making contact with the top of the press frame.

THE POWER PRESS

There are many types of power press: single acting, double-acting, triple-acting, etc., each being designed for different types of work such as blanking and piercing, bending, drawing, reverse drawing and so on. In each case energy

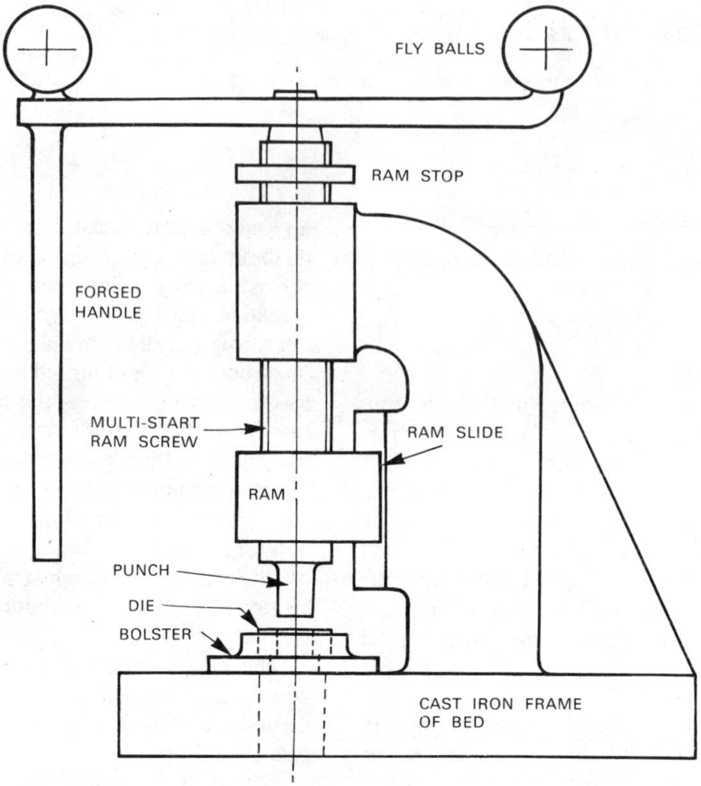

Fig. 3.1. Simple fly press with blanking tool set up. Guards and strippers are not shown

is transmitted from an electric motor, stored in a flywheel and delivered to the work by a tool through a mechanism which changes the angular motion of the flywheel to linear motion at the tool, just as the screw of the flypress changes the rotary motion of the handle and balls to linear motion at the tool. In this case, however, the mechanism is usually a connecting rod/crank linkage driving the ram.

In this work only the single acting press will be considered, a typical open frame press being shown in fig. 3.2. The flywheel is driven from an electric motor through multiple vee belts and energy is stored in it. The crank is driven from the flywheel by a clutch arranged to engage for one revolution when the operating pedal is depressed.

During the operation on the metal by the tool energy is expended and the flywheel slows down, speed being built up again and energy being re-stored before the next stroke.

SAFETY NOTE

If the energy contained in any type of press is expended on a human body the body suffers damage. Press tools must be guarded and if this is not possible during setting, the machine must be switched off while work goes on in the working zone of the machine.

Fig. 3.2. 450 kN power press. The rating (450 kN) is the force the press can be safely allowed to exert upon the work. (Reproduced by kind permission of the Cressex Machinery Co. Ltd.)

ADVANTAGES AND LIMITATIONS OF FLY AND POWER PRESSES

Basically the difference between fly and power presses is simply one of power available. However it is transmitted, the only power available in a flypress is that put into it by a human operator. In the case of the power press he only has to press a pedal and the power available is that of the motor driving the press.

The power press is much more rapid in action and with automatic feed of work to the tool the pedal can be locked down and speeds of 120 strokes/min are not uncommon. However, flypresses are much cheaper and lend themselves to the use of very simple tools.

Thus in general the power press is more expensive to buy, set up and run but is quicker and lends itself to high production rates. The flypress is cheaper to buy, set up and run but can only be used for slow running jobs at low production rates.

USE OF BOLSTERS

Specific Objective: *The student should know the reasons for the use of bolsters in press tools.*

Bolsters have already been mentioned as spreading the shock load of a press operation over the bed of a press. There are, however, other reasons for using them. Because of the nature of their work press tools must be hard and able to withstand shock loads. This calls for an expensive tool steel so the tool is made as small as is possible to do the job. The die is then mounted in a bolster of cheaper material from which it can be easily removed for maintenance and grinding. Further, if a number of dies are made to a standard design the same bolster can be used, thus reducing tooling costs.

BLANKING AND PIERCING OPERATIONS

Specific Objective: *The student should be able to distinguish between blanking and piercing.*

The most common work carried out on presses

Factor	Power-press	Fly-press
Force available	High	Low
Cost	High	Low
Setting-cost	High	Low
Tooling-cost	High	Low
Speed	Fast	Slow
Run-length	Long	Short

is blanking and piercing. These terms must be defined as to the casual onlooker the operations look the same. In the simplest terms:

Blanking produces a blank of metal for use – the sheet from which it was produced is scrap.

Piercing produces a hole of given size – the slug removed is scrap. During both blanking and piercing operations the punch passes into the die and it is necessary that there is clearance between them. This is called die clearance and the clearance *all round* is given by:

$$\text{Die clearance} = 0.1t$$

where t = thickness of metal being blanked (or pierced).

If one observes a blanking (or piercing) operation closely it is seen that the blank sticks in the die and the hole grips the punch. Thus in blanking, where the piece pressed out is required the *die is made to size and the clearance removed from the punch.* In piercing, where the hole is to be to size the *punch is made to size and the clearance added to the die.*

Consider a pair of punches and dies for producing a washer 50 mm outside diameter × 25 mm inside diameter in material 1 mm thick.

(a) For the outside diameter a *blanking* punch and die are required.

Blanking die diameter = 50 mm
Die clearance = 0.1 × 1 mm *all round*
Blanking punch diameter = 50 mm – 0.2 mm = 49.8 mm

(b) For the hole a *piercing* tool is required.

Punch diameter = 25 mm
Die clearance = 0.1 mm *all round*
Die diameter = 25.2 mm

Drawings of these tools are shown in fig. 3.3.

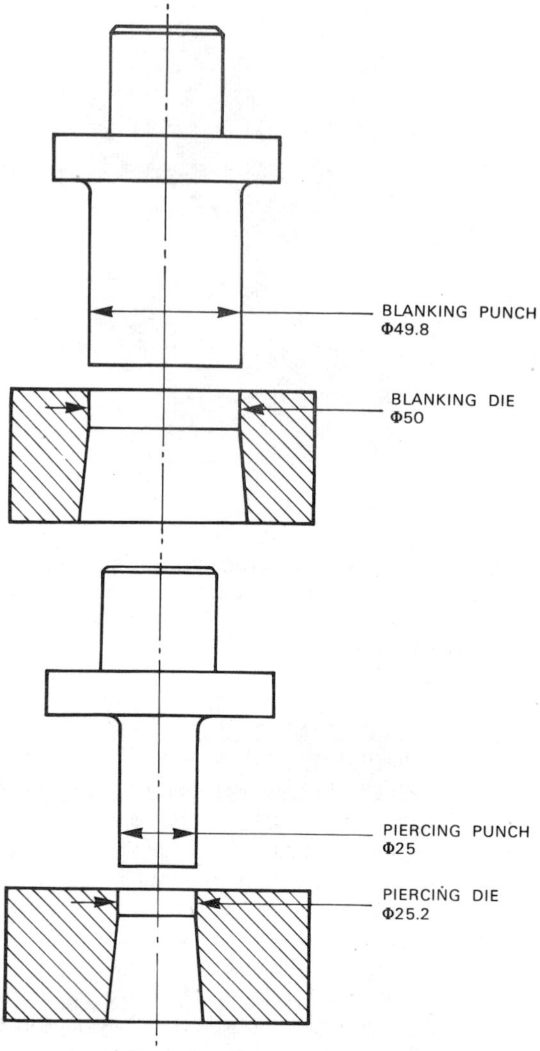

Fig. 3.3. Blanking and piercing tools. Note the application of clearance. In each case the shank and die outside diameters are the same to enable standard ram fittings and bolsters to be used

BLANKING PRESSURES

Specific Objective: *The student should be able to calculate the force required for a blanking operation.*

Again considering the washer of 50 mm outside diameter in mild steel of 1 mm thickness. In order to decide the size of press necessary to perform the operation it is necessary to be able to calculate the force required to press the blank out of the strip stock. It should be noted that this is a *shearing* operation as shown in fig. 3.4

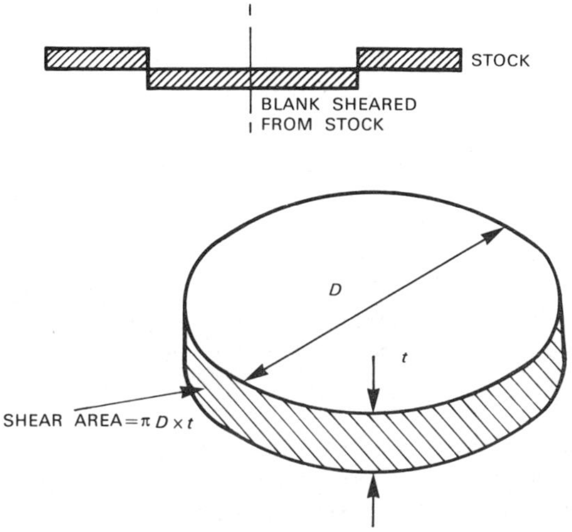

Fig. 3.4. Shearing effect in blanking and piercing

and it can be seen that the *shear area* is the circumference of the blank multiplied by the metal thickness.

In the case of a circular disc:

Shear area = Circumference × thickness
$= \pi D \times t$
$= \pi \times 50\text{ mm} \times 1\text{ mm}$
$= 157.1\text{ mm}^2$

If the average failing strength of the material in shear is 200 N/mm²

Blanking force = shear area × average shear strength
$= 157.1\text{ mm}^2 \times 200\text{ N/mm}^2$
$= 3\,1420\text{ N or }31.42\text{ KN}$

BLANKING OPERATIONS

Specific Objective: *The student should be able to specify and justify the order of procedure for blanking and piercing.*

If the Φ50 × Φ25 × 1 mm washer is again considered then the question must be asked:

Do we (a) produce the blank first and pierce the hole second,
or (b) pierce the hole first and blank out the complete washer second,
or (c) do them both at once.

If we consider the tooling required in each case we shall be led to a reasonable answer.

(a) *Blanking first*

Assuming, as is usual, that the washer is to be made from strip stock then the blanking tool must have the following features.

(i) A means of guiding the stock;
(ii) A means of obtaining the correct blank spacing in the stock;
(iii) A means of stripping the stock from the punch;
(iv) A means of collecting the blank.

The methods of designing these features into the tool are shown in fig. 3.5. The stock is guided over the die by means of a bridge piece which also strips the stock off the punch. Once the first hole is produced even spacing of subsequent holes is obtained by using a simple stop which the stock rides over moving forward (left to right) but which locates against the edge of the hole when the strip is pulled back. The blank simply falls through the die, bolster, and hole in the press bed into a work pan under the machine.

Having produced the blank it must now be pierced. This requires a separate tool having the following features:

(i) location for blank;

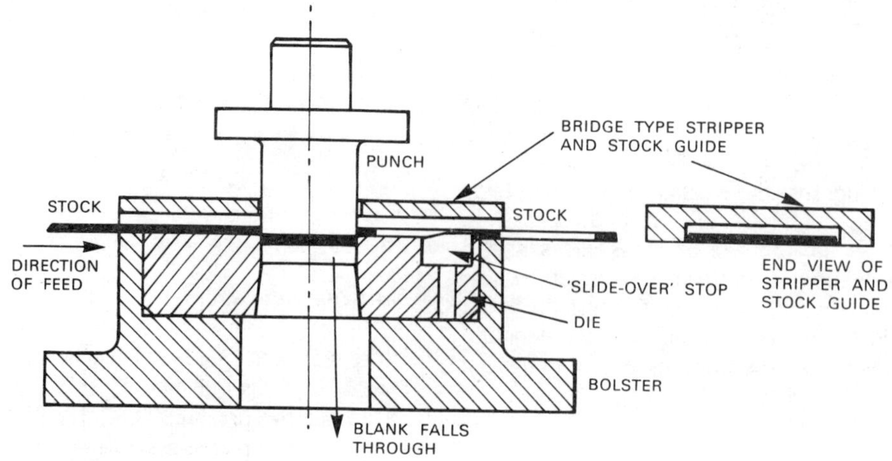

Fig. 3.5. Simple tool for blanking from strip stock showing stock guide stripper and feed stop

(ii) stripping of finished washer from punch;
(iii) disposal for centre pierced out from washer.

The features of such a tool are shown in fig. 3.6. The blank is located in a recess in the die, and in the design shown a stripper plate is fitted to the punch which is held by the stripper bolts shown and spring loaded downwards. As the punch enters the work the stripper plate is forced upwards against the springs and as the punch is raised the spring force behind the plate strips the washer off the punch and relocates it on the die. From here it must be removed by the operator before the next blank is placed in position.

Incidentally when tools of this nature have been needed quickly the author has successfully improvised strippers by cutting a hole in a block of hard rubber so that it fits on the punch. The block is a little longer than the punch and acts as a stripper. Its life is limited but it is easily replaceable.

(b) *Piercing first*

A similar tool to that shown in fig. 3.5 would be used for piercing at correct spacing along the strip and the strip then passed through another similar tool for blanking. In this case the blanking punch would have a bullet-shaped pilot to pick up the pierced hole and ensure concentricity.

Note that the pierced strips must be carefully stored and are more difficult to handle and locate than the blanks so, on balance:
For separate operations blank first.

(c) *Pierce and blank simultaneously*

The piercing and blanking operations are often mounted successively in one tool called a *follow-on-tool;* such a tool is shown in fig. 3.7. Here the piercing operation is done first and the blanking punch has a bullet-shaped pilot punch to locate in the pierced hole. A bridge-type stock guide is used which also strips the stock off the punches and, for hand fed tools a slide-over stop is used to locate the strip for each stroke of the press. With this type of tool it is also necessary to provide retractable feed stops for locating the strip when a new strip is started. The procedure is:

(i) Feed stop 1 in – feed stock to stop and pierce;
(ii) Feed stop 1 withdrawn – feed stop 2 in – feed stock to stop and blank first piece and pierce second;

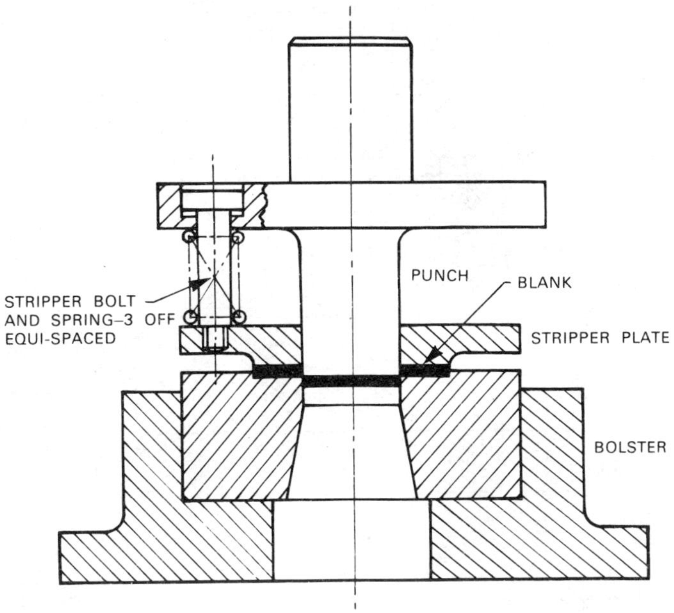

Fig. 3.6. Piercing tool for use after blanking. Note stripper arrangement and recess in die to locate blank

(iii) Feed stops both withdrawn – use slide-over stop in the normal way.

In this type of tool, accurate location for the punches and dies is obtained by mounting them in a die set or press tool pillar set as shown in fig. 3.7. These tools lend themselves to automatically fed presses where the stock is in rolls and is fed through the tool by rollers on either side of the press bed, while the punches and dies are disengaged. These are called roll-feed machines and are the subject of more specialist work than this.

In follow on tools pierce first.

BLANKING LAYOUTS

Specific Objective: *The student should be able to design blanking layouts for the most economic use of material.*

Blanking layouts are designed so the relative position of the successive operations in a strip give the most economic use of material, i.e. the maximum number of parts from the minimum amount of material. In designing a blank layout the following points must be considered.

1. Minimum allowable metal between operations; this is usually equal to the metal thickness;
2. Width of strip;
3. Subsequent operations; if a bending operation is to follow the blanking it is better if the bend is across the grain of the material, i.e. at right angles to the direction in which the strip was rolled. This is usually along the length of the strip.

Again considering our washer the only factor applying here is the strip width. If strip is assumed to be 52 mm wide the only layout is to have the blanks in line with a spacing of, say, 1

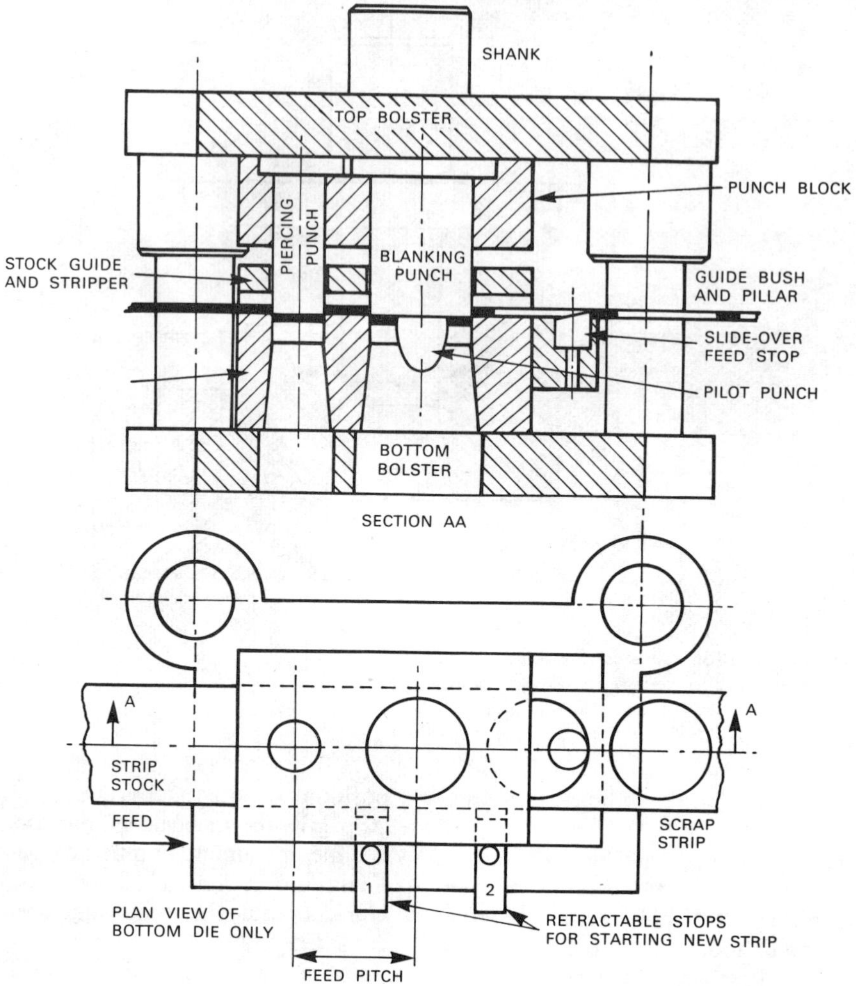

Fig. 3.7. Hand fed follow on tool for piercing and blanking washer

mm between the edges of successive holes as in fig. 3.8(a). Then:

$$\text{Area of stock/part} = 52\text{ mm} \times 51\text{ mm} = 2652\text{ mm}^2$$

$$\text{Area of part} = \frac{\pi D^2}{4} = \frac{\pi \times 50^2}{4}\text{ mm}^2 = 1964\text{ mm}^2$$

$$\text{Scrap area} = (2652 - 1964)\text{ mm}^2 = 688\text{ mm}^2$$

$$\%\text{ scrap} = \frac{688}{2652} \times 100\% = 25.94\%$$

Assume now we have stock 90 mm wide. Fig. 3.8(b) shows that two rows of blanks could be made, the holes being positioned so that the line AB is at 45° across the blank. Allowing 1

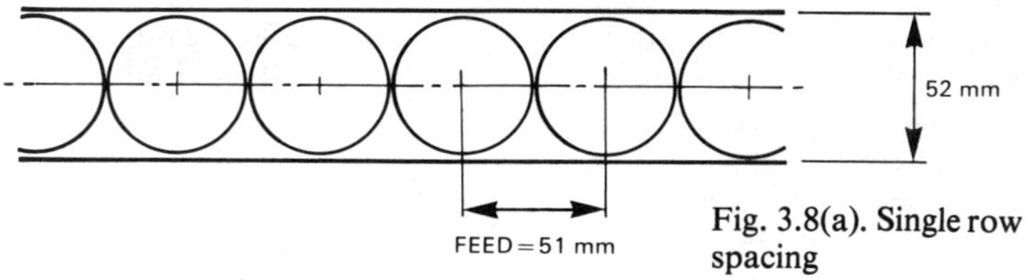

Fig. 3.8(a). Single row spacing

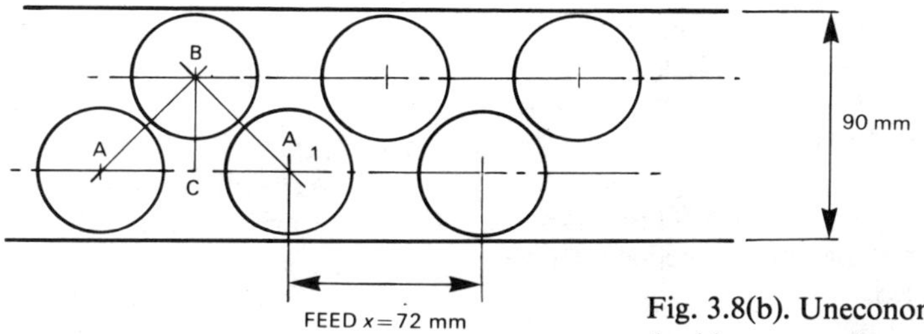

Fig. 3.8(b). Uneconomical double row spacing

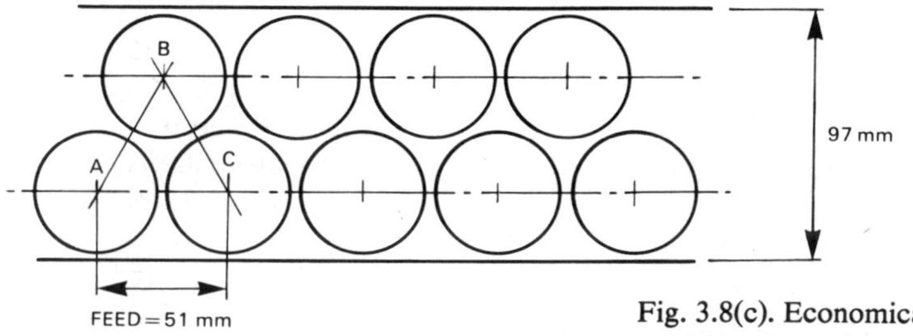

Fig. 3.8(c). Economical double row spacing

Fig. 3.8. Strip layouts for blanking 50 mm washer

mm of metal between the nearest adjoining hole.

$$AB = 51\ \text{mm}$$

$$AB^2 = AC^2 + BC^2$$

$$\text{as } AC = BC \quad AB^2 = 2BC^2$$

$$\frac{(51)^2}{2} = BC^2$$

$$BC = 36\ \text{mm}$$

$$\text{and as} \quad BC = AC = A_1C$$

$$\text{Strip feed length} = 2AC$$

$$x = 72\ \text{mm}$$

This length produces 2 blanks.

$$\text{Area of strip} = 72\ \text{mm} \times 90\ \text{mm}$$

$$= 6480\ \text{mm}^2$$

$$\text{Area of 2 blanks} = 42 \times 1964\ \text{mm}^2$$

$$= 3928\ \text{mm}^2$$

$$\text{Scrap area} = 6480 - 3928\ \text{mm}^2$$

$$= 2552\ \text{mm}^2$$

$$\%\ \text{scrap} = \frac{2552}{6480} \times 100\%$$

$$= 39.38\%\ \text{scrap.}$$

This is then much less economical than the single spaced method considered before.

Another approach is to assume that the rows of holes are to be spaced as close as is practical, as in the first case, and then to calculate the strip width to give the most economical spacing for two passes. This is shown in fig. 3.8(c). In this case three hole centres, A, B and C form an equilateral triangle at 51 mm centre distance.

$$\begin{aligned} BD &= AB \sin 60 \\ &= 51 \text{ mm} \times 0.866 \\ &= 44.166 \text{ mm} \\ \text{Strip width} &= BD + 1 \text{ blank diameter} \\ &\quad + 2 \text{ metal thicknesses} \\ &= 44.166 \text{ mm} + 50 \text{ mm} + 2 \text{ mm} \\ &= 96.166 \text{ mm} \end{aligned}$$

Assume that for safety a width of 97 mm is used with a feed of 51 mm then in one feed length two blanks are produced.

$$\begin{aligned} \text{Area of strip} &= 51 \text{ mm} \times 97 \text{ mm} \\ &= 4947 \text{ mm}^2 \\ \text{Area of 2 blanks} &= 2 \times 1964 \text{ mm}^2 \\ &= 3928 \text{ mm}^2 \\ \text{Scrap area} &= 4947 - 3928 \text{ mm}^2 \\ &= 1019 \text{ mm}^2 \\ \text{\% scrap} &= \frac{1019}{4947} \times 100\% \\ &= 20.5\% \end{aligned}$$

It is therefore more economical to double row the blanks as long as the spacing is as close as possible and this needs thinking through and calculating properly.

Note also that the strip should be turned *over* for the second pass, not end for end. Also a pair of retractable stops are required to ensure that the second row of blanks is correctly positioned relative to the first. Having produced the first hole in each row the normal slide-over type stop shown in fig. 3.5 is used to ensure correct spacing of the holes.

The strip layout for circular blanks is, of course, the simplest. With non-circular parts the shape must be considered and the blank layout chosen to give the most economic use of material. Consider the key tag shown in fig. 3.9:

In triangle ABC

$$\begin{aligned} AB &= 12 \text{ mm} - 6 \text{ mm} \\ AB &= 6 \text{ mm} \\ (BC)^2 &= (AC)^2 - (AB)^2 \\ (BC)^2 &= (15)^2 - (6)^2 \\ &= 225 - 36 \\ BC &= \sqrt{189} \\ BC &= 13.75 \text{ mm} \\ \sin \Phi &= \frac{AB}{AC} \\ &= \frac{6 \text{ mm}}{15 \text{ mm}} \\ &= 0.4 \\ \Phi &= 23°35' \text{ and } 2\Phi = 47°10' \end{aligned}$$

Area of part

= Area of sector $ADOD_1A$

+ area of sector ECE_1O_1E

+ 2 (area of rectangle BCED)

+ 2 (area of triangle ABC)

Area of sector $ADOD_1A$

$$= \pi \times (12)^2 \times \frac{(180° + 47°10')}{360°}$$

$$= \pi \times 144 \times \frac{227.166°}{360°}$$

Area of sector $ADOD_1A$

$= 285.46 \text{ mm}^2$

Area of sector ECE_1O_1E

$$= \pi \times (6)^2 \times \frac{(180° - 47°10')}{360°}$$

$$= \pi \times 36 \times \frac{132.83}{360°}$$

Area of sector ECE_1O_1E

$= 41.73 \text{ mm}^2$

As triangle ABC is right angled AB may be considered the base and BC the vertical height. Then:

$$\begin{aligned} \text{Area of triangle ABC} &= \frac{AB \times BC}{2} \\ &= \frac{6 \times 13.75}{2} \\ \text{Area of triangle ABC} &= 41.25 \text{ mm}^2 \\ \text{Area of rectangle BCED} &= BD \times BC \\ &= 6 \times 13.75 \\ \text{Area of rectangle BCED} &= 82.5 \text{ mm}^2 \end{aligned}$$

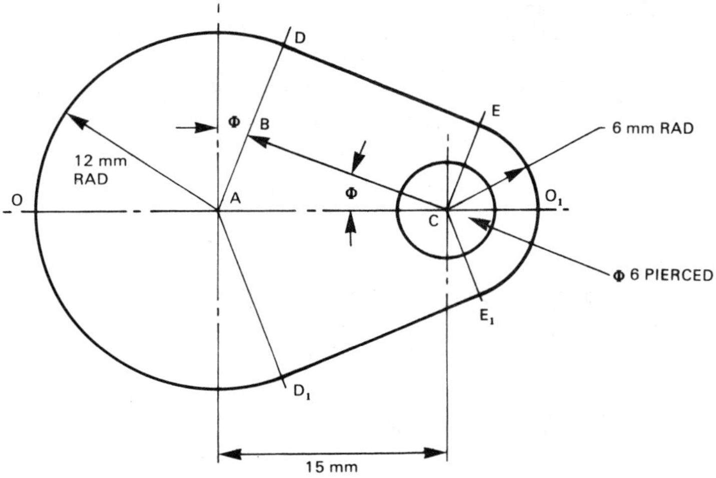

Fig. 3.9. Key tag to be produced by piercing and blanking from strip

Total area of key tag

$$= 285.46 + 41.73 + 2(41.25 + 82.5)$$
$$= 574.69\ \text{mm}^2$$

Having determined the area of the part it is now possible to calculate the percentage scrap for various strip layouts.

In fig. 3.10(a),

$$\text{Area of strip/part} = 34\ \text{mm} \times 26\ \text{mm}$$
$$= 884\ \text{mm}^2$$
$$\text{Scrap area/part} = (884 - 574.69)\ \text{mm}^2$$
$$= 309.31\ \text{mm}^2$$
$$\%\ \text{scrap} = \frac{309.31}{884} \times 100\%$$
$$= 35\%\ \text{(rounded)}$$

In layout 3.9(b),

$$\text{Area of strip/part} = 35 \times 25\ \text{mm}$$
$$= 875\ \text{mm}^2$$
$$\%\ \text{scrap} = \frac{(875 - 574.69)}{875} \times 100\%$$
$$= 34.32\%$$

It appears that using a wider strip and a shorter feed/stroke is more economical. The student should check whether this is generally true by considering say a rectangular blank of 20 mm × 10 mm with 1 mm clearance all around.

The strip layout in fig. 3.10(b) leads logically to a consideration of a double-pass layout using the same strip width as in fig. 3.10(c). In this particular case a complicated calculation can be simplified by drawing line AB connecting the centres of the 12 mm radii in adjoining points. Allowing, as in the other cases, 1 mm between the nearest points on the blanks it is seen that:

$$AB = 25\ \text{mm}$$
$$AC = 33\ \text{mm} - 24\ \text{mm}$$
$$= 9\ \text{mm}$$
$$\tfrac{1}{2}\ \text{feed/stroke} = BC$$
$$\text{and } BC = \sqrt{(25)^2 - (9)^2}$$
$$= \sqrt{544}$$
$$\tfrac{1}{2}\ \text{feed/stroke} = 23.32\ \text{mm}$$

Rounding this up to 23.5 mm gives:

$$\text{feed/stroke} = 47\ \text{mm}$$

$$\text{Area of strip/part} = 23.5\ \text{mm} \times 35\ \text{mm}$$
$$= 822.5\ \text{mm}^2$$
$$\text{Scrap area/part} = (822.5 - 574.69)\ \text{mm}^2$$
$$= 247.81\ \text{mm}^2$$
$$\%\ \text{scrap} = \frac{247.81}{822.5} \times 100\%$$
$$= 30.13\%$$

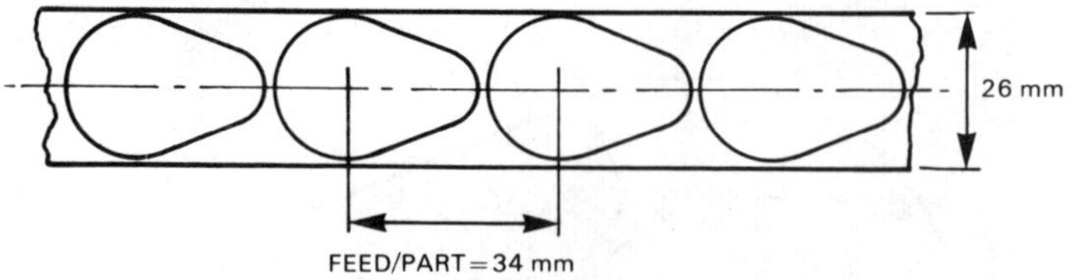

Fig. 3.10(a). Single row layout

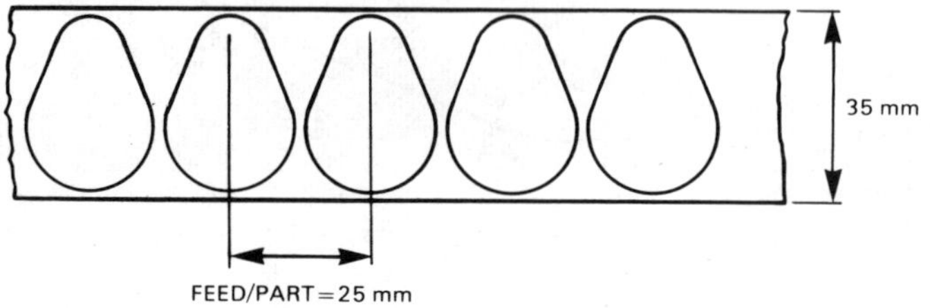

Fig. 3.10(b). Alternative single row layout

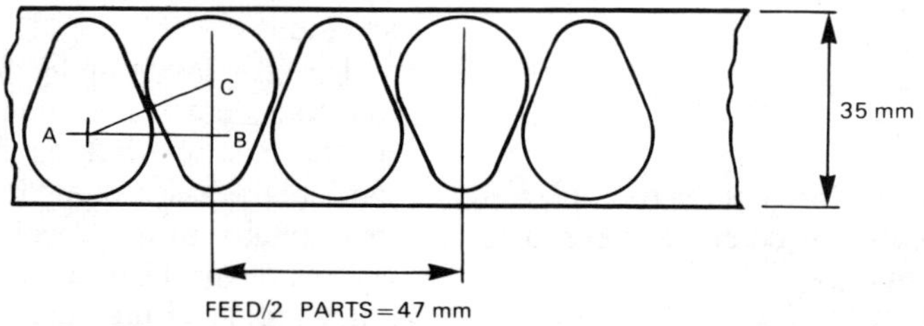

Fig. 3.10(c). Double row layout

Fig. 3.10. Various strip layouts for producing key-tag shown in fig. 3.9

As parts such as this are required in large numbers, or it would not be worth making a press tool, such attention to detail to reduce the scrap per part from 35% in case (a) to just over 30% in case (c) is well worth while.

BENDING OPERATIONS

Specific Objective: *The student should be able to calculate the raw material requirements for bending operations.*

Apart from the design of the tool for a bending operation it is first necessary for the tool designer to design the blank from which it is to be made and this involves the calculation of bend allowances, that is the amount of metal allowed for the bend.

If a simple india rubber is taken between the thumb and forefinger, bent almost double and examined while it is held bent three things become immediately apparent:

1. The rubber on the outside of the bend is stretched and in tension, its width being reduced

to compensate for the increase in length.

2. The rubber on the inside of the bend is in compression and its width increased to compensate for the reduction in length.

3. The shape of the cross-section of the rubber is therefore changed.

The same thing happens when metal is bent, a cross-section through a bend appearing as shown in fig. 3.11. As the outside is stretched and the inside compressed it follows that there

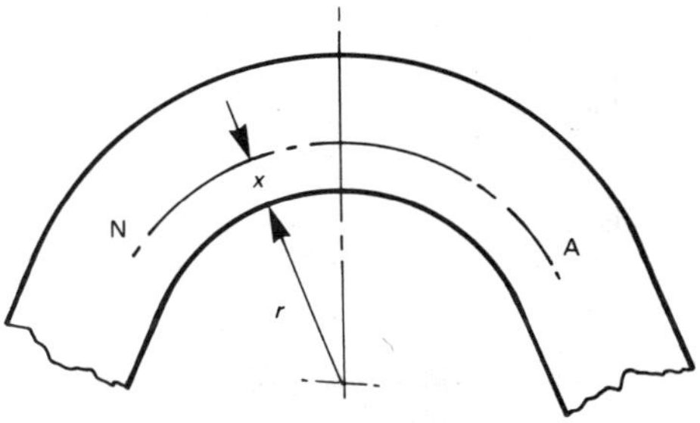

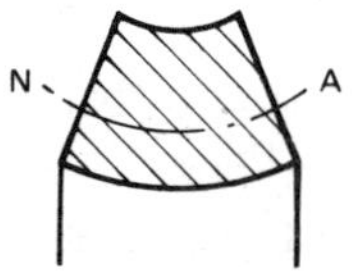

Fig. 3.11. Shift of neutral axis towards inside of bend (NA = neutral axis) [Reproduced from E. V. Crane, *Plastic Working in Presses* (John Wiley and Sons Inc.)]

is a plane in the bend, called the neutral axis, or NA in fig. 3.11, which is unstressed and whose length does not change. It is along this plane that the length of the bend must be calculated and, due to the change in section, the neutral axis is not at half metal thickness but has moved towards the *inside* of the bend.

Thus, for a bend of inside radius r in a piece of material of thickness t bent through an angle of $\theta°$ the bend allowance a will be given by

$$\text{Bend allowance, } a = 2\pi(r+x) \cdot \frac{\theta°}{360°}$$

The distance x from the inside of the bend to the neutral axis varies with the metal thickness and the bend radius. For mild steel over 1 mm in thickness a value for x of $0.4t$ is acceptable.

Thus for a component as shown in fig. 3.12, the developed length of the blank is given by:

$$L = 15\text{ mm} + a_1 + 15\text{ mm} + a_2 + 20\text{ mm} + a_3 + 12\text{ mm}$$

where a_1, a_2 and a_3 are the allowances for bends 1, 2, and 3 respectively.

$$\text{Allowance } a = 2\pi(r+x) \times \frac{\theta°}{360°}$$

In each case $r = 2$ mm

$$x = 0.4t = 0.8\text{ mm}$$

$$\therefore \quad a_1 = 2\pi(2+0.8)\frac{90°}{360°}\text{ mm}$$

$$a_1 = 4.4\text{ mm}$$

$$a_2 = a_3 = 2\pi(2+0.8)\frac{45°}{360°}\text{ mm} = 2.2\text{ mm}$$

$\therefore$ Blank length

$$L = 15 + 4.4 + 15 + 2.2 + 20 + 2.2 + 12\text{ mm} = 70.8\text{ mm}$$

In industry the most common angle of bend is 90° and many large firms produce a table showing the bend allowance for bends of different radius, in materials of various thicknesses. It must be realised that such tables are normally only for bends of 90° and basic

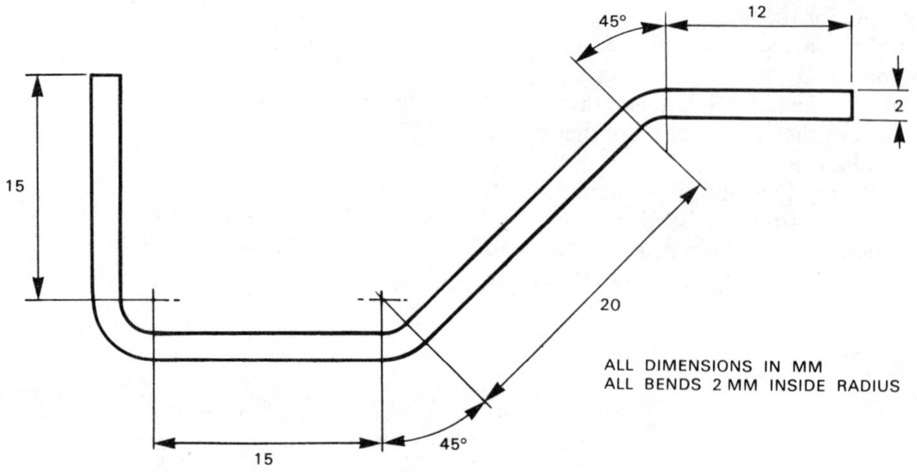

Fig. 3.12. Calculate the blank length for bending

principles as outlined above must be understood and used for other angles.

SPRINGBACK IN BENDING OPERATIONS

Metals are elastic materials and even if loaded beyond the elastic limit will return partly to their original shape when the load is removed. In a bending operation this elastic recovery is called 'springback', i.e. the metal, if bent through 90° will spring open and the bend will not be to the required angle.

There are two ways of overcoming springback; one is to overbend the metal, that is bend to a greater angle than is required so that the metal opens out to the correct angle. The amount of overbending required varies with the thickness, type and condition of the material, and the radius of the bend so that the amount necessary is largely a matter of judgement based on past experience.

The other method of overcoming springback is to hit home or bottom with the punch.

SAFETY NOTE

This requires careful setting and an avoidance of putting the next blank in the die without ensuring that the first is removed or a split die will result.

It is necessary to design the punch so that it only contacts in the bends. Too many bending tools are made to give contact over flat areas of metal which only squeezes these areas and does nothing for the bend. If the punch is cleared away as in fig. 3.13 then all the energy imparted by the press goes into the bend and is not dissipated on the flats. If now such a punch hits home on the bend it sets up a compressive stress across the metal thickness which eliminates the tensile stress in the outside fibres of the bend which is the main cause of spring back.

BENDING TOOLS

Specific Objective: *The student should be able to sketch, specify and justify the order of procedure for examples of simple bending.*

In many, indeed most cases, the sequence of operations envisaged by the untrained for bending a piece of metal to a given shape is the wrong one. Ask most people to suggest the order for bending the part shown in fig. 3.14

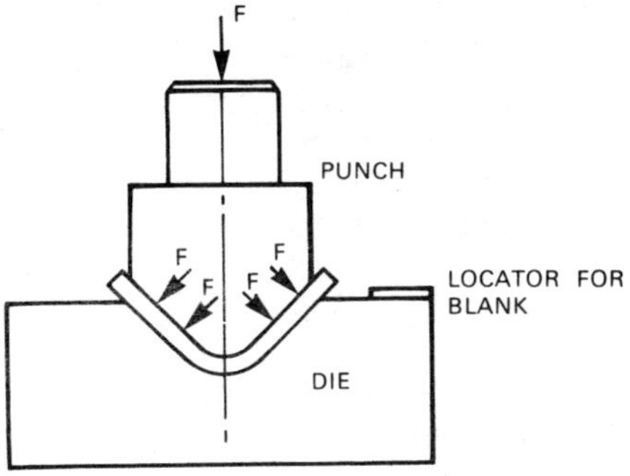

Fig. 3.13(a). Poor design of bending tool. The force 'F' is spread over the flat die faces [Reproduced from E. V. Crane, *Plastic Working in Presses* (John Wiley and Sons Inc.)]

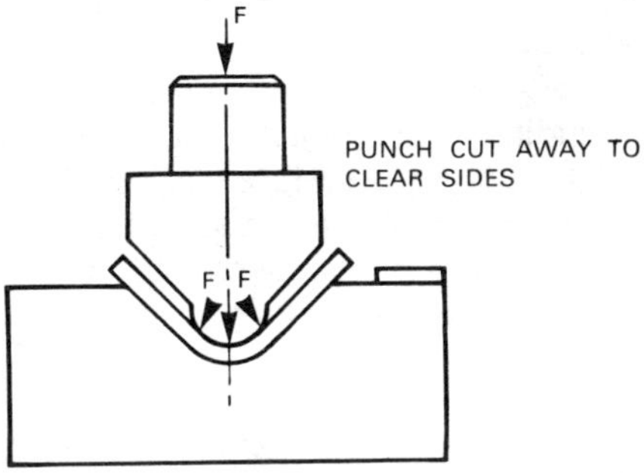

Fig. 3.13(b). Improved bending tool. The force 'F' is concentrated in the bend [Reproduced from E. V. Crane, *Plastic Working in Presses* (John Wiley and Sons Inc.)]

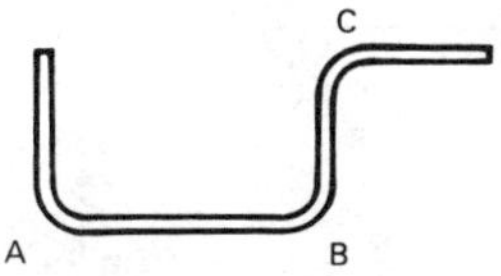

Fig. 3.14. Sheet metal bracket to be produced by bending in press-tools

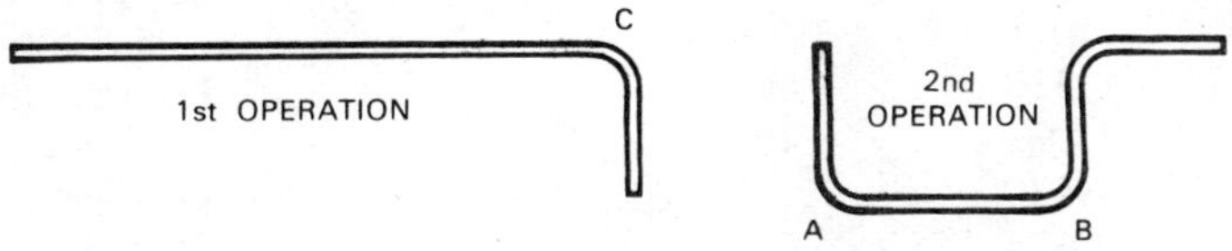

Fig. 3.15. Sequence of operations for part in fig. 3.14

they will suggest first making bends A and B simultaneously to produce a U shape and then turning the tag out by making bend C. A little thought shows that it is very difficult to make bend C in this way in a press – it can be done but it needs expensive tooling with a cam operated slide.

A far better way is to make bend C first, the sequence of operations being as shown in fig. 3.15 and the tools in fig. 3.16.

PROPERTIES OF MATERIALS FOR PRESSWORK

Specific Objective: *The student should be able to name the desirable properties of a material for blanking and forming.*

All presswork consists of some form of manipulating sheet metal which has been produced by rolling. As the final stages of rolling are usually carried out cold the material has two characteristics:

(a) It has become hardened due to cold working, and
(b) The grains of the material have become elongated in the direction of rolling giving the material directional properties, i.e. it can be manipulated better across the grain than along it.

Both of these effects can be eliminated by heat treatment of the material but annealing a large coil of steel without impairing its surface finish is expensive and therefore materials are available in a number of 'tempers' or degrees of hardness and the correct temper must be selected depending upon the amount of forming to be done and whether or not it is directional.

The table on page 80 is extracted from

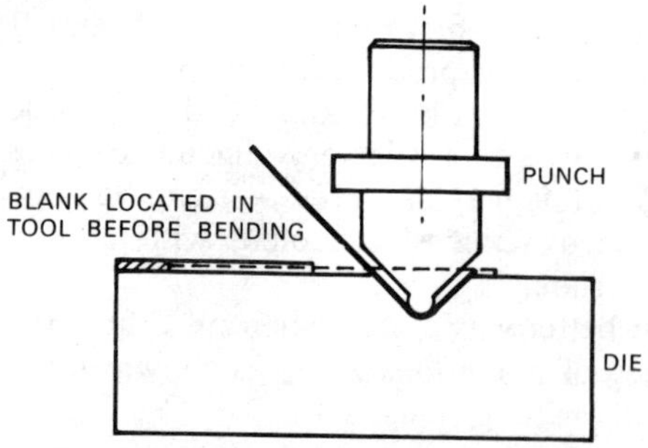

Fig. 3.16(a). First operation tool (bolster not shown)

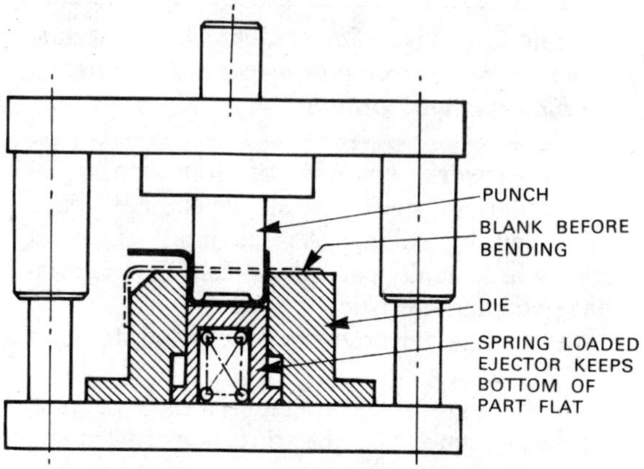

Fig. 3.16(b). Second operation tool built into die set

Fig. 3.16. Tools for bending bracket (fig. 3.14)

information published in *Plastic Working in Presses* by E. V. Crane (Wiley). Note how the so-called 'hardness' is characterised by tensile strength and ductility by % elongation. As the strength increases so the ductility decreases and the ability to be cold worked reduces.

Similar tables can be produced for other materials but in general it can be seen that the requirements for a material for it to be suitable for presswork depend on the amount of metal deformation to be produced. The factors to be considered may be summarized below.

1. As the degree of temper increases from 'very soft' to 'hard' so the ductility and the capacity of the material to be formed reduces.
2. As the ductility reduces so the direction of rolling becomes more important in bending operations.
3. The best condition for blanking is the 1/2 hard condition. Additionally the surface finish should be good as surface defects form the points of origin of tears and failures during forming operatings. Similarly slag inclusions during rolling can produce failures.

SAFETY NOTES

The Australian cricketer, Richie Benaud, once said that . . . you can always spot a wicketkeeper – he's the man walking down the

Table of presswork applications for sheet mild steel

Temper	Ultimate tensile strength MN/m^2	% Elongation	Application
Dead soft	276	35	Deep drawing
Soft	276	35	Bending, forming and shallow drawing
$\frac{1}{4}$ hard	310	17	18° bend across grain 90° bend along grain
$\frac{1}{2}$ hard	380	10	Best for blanking
Hard	415	—	Flat work only

street with his fingers sticking out at funny angles. Similarly you can often spot a press worker – he has no fingers, having had an accident either during the setting or operation of a press without taking adequate safety precautions. In order to keep his hands complete with fingers it is worthwhile for anybody working on presses to be able to recognise the dangers and guard against them.

It is interesting to note that there are over three times as many accidents annually, causing permanent disability, to handpress operators as to those operating foot-operated power presses. They nearly all consist of trapping the fingers between the punch and die, usually resulting in loss of one or more fingers. As most hand-presswork requires the operator to place the work in the tool the trouble is caused when the ram comes down unexpectedly before the hand is clear of the tool. This can be caused by:

1. Incorrect use of guards and safety devices.
2. The steep helix angle of the thread of a hand press plus looseness in the guides not providing enough friction to cause the ram to stay up until the press is operated.
3. Distraction of the operator's attention.

Accidents to press setters usually occur while they are working with their hands in the press area setting and adjusting the tool. Many such accidents could be avoided on power presses by switching the motor off while such adjustments are made. The author saw one such accident happen when a large spanner about 1 m long, used for adjusting the ram position, fell off the press table, bounced off the floor up under the operating pedal guard and down on the pedal of a power press. The motor was running and the ram came down leaving the setter with a thumb and half of a little finger only on one hand. This is most unusual but it could not have happened if the motor had been switched off.

Another cause of accidents to setters is in the handling of die sets out of the press. If the punch is lifted clear of the die on its pillars it will sometimes stay up. A slight knock or vibration will bring it down and if the hand is between the punch and die the results are disastrous. Straps should be provided to hold the punch up and they should always be used. If they are not provided a suitable block of wood between the two prevents the punch dropping. When carrying such die sets the straps should be used so that the bottom section of the tool does not fall due to the pillars sliding down the bushes. In any case to prevent such an accident *always* lift small die sets with one hand supporting the bottom section and the other holding the top.

Die sets are heavy, even small ones. In Level 1, *Workshop Processes and Materials*, safe methods of lifting were discussed, but it is most important not to try to lift beyond your capacity – get help or a hoist or both. It is easy to rupture yourself by lifting too heavy loads. Be modest and stay healthy.

The major causes of accidents may then be summarised as follows:

(a) *To operator*
- (i) Incorrect use of guards and safety devices;
- (ii) Hand press rams coming down by their own volition rather than under the operator's control;
- (iii) Distraction of the operator.

(b) *To setters*
- (i) Power press motor running while the setters hands are in the working area;
- (ii) Unsupported top die sliding down on the bottom die;
- (iii) Unsupported bottom die falling out of the top die on to the foot while carrying the tool;
- (iv) Rupture and back injuries due to lifting excessive loads.

TEC unit U76/056, around which this book is written, makes no mention of safety in presswork. The author believes, however, that writing on presswork should always make mention of hazards in this field. Space precludes a detailed description of the many types of press-guard available but an extremely detailed work on the subject is available. It is one of a series entitled Health and Safety at Work, booklet number 3 in the series being 'Safety Devices for

Hand and Foot Operated Presses'. The booklet is published by the Department of Employment through Her Majesty's Stationery Office and the author recommends it to all having an interest in presses and press-tools.

QUESTIONS

1. Both the fly-press and the power-press receive energy from a source, store it and transmit it to the work. For each type of press name the energy source, method of storage and mechanism of delivery of the energy. Compare the two machines from the aspects of initial cost, operating cost, tooling cost and force available.

2. Define blanking and piercing in the context of press-work. A rectangular part is required 15 mm × 10 mm with a 5 mm diameter hole in its centre and a 2 mm radius on each corner. The part is to be pressed from material 1.5 mm, thick whose mean shear strength is 250 N/mm^2:
   (a) make a dimensioned sketch of the part;
   (b) assuming that the blanking and piercing operations are to be carried out separately calculate the blanking force required;
   (c) calculate the dimensions of the blanking punch and die and piercing punch and die.

3. The part considered in question 2 is to be produced in separate tools in hand presses. State the order in which blanking and piercing would be carried out, sketch the tools and label important features.

4. The part considered in question 2 is to be produced in a follow-on type tool:
   (a) calculate the force required to operate the tool;
   (b) sketch the tool and label important features.

5. Assuming the part in question 2 is to be produced from strip with 3 mm metal all round calculate the percentage scrap produced when using:
   (a) 21 mm wide stock and
   (b) 16 mm wide stock.

6. The part shown in fig. 3.17 is to be made by blanking, piercing and bending from material 1.5 mm thick:

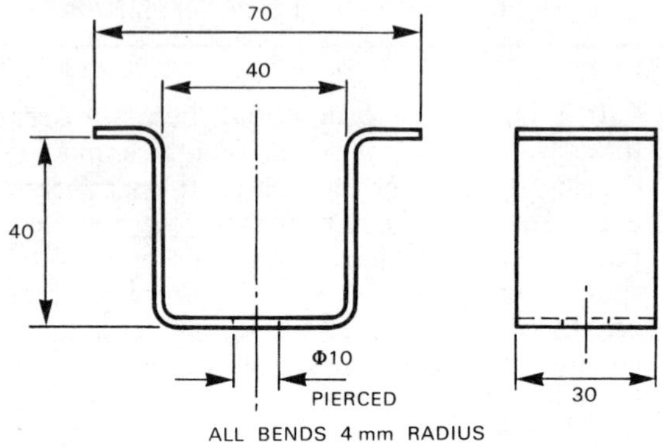

Fig. 3.17. Part to be blanked (cropped?) pierced and bent from 1.5 mm mild steel

   (a) make a dimensioned sketch of the blank, showing all calculations;
   (b) state the sequence of bending operations, sketching the part during each stage of manufacture;
   (c) sketch the bending tool(s).

7. In making the blank for the part shown in fig. 3.17 what factors, other than economics, may influence the choice of strip widths?

ANSWERS

2. (b) 23.354 kN
   (c) Blanking punch = 14.7 mm × 9.7 mm; blanking die = 15 mm × 10 mm
   Piercing punch = 5.00 mm diameter; piercing die = 5.3 mm diameter

5. (a) 46.31%
   (b) 49.10%.

4 Plastics forming

General Objective: *The student should appreciate the basic plastics forming methods of compression, transfer and injection moulding, and the forms of supply of ingredients for these processes.*

INTRODUCTION

In the Level 1 unit, *Workshop Processes and Materials,* considerable space was devoted to the nature and properties of plastics, but the working and forming methods described were those which could be carried out using the standard tools and equipment in a small general purpose jobbing shop, supplemented by inexpensive special items such as domestic ovens and modified electric soldering irons. However, the nature of plastics is such that they lend themselves to high production processes involving special moulding dies and machines whose high capital cost is spread over the large number of parts produced.

The most common processes used are compression moulding and transfer moulding for thermosetting plastics, and injection moulding for thermoplastic materials. In all of these processes the material is plasticised by the action of heat, or heat and pressure, and forced to flow under pressure until it conforms to the shape of a metal mould, after which it is allowed to solidify and is ejected from the mould.

It must be emphasised that these are by no means the only methods of plastic forming used in the plastics industry. Vacuum forming, blow moulding, extrusion and carding are other methods used depending upon the shape, size and form required of the finished product. It would be easy to say that processes to be discussed in this chapter produce the bulk of plastic parts manufactured but when the number of plastic bags in use is considered it makes this statement rather doubtful. However, it is probably true to say that the bulk of the parts used in engineering are made by the compression, transfer and injection moulding processes.

COMPRESSION MOULDING

Specific Objective: *The student should be able to describe with the aid of sketches the basic steps of the compression moulding process.*

Compression moulding is commonly used for moulding thermosetting plastics, the process being carried out in the following stages:

1. *Loading the mould*

With the two halves of the mould open the raw material is loaded into the mould. For a single impression mould the raw material may be in powder form and can be carefully weighed out, the operator weighing out the next load while that in the mould is curing. For simple moulds the load may be in the form of a tablet, one tablet being used per load. If the mould is more complex, pellets can be used, a fixed number of pellets constituting a mould load. In the case of very complex parts, the charge may be introduced to the mould as a preform, the powder having been cold-pressed to the approximate shape of the mould cavity before moulding, hence preform.

If powder is to be used for the moulding in the case of a multi-impression mould a loading board is used. This consists of a board having the same number of holes as the mould has impressions. The holes are slightly smaller than the impressions and spaced to align themselves with the impressions when the board is located on the mould. Sliding on this board is a similar board, the holes in which are again slightly smaller in diameter than those in the bottom board and whose depth has been calculated to hold the correct amount of powder. While the charge in the mould is curing, the operator slides the top board so that the holes are misaligned and fills the holes with powder. When the mouldings have cured and are ejected from the mould, the board is located on the mould, the top board slid so that the powder-filled holes locate over the holes in the bottom board, and the powder drops through into the mould.

2. *Pressing*
The mould is heated to cure the material and pressure is applied to cause the material to flow and take the form of the mould. After a short time the mould is opened to allow any gases to escape, closed and the pressure reapplied for a long enough period to allow curing to take place. This may be 1 to 5 minutes depending upon the material being moulded.

3. *Ejection*
When the time required for curing has elapsed the pressure is released, the mould opened and the part ejected. The mould can now be cleaned and the cycle repeated.

The presses used for compression moulding are usually of the vertical type, the mould cavity thus being horizontal to facilitate filling. The platens are heated so that the heat is conducted through to the dies to cure the material, heating usually being by steam or electricity. The temperature ranges from 100°C to 200°C depending upon the material being moulded. To obtain the pressures required the presses are hydraulic in operation and may be 'upstroke' or 'downstroke'.

In the case of the upstroke press the top platen is fixed and the lower platen moves up to it. This has the advantage that when the pressure is released the platens separate, rather than close, by gravity. This makes for safety in the case of hydraulic failure. The moving platen does not lend itself to automatic loading and therefore upstroke presses are largely used for hand operation.

Downstroke presses have a fixed lower platen, the hydraulic cylinder at the top moving the top platen down to close the dies and apply the moulding pressure. In this case hydraulic failure will allow the dies to close and it is necessary, during normal operation, to provide a positive means of raising the platen when the pressure is released. Downstroke presses are usually fitted with a lower cylinder to operate the ejector gear as shown in fig. 4.1.

TYPES OF MOULD

Consider the part, shown in fig. 4.2, to be moulded from a thermosetting plastic by compression moulding. The types of mould which may be used are positive moulds, semi-positive moulds and flash moulds.

1. *Positive moulds*
These are so called because the top part of the

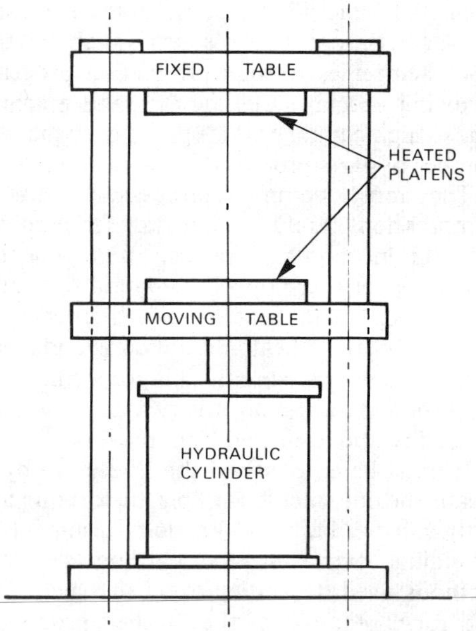

Fig. 4.1(a). Upstroke press for compression moulding

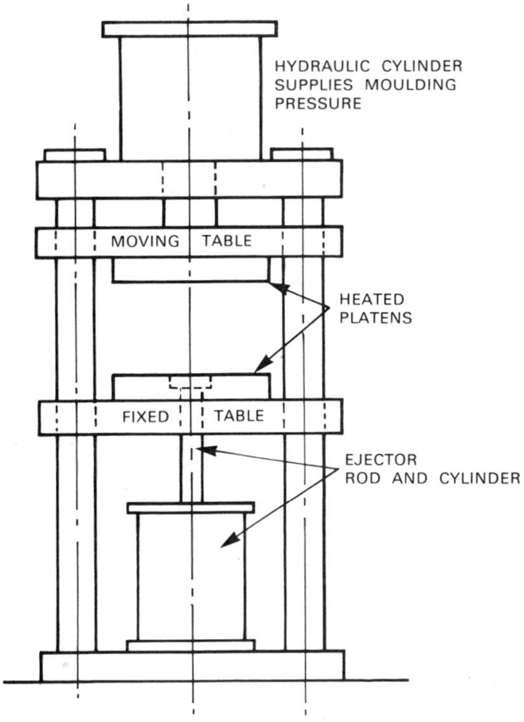

Fig. 4.1(b). Downstroke press for compression moulding

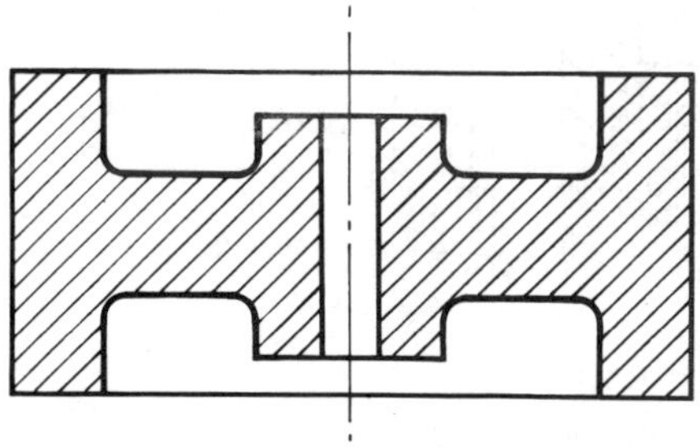

Fig. 4.2. Part to be moulded

mould, or top force, acts as a plunger and applies a positive location between the two halves of the mould. It suffers two disadvantages; the top mould being a close fit in the lower force, allows little or no provision for gas escape and excess material cannot escape. Thus the quantity of material introduced into the mould must be very accurately measured unless a variation in thickness can be allowed. It was found that as such moulds were used they became worn and allowed greater clearance between the plunger and the walls of the lower part of the mould. The restriction due to the small amount of clearance allowed the pressure to be maintained but at the same time allowed gasses to escape. It also allowed material to escape and form a flash and this gave rise to further variations in thickness so a pressure pad had to be introduced to limit the travel of the top force. This type of mould is little used with free flowing materials but is still used for moulding cloth or fabric filled types of thermosets. Such a mould is shown in fig. 4.3.

2. *Semi-positive moulds*
This is the most widely used type of compression mould and is of the form shown in fig. 4.4. Note that it still provides positive location between the two mould halves but also incorporates a land to limit the closing of the mould and thus control the thickness of the part produced. Excess material can escape into the flash space provided and control of the quantity of material introduced into the mould is not so critical.

3. *Flash type moulds*
These are the simplest and cheapest of the compression moulding tools but they have a major disadvantage and are little used. It can be seen from fig. 4.5 that the tool itself provides no location between top and bottom forces and all location must be provided by guide pins. Although guide pins are often used in other types of mould, particularly multi-impression moulds, this does rather defeat the low tooling cost objective. Further to this, if the moulding material is not evenly loaded into the die the side forces can result in bent core pins and guide pins. Uniform wall thicknesses are difficult to obtain.

Although it has been stressed that compression moulding is used for thermosetting plastics it is by no means limited to such

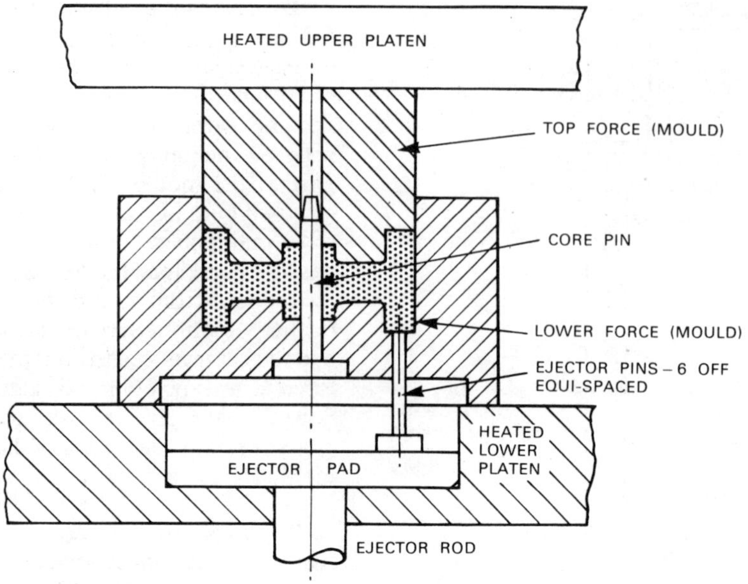

Fig. 4.3. Positive mould

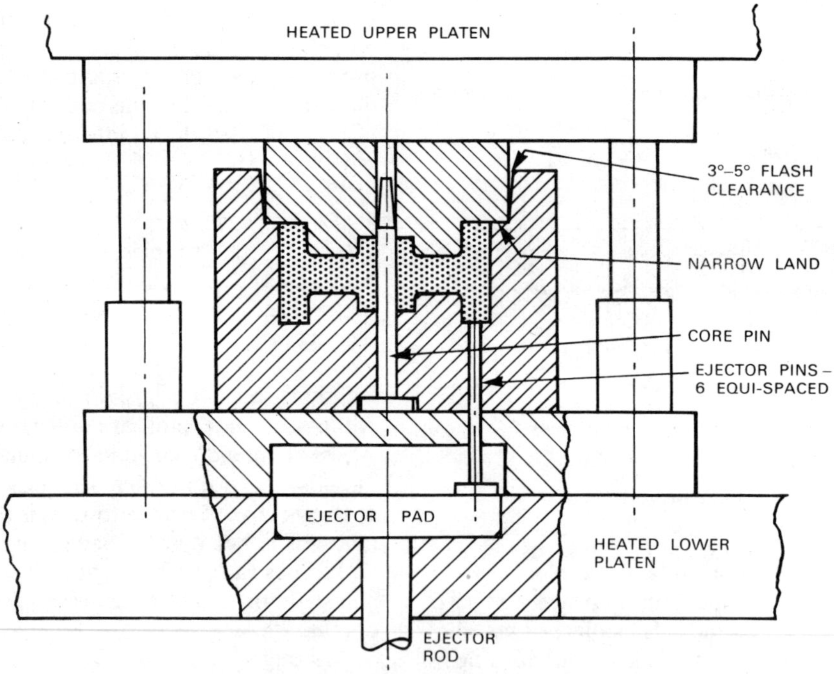

Fig. 4.4. Semi-positive tool with guide pillars and bushes using ejection method shown in fig. 4.1(b)

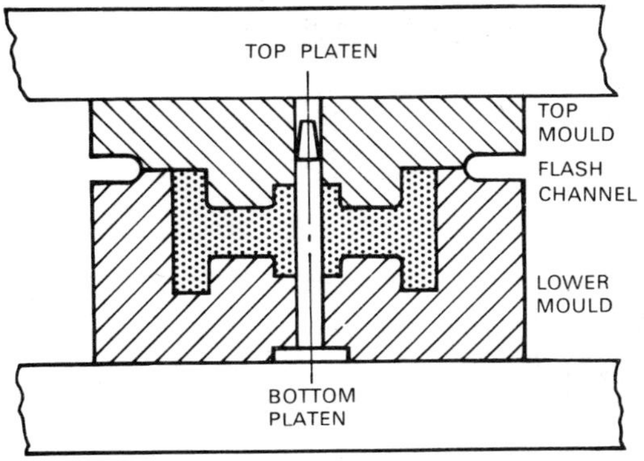

Fig. 4.5. Flash mould. Note that there is no location between top and bottom moulds. Uneven loading of powder could cause side thrusts which would bend the core pin

materials. Gramophone records are made from thermoplastics but are compression moulded and the quality of sound reproduction from modern recordings is a tribute to the ability of the process to reproduce the finest features in the dies on the parts produced from them.

TRANSFER MOULDING

Specific Objective: *The student should be able to describe with the aid of sketches the basic steps of the transfer moulding process.*

One of the main problems with compression moulding is that the material is introduced into the mould in the solid state and has to plasticise and flow within the mould. This requires very high pressures and even so it is difficult to get the material to flow into intricate parts of complex moulds. In the transfer moulding process, the material is plasticised in a chamber outside the mould by heating and forced through an orifice into the mould by the action of a ram. Being fluid when it enters the mould, it fills intricate shapes more easily and is less likely to displace inserts and core pins than is the pressure transmitted through a powder before it is properly plasticised, as in a compression mould.

There is one disadvantage to transfer moulding. It is impossible to completely empty the heating chamber and the material left in the chamber, and in the sprue, the channel leading from the chamber to the mould, having been heated under pressure, will have cured along with the material in the mould. Being a thermosetting material it follows that this material must be extracted and discarded as scrap. Thus the wastage in material is rather higher than with compression moulding.

The method of removing this cull, as the waste material is called, is to make the nozzle tapered, having its smallest diameter where it enters the mould. As the dies separate, the cured material in the sprue breaks off at this point, but is still attached to the ram which has a dovetail cut in it. When the ram is withdrawn the cull can be removed and the pot, or heating chamber, recharged with fresh material.

From this description of the transfer moulding process, a tool for which is shown in fig. 4.6, it can be seen that, in addition to the heated platens on which the moulds are mounted there must also be a separate pot in which the material is heated and a ram to transfer it from the pot to the mould under pressure. Fig. 4.7 shows an upstroke type press with a middle platen carrying the top half of the mould and the pot, and an upper platen carrying the cylinder and ram for transferring the material in its plasticised state to the closed mould.

Transfer moulds are often made to be used on a simple press, the transfer ram and pot being incorporated in the tool and the force applied by the ram on the top mould through the material keeping the moulds closed. When the material plasticises and enters the mould under pressure this pressure tends to separate the mould halves. As

$$\text{Force (kN)} = \text{pressure (kN/m}^2) \times \text{area (m}^2)$$

and it is necessary for the closing force to be greater than that tending to open the mould, it follows that the ram area must always be greater than the projected area of the moulds and runners.

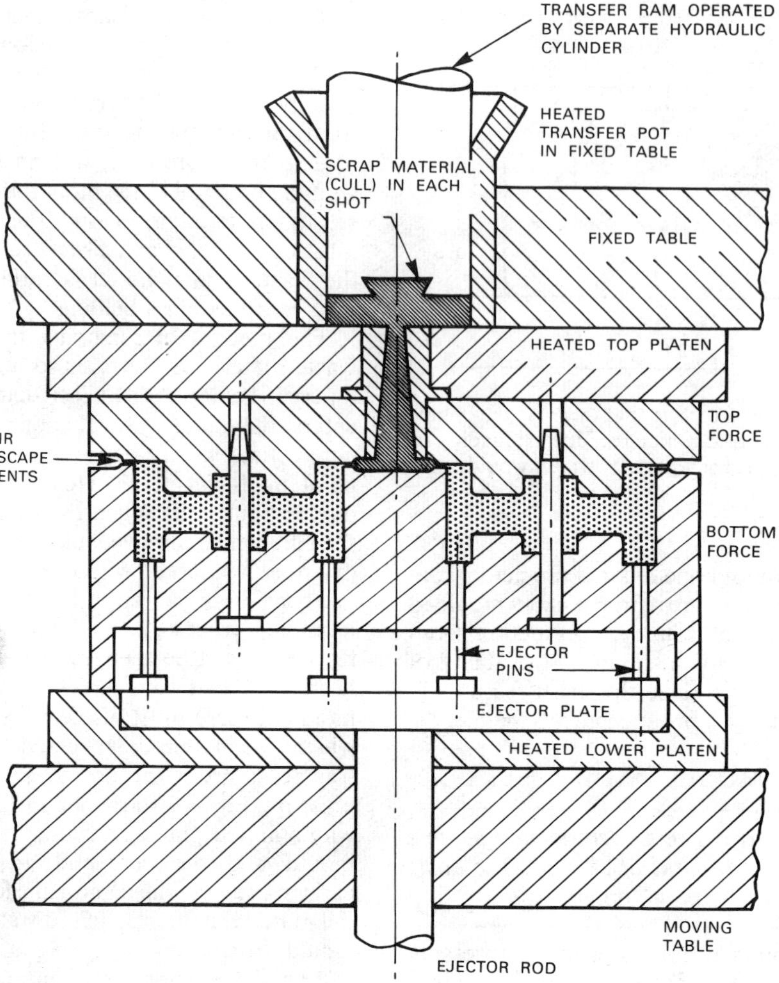

Fig. 4.6. Multi-impression transfer tool for use on press of the type shown in fig. 4.7

Note also that in the transfer moulding tool allowance must be made for displaced air to be expelled from the mould as the plasticised material is forced in. The air escape vents in fig. 4.6 are exaggerated and should be grooves only about 0.01 mm deep, the number required depending upon the mould size and material being moulded.

Bearing in mind the wastage, greater complexity and cost of the tool it follows that transfer moulding will only be used where its ability to produce complex mouldings outweighs the disadvantages. However, as many plastic parts tend to be complex it is a widely used process.

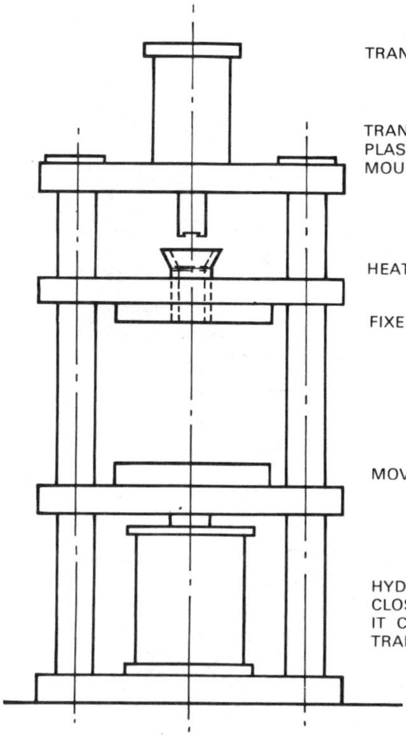

Fig. 4.7. Vertical transfer press

INJECTION MOULDING

Specific Objective: *The student should be able to describe with the aid of sketches the basic steps of the injection moulding process.*

Injection moulding is similar to transfer moulding in that the material is preheated in a cylinder and injected into the mould under pressure exerted by a ram. It differs from the transfer process in that it is used for moulding *thermoplastic* materials which, it should be recalled, plasticise when heated and solidify when cooled. Thus the material left in the cylinder is not wasted; it remains in the plastic state and is used on the next component. Further, the waste material from the mould can be broken up and returned to the hopper for re-use.

Fig. 4.8 shows in diagrammatic form the hopper, ram and nozzle of an injection moulding machine. Note the heating coil to plasticise the material and the 'torpedo' which forces the material to the outside of the chamber to present a shorter heating path. Without the torpedo, heat would have to be conducted through the thickness of the plastic to ensure that the material in the centre was plasticised. Plastic materials are not the best conductors of heat, hence the torpedo, or spreader as it is sometimes called.

A tool for injection moulding our component is shown in fig. 4.9. Note that, like the transfer tool in fig. 4.6 it is a multi-impression mould to produce a number of parts at each stroke of the press. A major difference is that in this case the platens are water cooled, rather than heated, in

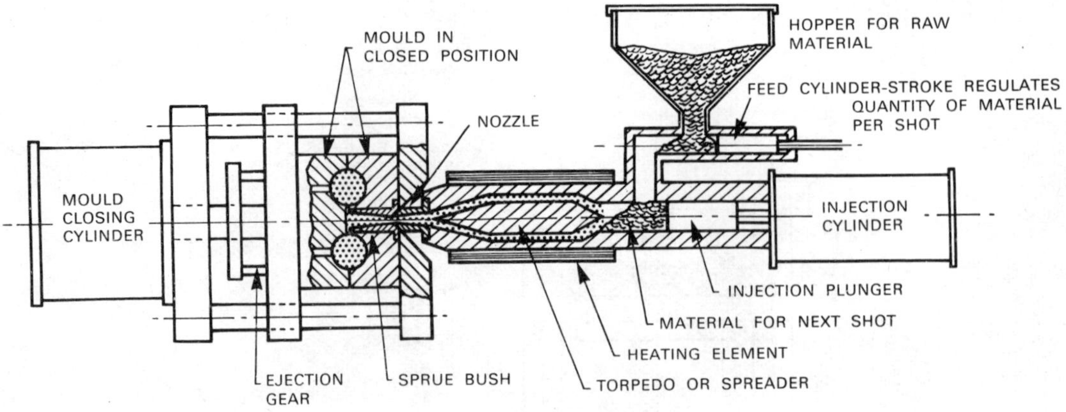

Fig. 4.8. Injection moulding machine

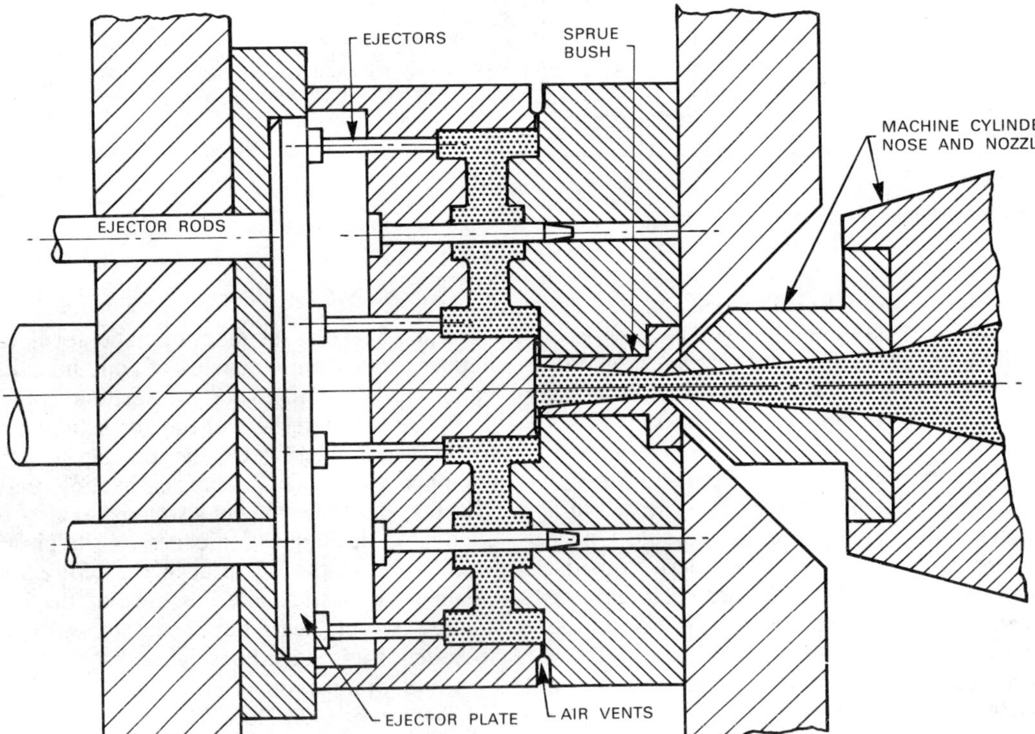

Fig. 4.9. Multi impression injection moulding tool for use on the type of machine shown in fig. 4.8

order to facilitate solidification of the material and reduce the cycle time for the process.

A comparison of these three methods of moulding plastic materials is tabulated below.

accurately locating the insert. In this case it is easy – the core pin could be used. In other cases special provision must be made for location of inserts.

Moulding method	Moulding material	Features
Compression	Thermosets	Low cost tooling; heated mould; difficult for intricate parts.
Transfer	Thermosets	Higher cost tooling; heated mould; good for intricate parts; material wastage high.
Injection	Thermoplastics	High cost tooling; cooled moulds; no material wastage.

THE USE OF INSERTS

Specific Objective: *The student should be able to explain where inserts may be used in simple components.*

Consider again the part shown in fig. 4.2. It may be that this is a pulley or wheel to be moulded from a material which would rapidly wear the shaft on which it is to run. Again, plastics are not the best materials in which to mould threads, particularly if the thread is likely to be frequently unscrewed and retightened. The thread can be easily broken by excessive tightening and the material is likely to crumble.

These are two cases where inserts can be used to advantage. Inserts are metal parts moulded into the plastic during manufacture, the metal being chosen to have suitable properties for the function that the part is to be called upon to carry out in service.

In the case of the part we are considering let us assume that an insert in the form of a brass bush is to be used. The following factors must be considered:

1. LOCATION

Provision must be made in the die for

2. PREVENTION OF ROTATION

Usually the insert must not rotate in the part into which it is moulded. This can be prevented by making it from square or hexagonal stock but a coarse knurl is often cheaper and as effective.

3. PREVENTION OF WITHDRAWAL

If an end load is to be applied to the insert it must not pull out of the plastic moulding. This can usually be prevented by machining one or more undercuts in the insert.

Thus a bush to be used as an insert for the component we are considering would appear as in fig. 4.10.

SAFETY NOTES

Specific Objective: The student should be able to name the sources of danger in using plastic moulding machines and the relevant precautions.

Plastics are moulded hot in machines working at high pressures. Therein lie the dangers in plastic moulding machinery and an analysis of the dangers enables the hazards to be avoided. These are:

1. Parts of the machine are hot; do not touch.

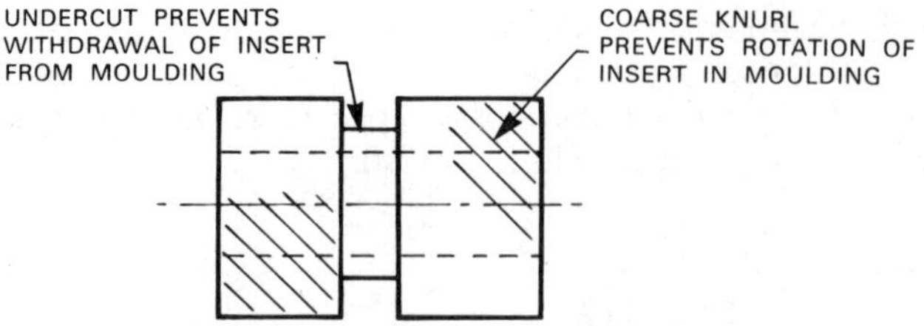

Fig. 4.10. Insert of the type which could be used as a bush in a plastic component

2. The plastic is hot and under pressure. If the two halves of the mould are not quite closed and pressure is applied a dangerous hot stream of plastic will squirt out. If this hits the face it can cause severe damage, particularly to the eyes. Wear safety glasses and stand clear of the split line when the press is operated.
3. The mould halves come together under great pressure; keep your hands clear when the press is being operated.
4. In the cases of transfer and injection moulding the mould halves are subject to internal pressure. The die closing pressure must be greater than the transfer or injection pressure or the dies may open. If a hand operated press is used where the dies are closed mechanically a mechanical locking device is provided and must be used.

These comments on safety apply particularly to machine setters who, by the nature of their job, have to work on occasion without the safeguards provided for the operator. When modern presses are set up and in operation there are interlock devices which prevent the press being operated without the guard in position or if the mould is not closed. In addition the operating controls are usually placed so that the operator cannot be in the danger zone when he is operating the machine. However, he should always use the guards provided – it is an offence not to use them under the Health and Safety at Work Act, wear safety glasses and observe the correct operating procedures when running any type of plastics moulding press.

QUESTIONS

1. State the forms of supply of the materials for each of the following operations, explaining in each case how the correct quantity of materials is arrived at:
   (a) simple compression mould;
   (b) more complex compression mould;
   (c) transfer mould;
   (d) injection mould.
2. A compression mould with a six impression die is to be made, each impression having a maximum diameter of 15 mm and requiring a volume of 1.5 cm^3 of moulding powder. If the moulds are set on a pitch circle diameter of 35 mm design a suitable loading tray and explain how it would be used.
3. The component shown in fig. 4.11 is to be moulded by compression moulding in a four

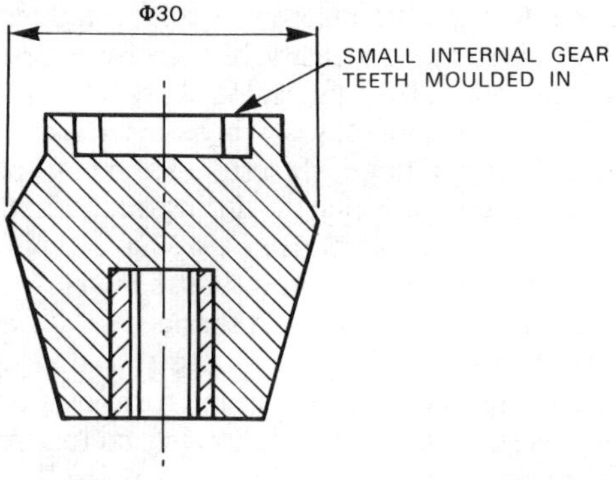

Fig. 4.11. Control knob moulded in plastic with a brass threaded insert. The diameter is given as an indication of overall size

impression mould, the threaded bush shown being an insert. Sketch one impression of a suitable mould showing how the insert would be located in the mould. Hint: it may be quicker to screw the insert on a loose piece to locate it in the mould.

4. State what factors would influence the choice of process for the part shown in fig. 4.11, between transfer moulding and injection moulding. There is a short answer to this question.

5. Assuming that the insert shown in fig. 4.11 is 10 mm outside diameter by 15 mm long make a detailed drawing of it. You need not detail the thread.

6. The recess in the end of the part shown in fig. 4.11 is to have internal gear teeth moulded in. The teeth are approximately 2 mm deep and 1.5 mm thick at the pitch line. If the part is to be made in a thermosetting plastic state which moulding method should be used for making the part. Give reasons for your choice.

7. An injection moulding press operator was hit by a stream of hot plastic from the mould joint line of his own press. At the enquiry it was established that:

(a) The press was fitted with a guard which was not in place.
(b) The interlock switch preventing the machine operating until the mould was properly closed had been tampered with by the operator – it made the job easier!
(c) The micro-switches which prevent the machine operating when the mould is not properly closed were faulty and had not been tampered with by the operator.

The operator claimed that management was at fault because, although he was not using the guard, the micro-switch system should have protected him. Management claimed that the operator was at fault because if the guard was in place he would not have been hurt.

Who do you think was at fault? Give reasons for your conclusions.

8. The part shown in fig. 4.11 is to be transfer moulded in a machine which does not have a separate cylinder to operate the transfer ram. The ram and pot must therefore be incorporated in the die. If the die has four impressions calculate the smallest allowable diameter of the transfer ram. Explain why this diameter is necessary.

9. What is the main difference between the moulds used for transfer moulding and injection moulding? Explain why a small dovetail is necessary in a transfer ram and not in the ram of an injection machine. Why is there less wastage in injection moulding than in transfer moulding?

ANSWER

8. Assuming ram force is to be twice the force tending to open the dies, ram diameter = 85 mm (rounded).

5 Metal cutting

General Objective: *The student should recognise the principles of metal cutting with single point tools.*

In Level 1 we considered the metal cutting process from an essentially practical standpoint and showed that for metal cutting to take place:

1. The tool must have adequate but not excessive clearance[1] and the clearance angle depends largely on the geometry of the workpiece.
2. The tool must have a rake[1] angle which depends largely on the material being cut.
3. The machine must provide the tool motions and the forces necessary to cause a shearing action to take place and the chip to be removed.

In this chapter the mechanics of metal removal will be examined more closely and some of the more sophisticated types of tool available to the engineer will be discussed.

THE MECHANICS OF METAL CUTTING

A discussion of the mechanics of metal cutting requires an examination of the forces exerted by the tool on the work and vice versa. For purposes of simplicity, only *orthogonal cutting* will be considered in detail, rather than the more usual oblique cutting adopted in practice.

Orthogonal cutting is a particular case where only two forces are measured and the cutting edge is at right angles to the direction of feed. A typical case is that of machining the end of a tube as shown in fig. 5.1(a) and the forces which can be measured are:

1. Tangential force F_t, tangential to the work surface. In association with the cutting speed this is the main power-consuming force.
2. Axial force F_a, parallel to the work axis and resisting the feed motion. In association with the feed motion this also consumes power, but as the feed rate is usually very slow compared with the cutting speed, the power consumed by F_a is normally very small.

If the cutting edge is not at right angles to the direction of feed, as is more usual, a third force:

3. Radial force F_r is measured, which tries to force the tool out of the work and is resisted, in the case of a lathe, by friction in the cross slide and its lead screw. As no motion is associated with F_r it does not consume power but influences F_a.

The three-force, or oblique, cutting system, is shown in fig. 5.1(b).

MEASUREMENT OF TOOL FORCES

It has been emphasized above that the forces F_t and F_a can be *measured*. They are chosen because they are at right angles and can be easily resolved, using a triangle of forces, to find the *resultant* force R which is, in fact, the only real force acting. It is convenient simply to measure these forces, using a tool-force dynamometer, determine R and resolve R into pairs of forces at right angles to each other which are of interest.

Many forms of dynamometer are available, all working on the same basic principle. Due to the forces applied, the tool deflects, and if its deflections are measured they can be related to the forces which produce them. The differences in dynamometers are largely in the methods of measuring these deflections.

The instrument shown in fig. 5.2(a) carries the tool in a holder mounted on the front of a disc or diaphragm which is clamped to the main

[1] It is assumed that at this stage the student is familiar with the definitions of the terms used.

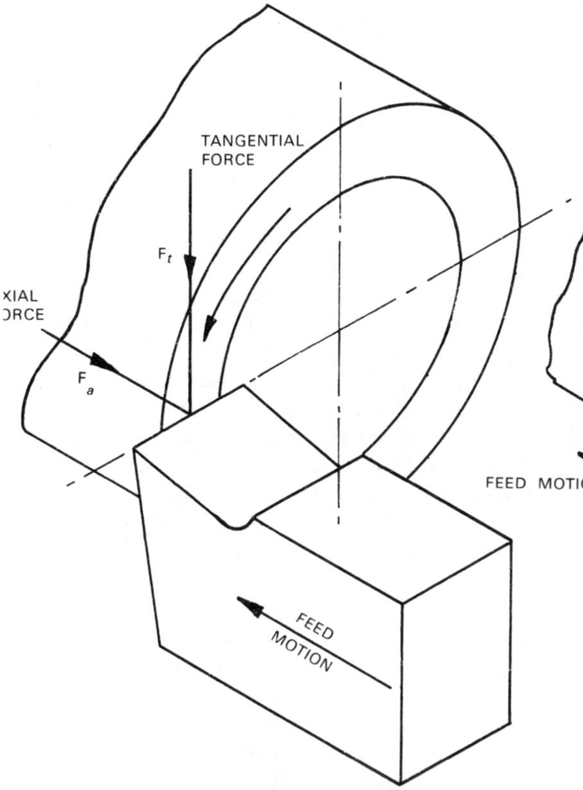

Fig. 5.1(a). Orthogonal cutting – two-force system

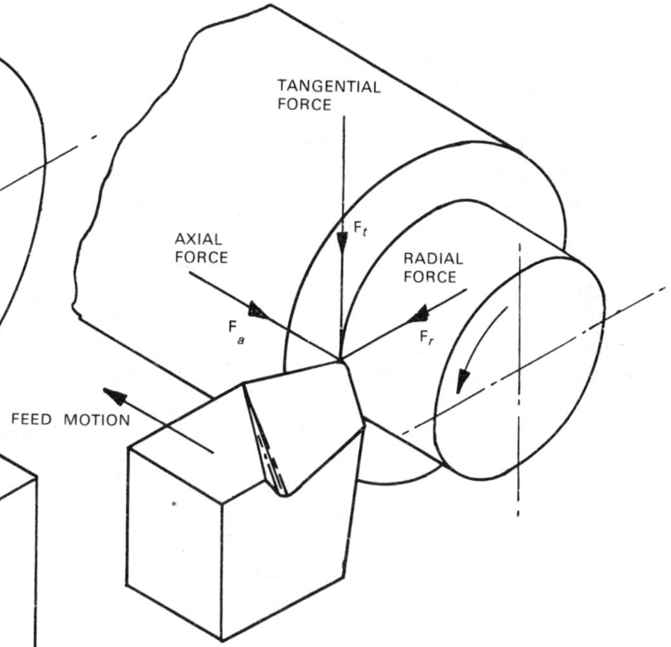

Fig. 5.1(b). Oblique cutting – three-force system

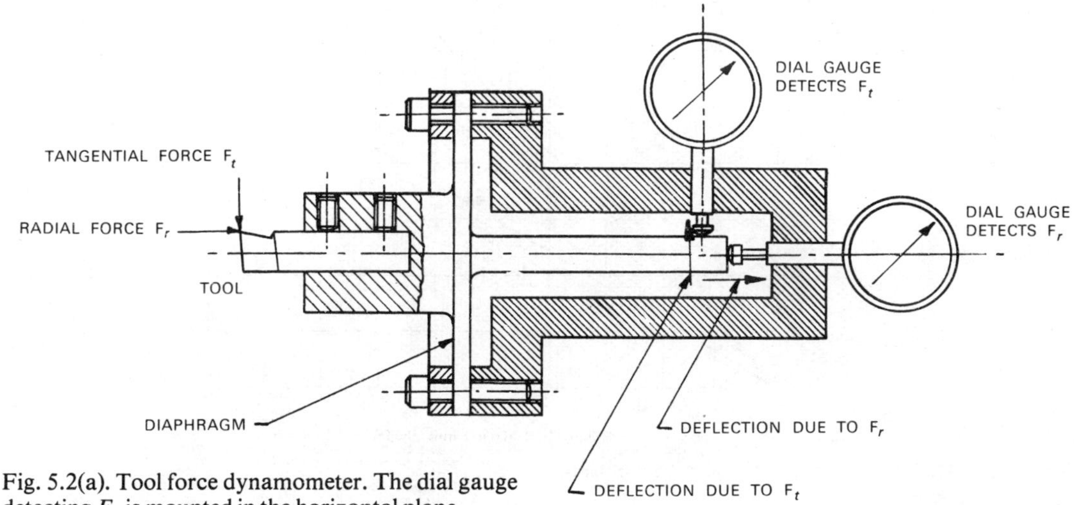

Fig. 5.2(a). Tool force dynamometer. The dial gauge detecting F_a is mounted in the horizontal plane

body of the dynamometer. The back of the diaphragm supports a bar against which dial gauges bear to detect the movements of the bar caused by the tool forces deflecting the diaphragm. A third dial gauge (not shown), operating in the horizontal plane, can be used to measure the third force, F_a if necessary.

The tool is first replaced by an adaptor carrying a ball which is then loaded by known dead weights and a graph or calibration chart is made of load against deflection as shown in fig. 5.2(b). In operation, the tool should project a fixed amount so that its point is in the position of the ball centre. Its deflections are noted during cutting and by referring these deflections back to the calibration chart the forces involved can be determined.

SIMPLE ANALYSIS OF FORCES IN ORTHOGONAL CUTTING

If we use a two-force dynamometer to measure the forces F_t and F_a, we can determine by means of a triangle of forces the resultant force R exerted by the work on the tool, as shown in fig. 5.3.

Note: In all the force diagrams the tool is shown as if mounted on a shaping machine, i.e. moving horizontally from right to left.

A simple way of finding any two components of R at right angles to each other is based on the fact that the angle subtended at the periphery of a semi-circle from its diameter is always a right angle. If the force F parallel to the rake face and the force N normal to the rake face are required, we extend the rake face to meet the circle and from this point draw to the other end of R (the diameter). These two vectors to scale represent forces F and N.

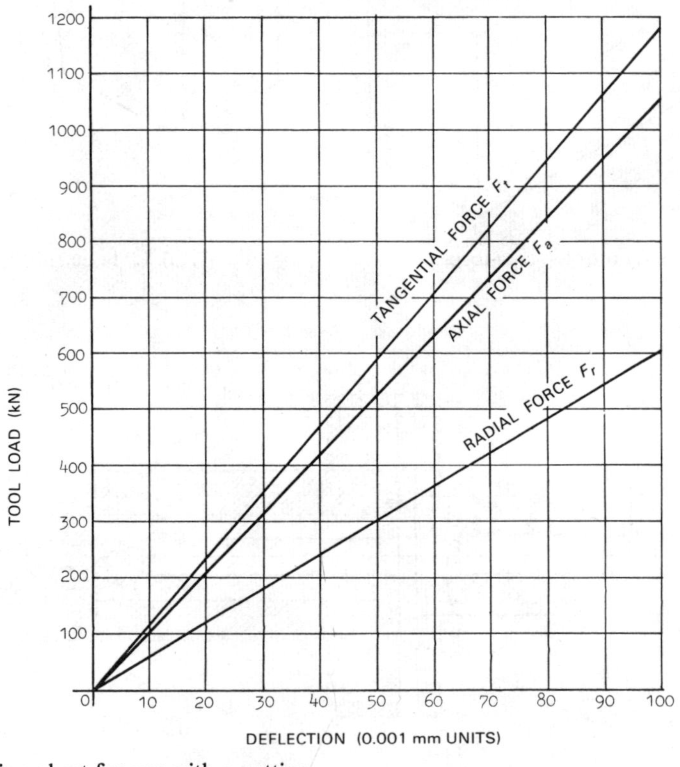

Fig. 5.2(b). Calibration chart for use with a cutting tool dynamometer

Fig. 5.3. Vector diagram of the forces on a cutting tool (orthogonal cutting):
R is the applied force due to the cutting action
F_t and F_a are the two forces most convenient to measure
F is the chip friction force
N is the force normal to the rake face

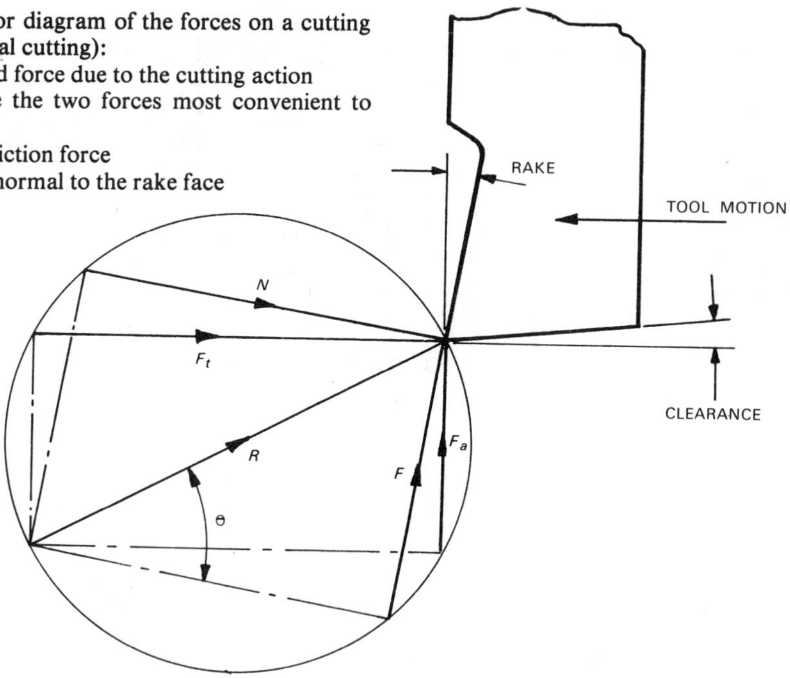

The force F is equal and opposite to the force of friction which must be overcome to move the chips along the rake face and N is the force, normal to the rake face, producing that friction.

The ratio $\frac{F}{N}$ is the coefficient of friction

$$\tan\theta = \frac{F}{N} = \mu$$

What is the effect of changing the coefficient of friction? Let us assume that the normal force N remains constant but, due to a reduced coefficient of friction, F is reduced. We can now reverse the above procedure and use the friction force F and the normal force N to find the resultant, and hence find new values for F_t and F_a, as shown in fig. 5.4. We find that both F_t and F_a are reduced because the resultant R is reduced. Recalling that F_t is the power-consuming force it follows that a reduced friction coefficient between the chip and the tool rake face reduces the power consumption and the heat generated. This, then, produces a better cutting process and higher rates of metal removal may be employed.

A similar analysis can be used for examining the effect of an increase in rake angle but before we do this we must consider the effect of the rake angle on the angle at which the metal shears ahead of the tool. This is shown in fig. 5.5(a) and (b). The metal may be considered to shear along a line slightly steeper than the direction of N, the normal to the rake face. If the rake angle is increased, the angle at which shearing occurs is also increased, thus reducing the length of the shear line.[1] As the length of

[1] This is a simplified version of the work carried out by two men, Ernst and Merchant of the Cincinnati Milling Machine Co. Later researchers hold that shearing is not along a line but that a zone of deformation occurs ahead of the tool. For purposes of simplicity we will assume that shearing takes place along a line.

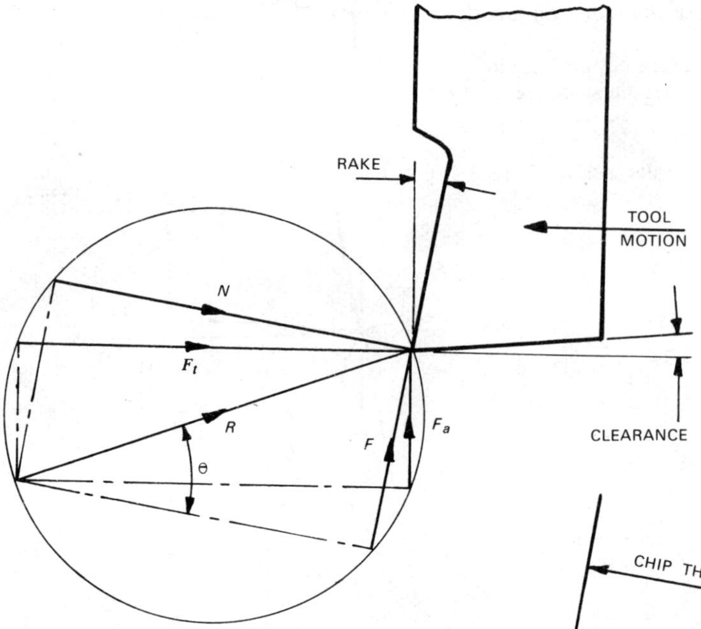

Fig. 5.4. Effect of reduced friction force *F*, *N* remaining the same. F_t and F_a, and hence *R*, are all reduced

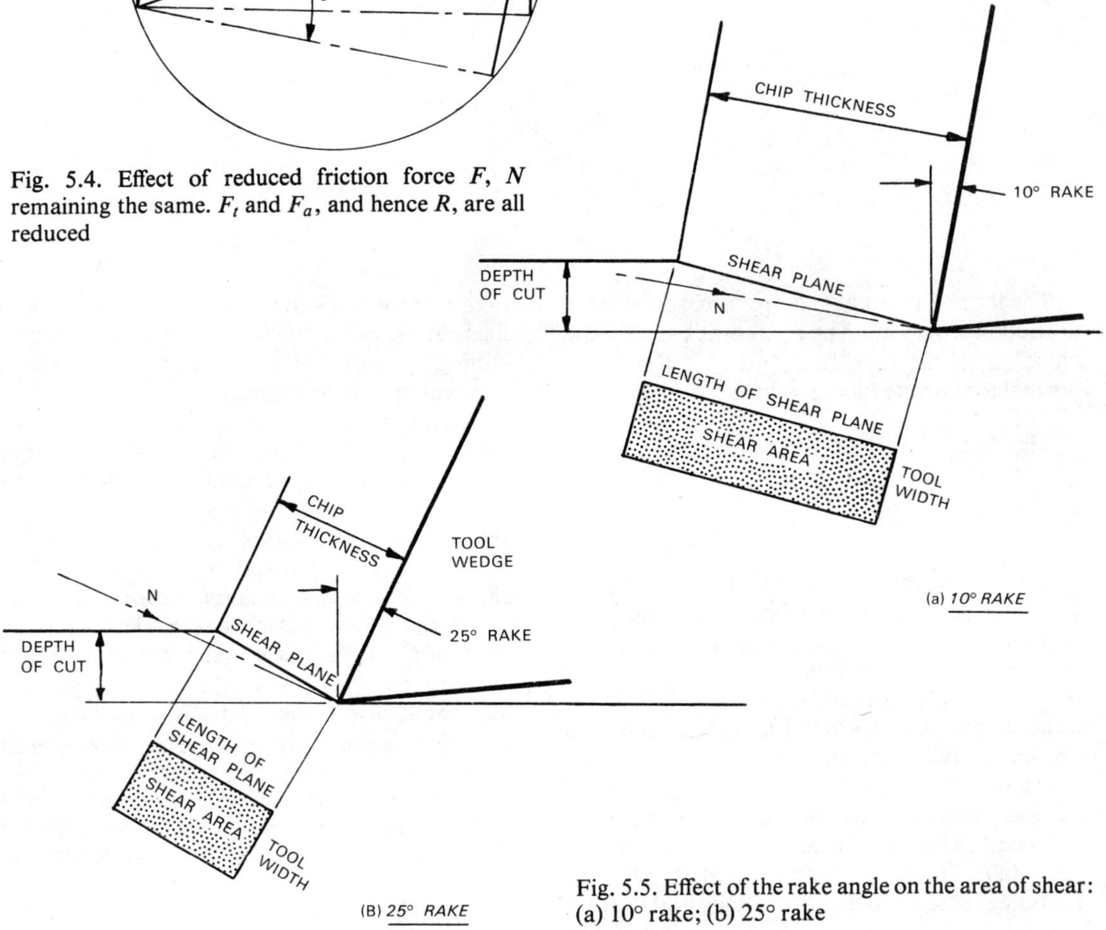

Fig. 5.5. Effect of the rake angle on the area of shear: (a) 10° rake; (b) 25° rake

Fig. 5.6. Effect of increased rake angle. R is reduced and therefore all other forces are reduced, as well as their proportions being changed

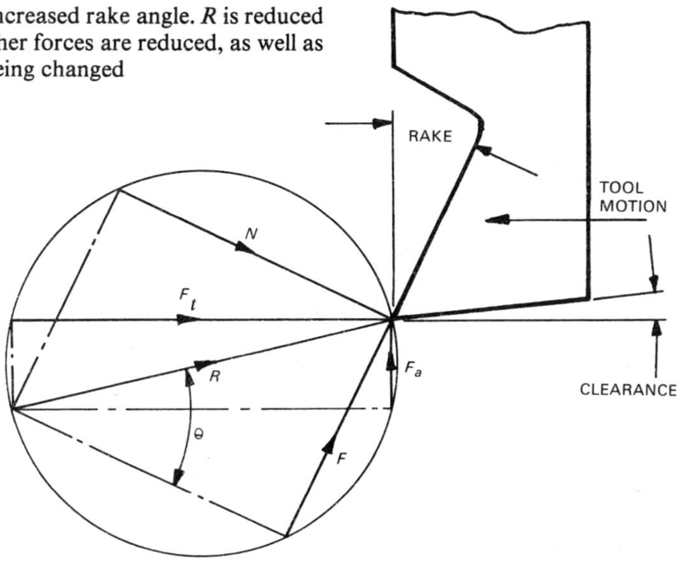

the shear line is reduced the shear area is correspondingly reduced and since

Shear force = Shear strength × shear area

the shear force is reduced. Thus the forces necessary to produce cutting and the resulting force F are reduced (note that the ratio F/N remains the same). Using the circle method we can again draw in the forces, as in fig. 5.6, which shows that F_t and F_a are reduced. As F_t is the power-consuming force, the power consumption and heat generated will both be less.

It would appear then that the greater the rake angle the better will be the cutting conditions. While this is true up to a point it must be noted that as we increase the rake angle the force N, normal to the rake face, is operating in such a direction that it is likely to break off the tool point and that, for a given depth of cut, as the strength of the work material increases so does N increase. Furthermore, as the rake angle is increased the tool is weakened and therefore if the rake angle is made to steep and a tough material is being cut, early failure of the tool will result through the tool tip breaking off.

CHIP FORMATION IN METAL CUTTING

There are three fundamental types of chip produced in the metal cutting process, as shown in fig. 5.7:

(a) the continuous or ribbon-type chip
(b) the continuous chip with a built-up edge
(c) the discontinuous chip.

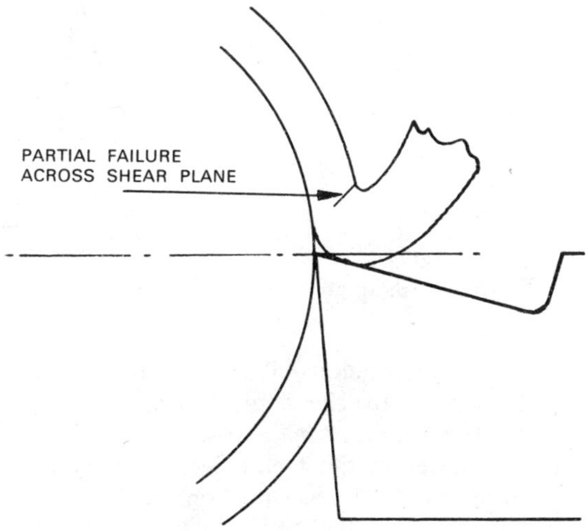

Fig. 5.7(a). Continuous chip

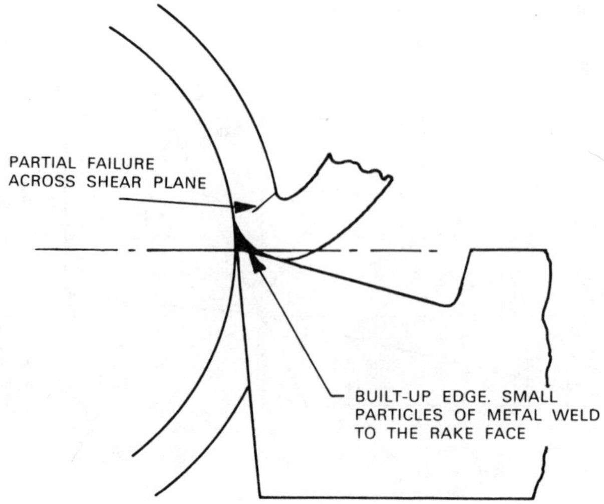

Fig. 5.7(b). Continuous chip with built-up edge

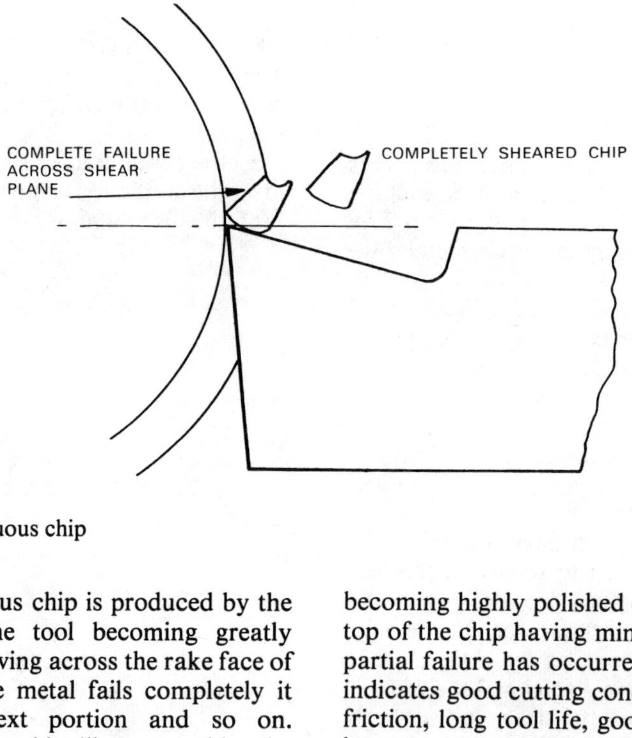

Fig. 5.7(c). Discontinuous chip

(a) The continuous chip is produced by the metal ahead of the tool becoming greatly compressed and flowing across the rake face of the tool. Before the metal fails completely it compresses the next portion and so on. Examination of the chip illustrates this, the underside in contact with the rake face becoming highly polished or burnished and the top of the chip having minute serrations where partial failure has occurred. This form of chip indicates good cutting conditions with low chip friction, long tool life, good surface finish and low power consumption. It usually occurs when cutting a ductile material at high speed and fine feed, with a large rake angle, polished tool face and good lubrication.

(b) The continuous chip with the built-up edge is formed in a similar manner but due to intense pressure and high temperature small pieces of work metal become welded to the tool tip. Due to the extreme pressure this weldment becomes work-hardened and brittle so that it eventually breaks off and is carried away by the chip over the tool. Being very hard it is also extremely abrasive, and causes rapid tool wear since the process of build-up and breakdown is repeated with very high frequency. The built-up edge usually occurs when a ductile material is being cut under conditions causing extreme heat and pressure, conditions such as poor tool-face finish, poor lubrication and too small a rake angle.

(c) The discontinuous chip occurs when the force involved in bending the chip causes complete breakdown of the shear zone and individual segments of metal come away separately. This can be due to the material itself being discontinuous, e.g. grey cast iron, where the discontinuities caused by the graphite flakes break the structure of the metal, or can occur with a material such as brass whose shear strength is relatively low compared with the force required to bend it. With these materials a good surface finish results and the forces involved, and hence the power consumption, are low.

This type of chip can also occur with a ductile material if the cutting speed is too low and the rake angle too small. It is often induced deliberately in repetitive operations to give easy separation of the swarf from the work and to make the swarf itself easy to remove in large quantities. A step, called a *chip breaker*, is ground behind the cutting edge, as shown in fig. 5.8, which causes excessive bending of the chip and hence premature fracture. It should be noted that use of a chip breaker produces larger cutting forces and hence higher power consumption but these disadvantages are outweighed by simplified swarf clearance.

PRACTICAL EFFECTS OF VARIATIONS IN THE CUTTING PROCESS

Specific Objective: *The student should be able to state the effects of variations in speed, feed, depth of cut and tool geometry on cutting force, power consumption and tool life.*

In considering the effects of variations in the cutting process on the force required and power

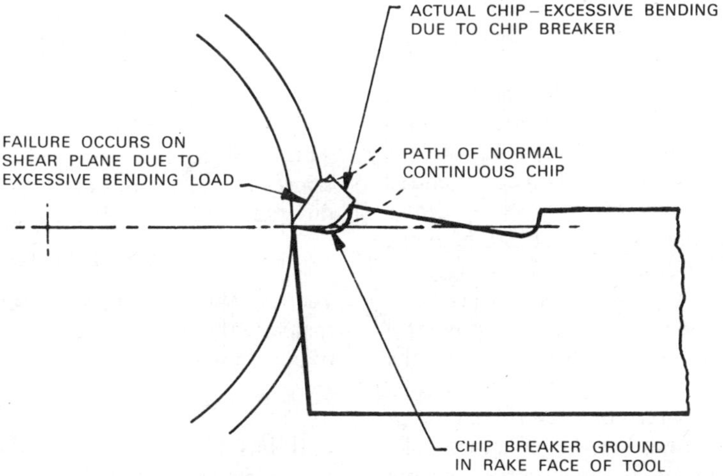

Fig. 5.8. Chip breaker

consumed in the process there are two facts which must be taken into account:

1. The force required depends on the shear strength of the material and the area being sheared.
2. Power is the product of force and velocity, in this case largely the tangential force, F_t, and the cutting speed.

As an increase in cutting speed does not change the shear area there should be little change in the cutting force with speed. This is verified by experimental work on metal cutting in which measurement of cutting forces at different speeds shows that, starting with very slow speeds the force is high. As the speed increases the force drops sharply and then remains almost constant, increasing only slightly as the speed goes up. These variations are due to the changes that occur in the friction characteristics with the cutting speed and once the initial condition has stabilised there is little change. However, the power required, being proportional to the speed, increases almost directly with the cutting speed.

If the feed rate is increased in mm/rev the shear area also increases, as it does with an increased depth of cut. It follows that the cutting force, and therefore the power will both increase if the feed rate or depth of cut are increased.

We have shown that an increase in the rake angle decreases the shear area and therefore the force and power required are both reduced. The amount the rake angle can be increased is limited, however, as the tool becomes weaker if the rake angle is increased. If the rake angle is too large the tool will break and recommended rake angles are therefore a compromise depending upon tool material strength, work material strength and efficiency of cutting.

Practical tests can be carried out on a lathe to verify these statements. A robust powerful machine is needed, with a wattmeter in the power circuit to measure power requirements, and the variables changed one at a time. Except for the tests to determine the effect of rake angle a tool having recommended cutting angles should be used and the rake surface lightly stoned between each test. Also the machine should be run with the feed engaged and the power noted *without* cutting before the test is carried out. This will give an indication of the power required to drive the machine and as this will vary with the different conditions it should be subtracted from the power measured during cutting to give, as far as possible, a measurement of power at the tool point.

Test 1 With feed rate and depth of cut constant, measure the power required at the speeds available on the machine. Plot graphs of:

(a) Power (kW) *v* Cutting speed (metres/min)

(b) Specific metal removal rate (mm³/min kW) *v* Cutting speed.

Note. The specific metal removal rate is a measure of the amount of material which can be removed per minute for the expenditure of 1 kW of power. If the feed rate is l mm/rev and the depth of cut d mm then from a bar of diameter D mm the volume, V, of metal removed per revolution is:

$$V = \frac{\pi}{4}(D^2 - d^2) \times l \text{ mm}^3$$

If the cutting speed is N rev/min then volume removed/min is

$$\text{Vol/minute} = \frac{\pi}{4}(D^2 - d^2) \times l \times N \text{ mm}^3\text{/min}$$

and specific metal removal rate

$$= \frac{\pi}{4}\frac{(D^2 - d^2)}{\text{kW}} \times l \times N \text{ mm}^3\text{/min kW}$$

Test 2 With cutting speed and depth of cut constant, measure the power required at different feed rates. Plot a graph of:

Power (kW) *v* feed rate (mm/rev)

Test 3 With cutting speed and feed constant measure the power required for different depths of cut. Plot a graph of:

Power required *v* depth of cut (mm)

It has been suggested that an estimate of power at the tool can be obtained by running the machine to measure the power required under no-load conditions before the test is carried out. This is only an approximation because the frictional forces set up in the slideways, gears, and bearings of the machine will themselves increase when the cutting forces are applied. Although the tests give an indication of how changes effect the machine as a system, a more precise method of determining the effects of the changes on the tool itself is to use a tool dynamometer as shown on p. 95 and measure the forces on the cutting tool. In this case graphs can be plotted of the three forces, F_t, F_a, and F_r, against speed, feed, and depth of cut.

Similar tests can be carried out with tools of different geometry to determine the effects of changes in rake and plan approach angles on the forces involved in the cutting process. The effect of the rake angle has already been discussed. If the radial and axial forces are considered it will be appreciated that both are in the horizontal plane and could be resolved into a single force. It is the author's opinion that this force varies little and it is the ratio between the two horizontal forces which changes as the plan approach angle is changed. The main effect of changing the plan approach angle is to change the tool life as, for a given depth of cut, the length of tool edge cutting the work increases from a minimum with a knife tool as the angle increases. Thus the work done by the tool is spread over a greater length of cutting edge and the load per unit length of edge is decreased. This means that the greater the plan approach angle the longer will be the tool life. This, however, is only true up to the point where shattering commences but as a general rule the plan approach angle should be as large as is practically possible.

TOOL FAILURE AND TOOL LIFE

Under modern cutting conditions a tool usually fails in one of two ways:

(a) A land is worn on the clearance face, thus destroying the clearance angle and giving rise to rubbing, as shown in fig. 5.9(a).

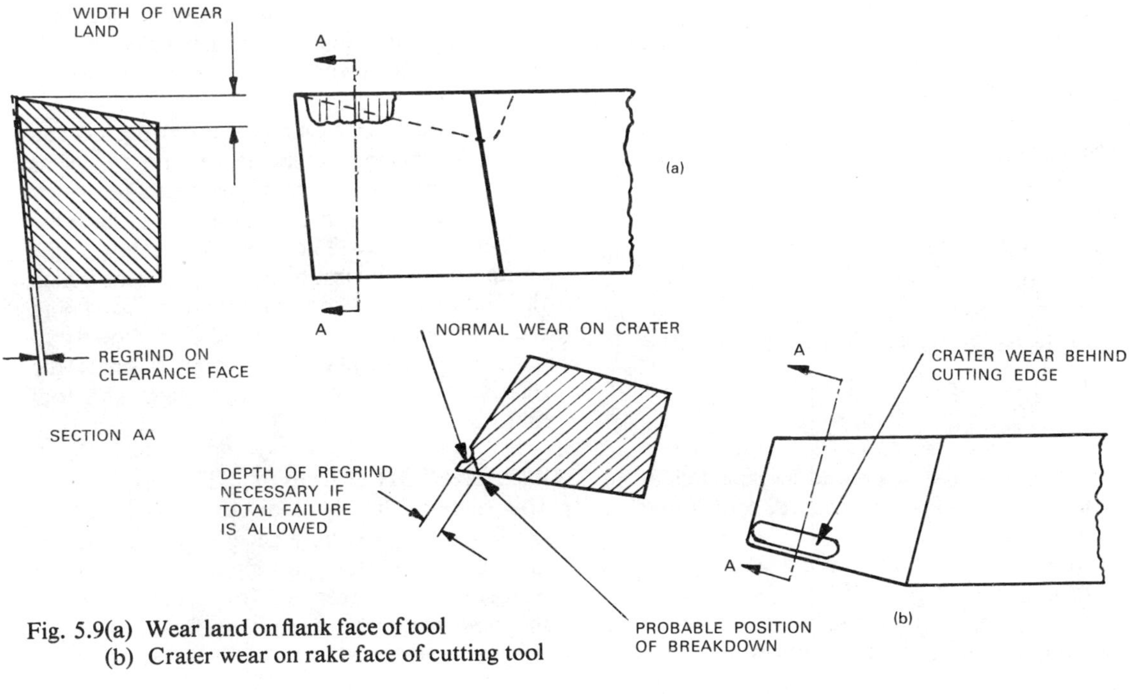

Fig. 5.9(a) Wear land on flank face of tool
(b) Crater wear on rake face of cutting tool

(b) A crater is worn in the rake face behind the tool edge.

If the crater is allowed to develop, the cutting condition changes to one of high positive rake, giving rise to weakness and a calamitous tool failure, as shown in fig. 5.9(b).

Either of these conditions can demand an excessive amount of tool regrinding and in the case of cemented-carbide tipped tools is an expensive business. A tool is therefore considered to have failed when either:

(a) the wear land exceeds 0.25 mm

or

(b) at the onset of cratering.

Modern production methods require that the tool life should be known and ideally last longer than one work shift, but a complete new set of pre-set tools is kept ready to be fitted into the machine with the minimum of difficulty. At the end of the shift the old tools are removed *before* they have failed, and are replaced by the set of new ones, for the next shift. This avoids replacing individual tools at irregular intervals, with a consequently greater machine 'down time'.

The tool life under the known cutting conditions can be found by testing under these conditions. It is fortunate that there is a mathematical relationship between the tool life T in minutes and the cutting speed V in metres per minute. The relationship is given by

$$VT^n = C$$

where n and C are constants depending on the conditions, i.e. depth of cut, feed rate, coolant, etc.

This expression can be transposed to:

$$T = \sqrt[n]{\frac{C}{V}}$$

and it follows that, as would be expected, as the cutting speed, V, is increased, tool life, T, is reduced.

It must be noted that n and c vary with different operating conditions, but they can be found by experiment and used to calculate a cutting speed for an operation which gives an optimum tool life. This technique is discussed more fully in the following book for Level 3.

CUTTING TOOL MATERIALS

Specific Objective. *The student should be able to compare the advantages and limitations of high speed tool steels, cemented carbides, ceramics, and diamonds.*

The development of cutting tool materials has been essentially a search for materials able to withstand the high temperatures generated at high cutting speeds. Ideally, a cutting tool should have the following properties:

(a) hardness to resist abrasion

(b) the ability to retain this hardness at high temperatures

(c) toughness to withstand any shock loads imposed, particularly during an interrupted cut

(d) the ability to provide a good surface finish to reduce the chip/tool friction.

In practice it has been found impossible to arrange for a given material to have all these properties and research has concentrated on (b), the hot hardness, other means having been found to overcome the deficiencies of the material so far as the other desirable properties are concerned.

The hardness of different tool materials can be plotted against temperature on a graph, as shown in fig. 5.10, across which can be drawn the line of the minimum hardness required for efficient cutting. The line then shows the comparative maximum temperatures at which the different tool materials can be used and although the graph cannot be used directly to determine cutting speeds it does indicate the relationship between cutting speeds and tool materials.

PLAIN HIGH-CARBON STEEL

This material has been discussed at length in Level 1. Here it is only necessary to point out that the steel has the property of softening as its temperature is increased. This process, known as *tempering* is most useful in many applications but is a limitation as far as metal

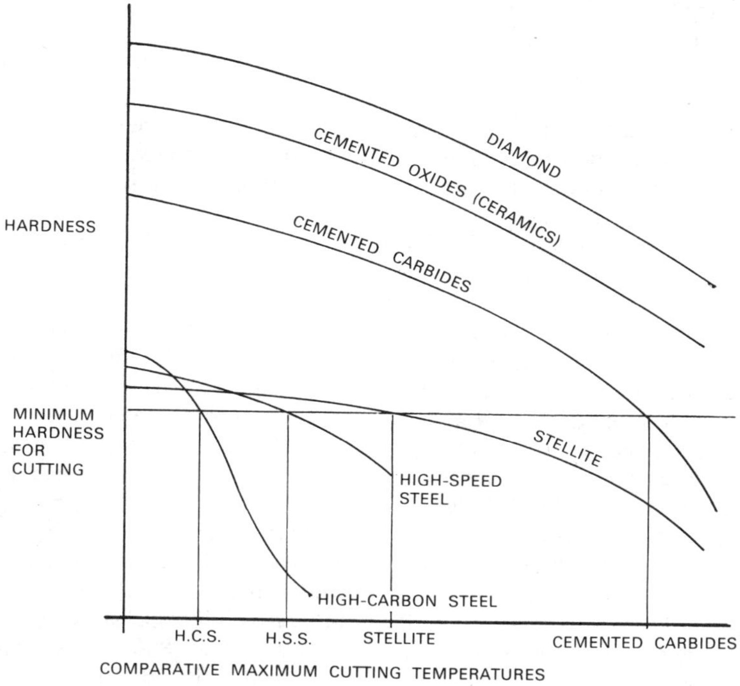

Fig. 5.10. Comparative values of maximum cutting temperatures for various tool materials

cutting is concerned. Cutting speeds with high-carbon steel tools are about one third of those with high-speed steel tools and the usefulness of high-carbon steel as a cutting tool material is largely limited to hand tools and other cutting tools operating at low cutting speeds.

HIGH-SPEED STEEL

Again, this material has been discussed at length in Level 1. Fig. 5.10 shows that its hardness does not drop below that required for cutting until a much higher temperature than that for plain carbon steel has been reached. This ability to remain hard at high temperatures is largely due to the inclusion of tungsten as an alloying element. The commonest high-speed steel is known as 18–4–2 H.S.S., these being the percentages of tungsten, chromium and vanadium present in a steel containing 0.7% carbon. The tungsten and chromium, in association with the carbon, form very hard carbides which themselves associate to form double carbides of high stability at high tempertures, hence the hot hardness. The vanadium has the effect of refining the grain structure and improving the shock resistance of the material.

High-speed steel is still probably the most common cutting tool material for use in machine tools. It is an extremely good material in the medium speed ranges, the cutting speed for its use on low carbon steel work material being approximately 30 m/min. Regrinding is a simple operation; cooling of the tool during grinding is necessary but not so critical as with plain carbon steel tools.

STELLITE

This is the proprietary name of a non-ferrous cast alloy of cobalt, chromium and tungsten, containing iron only as an impurity. It is hard in the as-cast condition and cannot be further hardened or softened by heat treatment. Fig. 5.10 shows that it retains its hardness up to higher temperatures than high-speed steel and it will in fact cut when red hot. It is also tough enough to be used as a tool bit or drill without having to be brazed as a tip on a tough steel shank. Stellite as a cutting tool material has been rather overshadowed by the cemented carbides but due to its extreme hot hardness it is said to have the ability to 'drill holes in hardened steel'. Actually this is not strictly true. Holes can be made in hardened material because this is locally softened by the heat generated by the drill friction. The work is clamped to the machine table, the drill run at high speed and pressed against the work until both work and drill become hot. This softens the work locally but not the drill which, although it may be glowing a dull red, proceeds to cut as soon as the work is soft enough. Obviously no coolant must be used in this process.

Apart from the specialised application outlined above, the main use of stellite is now not in the metal cutting field. It can also be deposited on steel as a hard facing by a process similar to welding and is thus used for repairing tools which do not cut but are subject to friction and wear at high temperatures. A typical example is a pressure die-casting tool where the hot metal moving at high speeds erodes the die, particularly at the point of entry. This wear is made up by the deposition of stellite. Stellite can only be finished by grinding, of course, since it cannot be softened to enable it to be cut by conventional tools.

CEMENTED CARBIDES

Carbides of certain metals, notably tungsten and boron, are extremely hard but are not in themselves tough enough to stand up to the rigours of the metal cutting process. They are therefore bonded together by another material just as bricks are cemented together by mortar, hence the term cemented carbides. In this case the bonding material is the metal cobalt. The tungsten carbide, or boron carbide, is milled to a fine powder and mixed in the desired proportions with powdered cobalt. The mixture is then placed in a mould and heated under pressure to form a slug of the material. At this stage it is crumbly or friable and is in what is known as the 'biscuit' form. It can, however, be handled and is then heated to a very high temperature in the presence of hydrogen. The hydrogen atmosphere prevents damage by atmospheric oxygen and the now familiar cemented-carbide tool tip is produced. Fig. 5.10 gives some indication of its hardness values, which may be as high as Rockwell 85C. Increases in temperature reduce the hardness slowly but at 800°C the hardness is about Rockwell 65C, harder than fully-hardened high-speed steel at room temperature.

For many years cemented carbides were used as tips brazed to a tough medium carbon steel shank but later developments include the clamped-tip tool discussed in detail in a subsequent section.

As a general guide, the cutting speed for a given material using cemented-carbide tools is about three times that for high-speed steel with the same material. Certain precautions must be observed in the use of cemented carbide tools, notably:

(a) The machine must be rigid and free from vibration;

(b) The power and speeds available must be adequate to allow the tool to be used at high rates of metal removal;

(c) The tool must not be allowed to rub and it is better to withdraw the tool while the feed is still engaged rather than disengage the feed first;

(d) It is well worth while stoning the rake

face of the tool to a high degree of surface finish.

Regarding (a) and (b), it is remarkable how many tools of this type are used on unsuitable machines because it is the done thing to use carbide-tipped tools. They are expensive to produce and maintain and should not be used where high-speed steel would do equally well. Regarding cutting speed, it will be found that if the cutting speed is too low a good but dull matt surface finish is produced. If the speed is raised, a point will be reached where the finish is equally good but brilliant and shining. The speed is then about right.

CERAMIC TOOLS

The word 'ceramic' is derived from a Greek word meaning 'of the earth' and referred originally to earthenware products of the potter. With the growth of the electrical industry the same processes were used for electrical insulators[1] and as these developed it was realised that certain of the materials had the basic properties required for cutting tools, i.e. hardness and the ability to retain that hardness at high temperatures.

The most common ceramic material used for cutting tools is aluminium oxide, Al_2O_3. This is an abrasive or cutting constituent of grinding wheels but for the conventional metal cutting process a much denser form is required. The grit is milled to a very fine powder, mixed with very small amounts of bonding material and pressed into shape in moulds. The slugs produced are then fired in kilns at extremely high temperatures to produce the tool tips used for cutting. They are so hard as to be almost unmachineable and are finished by light grinding with a diamond wheel.

The material is weak in bending strength and extremely brittle, so must consequently be well supported. The introduction of these ceramic tips led to the use of clamped throwaway-tip tools, since regrinding is extremely difficult and they cannot be brazed to tool steel shanks.

These tips are particularly well suited to finish-turning at very high speeds with low feed rates, the cutting speed being of the order of ten times that used with high-speed steel. They particularly lend themselves to negative-rake cutting.

DIAMOND TOOLS

Being the hardest material available to man diamonds were used for cutting tool purposes before either cemented carbides or ceramic materials, both of which are man made. Diamonds occur in nature and were therefore available once the techniques for their use had been developed.

The diamond is usually clamped in a special toolholder, the cutting edge profile being either ground to an arc of a circle or as a number of very small facets. In either case, when the edge is worn it is rotated to bring a fresh facet or length of arc into position.

Diamond tools are essentially finishing tools, the maximum depth of cut being 0.15 mm and the feed rate an absolute maximum of 0.05 mm/rev. As these tools are used to give much finer finishes than can normally be obtained by turning feed rates of 0.015 mm/rev. are much more common, with speeds up to 650 m/min.

These tools can be used to give very long life and a remarkable finish when machining materials which can normally be easily machined or for machining abrasive materials such as plastics where their life can be measured in months rather than minutes as with other tool materials. As with most very hard materials, diamonds are extremely brittle and easily damaged if subject to shock. For this reason the machine must always be running when the tool is in contact with the work and the tool should not be allowed to rub. Vibration of any type will cause early breakdown of the tools and specially designed machines are used in which gear drives are eliminated.

Apart from machining abrasive materials the main purpose for which diamond machining is used is to give very high accuracy and surface finish on bearings and aluminium alloy pistons.

[1] A spark plug manufacturer kindly provided the author with some specially formed sticks of his insulator material which successfully cut mild steel at 300 m/min surface speed.

NEGATIVE RAKE CUTTING

Specific Objective: *The student should be able to compare the relative merits of positive and negative rake angles.*

We have seen that the modern tool materials, cemented carbides and ceramics, have the basic properties of abrasion resistance and hot hardness, but are extremely brittle and will not withstand shock or bending loads well. They must be well supported and designed so that the applied loads do not give rise to conditions which cause early failure.

Fig. 5.11(a) for positive-rake cutting shows that the force N normal to the rake face acts towards the unsupported part of the tool and tends to break off the tip.

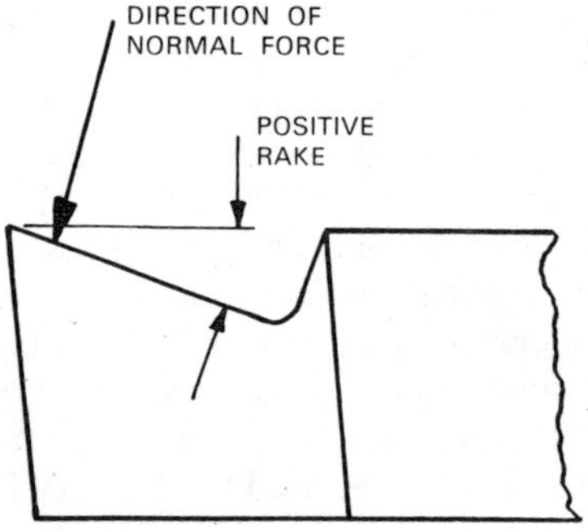

Fig. 5.11.(a) Positive rake cutting tool

If the rake angle is made negative as in fig. 5.11(b), the normal force N is inclined backwards into the body of the tool and the tip is in direct compression, which is a form of loading it is better equipped and supported to withstand.

Due to the much higher cutting speeds used with these tool materials, conditions in the cutting zone are different from those in conventional cutting. The action is so rapid that the heat generated by chip friction does not have time to pass into the work or the tool and almost all the heat is concentrated in the chip. The chip becomes much more plastic and a thin continuous chip results from a rake angle which would undoubtedly give a thicker discontinuous chip at lower speeds.

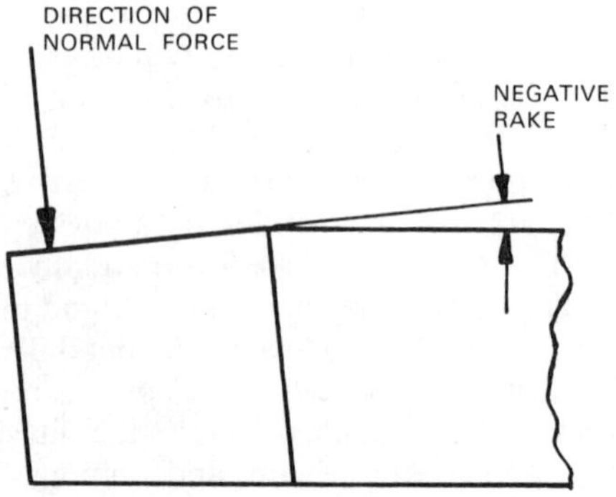

Fig. 5.11(b). Negative rake cutting tool

CLAMPED-TIP TOOLS

If the tool used has a negative rake angle equal to the clearance angle, the angle between the rake and clearance faces is 90°. Thus a square slug of material with a suitable nose radius on each corner can be used directly as the cutting medium. Such a tool is shown in fig. 5.12. The tip is clamped in position, well supported to resist the cutting forces and with a minimum amount of tip overhang. The shank is so designed that the tip is inclined about an axis along the cutting edge, this inclination providing rake· and clearance, and the cutting edge is thus at constant height over the whole depth of cut.

When the edge is worn the tip is turned through 90° to bring a new edge into action. When all four top edges are worn the tip is turned over to reveal four new edges, giving a total of eight in all. When all the edges are worn the tip is disposed of, hence the name 'throw-away' tip.

A similar development has taken place with milling cutters. Until recently, inserted-tooth face mills had a series of brazed tips set into the body. When the tips were worn, the regrinding and resetting of the teeth were critical to get all the tip points at the same height. Modern cutters enploy throw-away tips which are

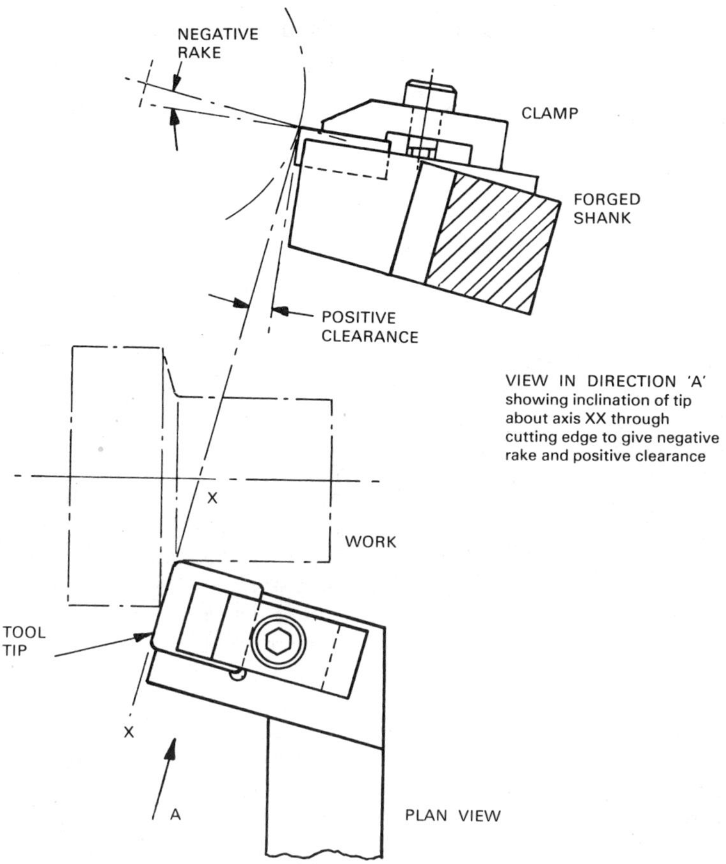

Fig. 5.12. Throw-away type clamped tip tool

accurately located in the body and, when replaced, position themselves accurately.

CUTTING FLUIDS

General Objective: *The student should appreciate the roles of cutting fluids.*

In the metal cutting process heat is generated in four places by two different methods of converting mechanical energy into heat.

Specific Objective: *The student should be able to state four characteristics of a cutting fluid, including cooling and lubricating properties.*

(a) *Frictional heat generation* In the section on tool failure and tool life (p. 103) it was shown that wear due to friction takes place on both the rake face and the flank face of the tool. This wear is due to friction and the energy used in overcoming friction is eventually converted to heat. Observation of a non-cooled chip or inadvertently touching a tool just after a cut has been taken emphasises the point.

(b) *Heat generated in metal deformation* Although not so obvious a little thought shows two zones in which the metal is deformed. Immediately ahead of the tool cutting edge deformation and shearing takes place and bending of the clip occurs as it curls over the rake face. Energy is required to cause these deformations and this mechanical energy is ultimately converted to heat. These points of heat generation are shown in fig. 5.13.

Although the heat generated by friction cannot be entirely eliminated it can be reduced by providing good lubrication between the tool faces and the work to reduce the coefficient of friction. The remaining frictional heat and all of the heat generated by metal deformation must be carried away by the cutting fluid in order to cool the tool so that it retains its hardness.

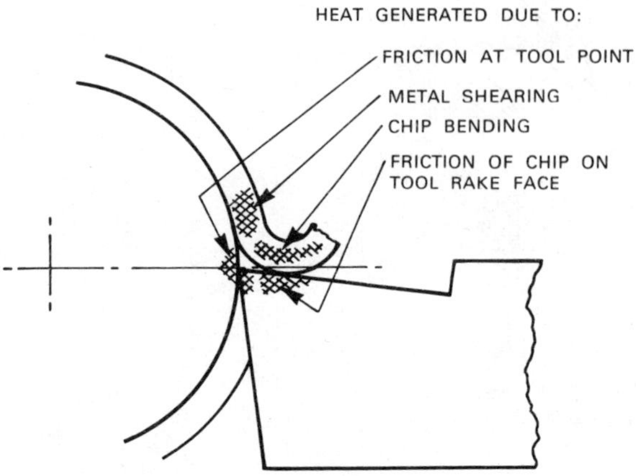

Fig. 5.13. Heat generated in metal cutting

As the chip passes over the rake face of the tool the local pressures are extremely high and the two surfaces, rubbing against each other under such pressure become chemically clean. Heat, pressure, and chemical cleanliness are ideal conditions for pressure welding. If particles of metal do weld to the tool face they form a built up edge (fig. 5.7(b)) which breaks off and is carried away over the tool faces. This process is repeated at high frequency and, as the particles forming the built up edge are work hardened they are abrasive and rapidly accelerate tool wear and breakdown. A further function of the cutting fluid is therefore to prevent local welding of metal particles to the tool.

Apart from cooling the tool the fluid must also cool the work in order to minimise heat distortion and it is also useful in flushing swarf away to prevent clogging of the work zone.

The cutting fluid must therefore be:

1. A good lubricant to minimise heat generation.
2. A good coolant to carry away heat formed during cutting.
3. A contaminating agent to prevent welding at the chip/tool interface. Additionally the fluid must be chosen so that it:

1. Prevents rusting of the machine slideways and working surfaces.
2. Does not cause staining of the machine or workpiece by chemical action.
3. Does not cause smoke or fog when heated during cutting.
4. Does not smell objectionably or decompose.
5. Does not lose its physical properties.
6. Does not infect or otherwise harm the operator.

The development of cutting fluids is really the development of liquids having these properties. Fluids available are:

1. WATER

Water has most of the properties of a coolant. It has a high specific heat, about twice that of most oils, its thermal conductivity is about four times that of oil, and its latent heat of vaporisation about seven times that of oil. Unfortunately it has numerous disadvantages. It causes rusting of ferrous parts, both work and machines, it is a poor lubricant and offers no protection against local welding. Strangely enough it is not a good wetting agent in that it forms droplets and does not form an even film over the work and tool surfaces (pour too little water to cover the bottom into a clean saucepan and see if it spreads all over the surface). Its wetting properties can be improved by the addition of soda and it is then less corrosive to ferrous materials but a soda solution should never be brought into contact with aluminium.

2. SOLUBLE OILS

Oil will not dissolve in water but it can be made to form an emulsion by the addition of agents which disperse the oil throughout the water as tiny droplets. These materials should then really be called *emulsifying oils* rather than soluble oils. They have very good cooling properties, probably better than water as they have better wetting properties. They do not cause rusting if properly used, but it is not advisable to leave a film of the coolant lying on machine slideways over a long period – clean them off when work is finished. They have some lubricating properties but no antiwelding action.

Soluble oils are best suited to light cutting operations on general purpose machines where high rates of metal removal are not of prime importance.

3. STRAIGHT CUTTING OILS

The basis of all straight cutting oils is mineral oil of a suitable viscosity. Its main advantage over soluble oil is its improved lubricating property but it will only maintain a lubricating film at relatively low pressures. This is improved by the addition of fatty oils such as rape oil and lard oil, these fatty oil blends being the most common of the straight cutting oils.

Fatty oils are rarely used alone as they tend to decompose and form sludge but they blend well with mineral oils which inhibit the decomposition.

4. EXTREME PRESSURE (E.P.) OILS

It is under heavy cutting conditions that the extreme pressures are produced which cause welding of particles of metal to the tool face. It has been found that the addition of flowers of sulphur causes a slight contamination of the tool surface which prevents this welding taking place. Thus the first of the E.P. oils was sulphurised mineral oil. However, sulphur can cause staining of yellow metals and some oils have been developed containing sulphur in such a way that no staining occurs.

Chlorine additive also give E.P. properties and oils containing chlorine in a suitable form are very good for use on alloy steels containing chromium, the stainless steels. Later developments include oils containing both sulphur and chlorine suitably combined to give multi-purpose E.P. oils.

5. PARAFFIN

The main use of paraffin as a cutting fluid is in the machining of aluminium alloys. It is not normally used neat but with the addition of up to 50% mineral oil.

SELECTION OF CUTTING FLUIDS

Specific Objective: *The student should be able to select and give reasons for using a suitable cutting fluid for a given operation.*

In order to select a suitable cutting fluid for a given job several factors have to be considered. These are:

1. Cutting tool material.
2. Workpiece material.
3. Operation being performed.
4. Machine tool in use.

1. CUTTING TOOL MATERIAL

As far as coolant application is concerned cutting tool materials can be divided into three categories.

(a) Plain carbon steels
(b) High speed steels
(c) Stellite, cemented carbides and diamond.

Years ago plain carbon steels were used for all cutting purposes. They have now been superseded by groups (b) and (c) in most cases and are currently used only for cutting at low speed with low metal removal rates. If, however, they are used for more arduous operations it is essential that they are well cooled since they start to lose their hardness at low temperatures. A copious supply of soluble oil is therefore used.

The materials in group (c) stay hard when hot and their composition tends to prevent chipwelding and the formation of a built up edge. If any coolant is necessary it is normally to cool the work to prevent distortion due to heat build up. However, in well designed tools of this type most of the heat goes away in the swarf and no coolant is necessary.

The main difficulty is in choosing a coolant for use with group (b), the high speed steels and here the other factors are an important consideration.

2. WORKPIECE MATERIAL

Certain work materials present particular problems and are dealt with individually. Other materials must be considered in conjunction with the operation being performed so that an overall assessment of the arduousness of the operation is made in selecting the coolant.

Materials are considered separately are as follows.

(a) *Cast iron*

This material breaks up, on being cut, into small chips and carbon dust. It is normally cut dry but soluble oil may be used to wash away the chips and lay the dust which can form a sludge in the coolant sump which may have to be cleaned out. On balance cast iron is better cut dry.

(b) *Yellow metals*

These again can be cut dry. If a coolant is used it *must not contain sulphur additives* or staining of the work will result. In any case such oils are not usually necessary and a soluble oil is normally suitable.

(c) *Aluminium alloys*

Most aluminium alloys cut easily and a soluble oil is suitable. Again oils containing sulphur should not be used or staining will result.

(d) *Magnesium alloys*

If magnesium in finely divided form is dropped in water it reacts and gives off hydrogen. This increases considerably the existing fire hazard present in machining magnesium and it follows that *no water-based coolant or soluble oil must be used* on magnesium. In addition sulphur and chlorine compounds tend to attack the metal and it is fortunate that it is an easily machined material which is normally cut dry.

(e) *Bearing metals*

Most bearing metals can be machined dry but if a cutting fluid is used to give the required finish soluble oils and E.P. oils should be avoided since the metal become slightly impregnated with the oil and trouble may arise in service. If a fluid is necessary a straight mineral oil should be used.

For all other cases the actual machining operation as well as the material to be cut must be considered. Any operation where the tool loads are high is likely to require an E.P. oil and therefore most broaching, tapping, and gear cutting operations require this type of oil. On the other hand toolroom turning operations, where the cut is continuous and not interrupted, and usually not of such an arduous nature, can generally be satisfactorily carried out using soluble oils. If a matrix is made up as in fig. 5.14 having the type of operation on one axis and the material on the other then a guide can be obtained as to the type of cutting fluid needed. It must be noted that this chart is really a 'basis for negotiation', as a really heavy turning operation with a fast feed rate and deep cut could move up the scale to the 'deep drilling' position, and light end milling could go down the scale to the turning position. Each case must be considered on its merits.

MACHINE TOOL IN USE

In some cases the actual type of machine may influence the choice of coolant. Where the work area is enclosed and flooded with a cutting fluid from a number of jets it can very easily seep into the bearings and under gearbox covers. Soluble oil could easily cause damage to such machines and this is an additional reason for using straight cutting oil on expensive machines such as multi-spindle automatics, gear cutting machines, etc. A slight seepage of these oils will do little or no harm to the machine tool.

APPLICATION OF CUTTING FLUIDS

Specific Objectives: *The student should be able to list and give reasons for using the most common methods of applying cutting fluid to given cutting operations.*

The method of applying cutting fluid depends again on the particular operation. Methods

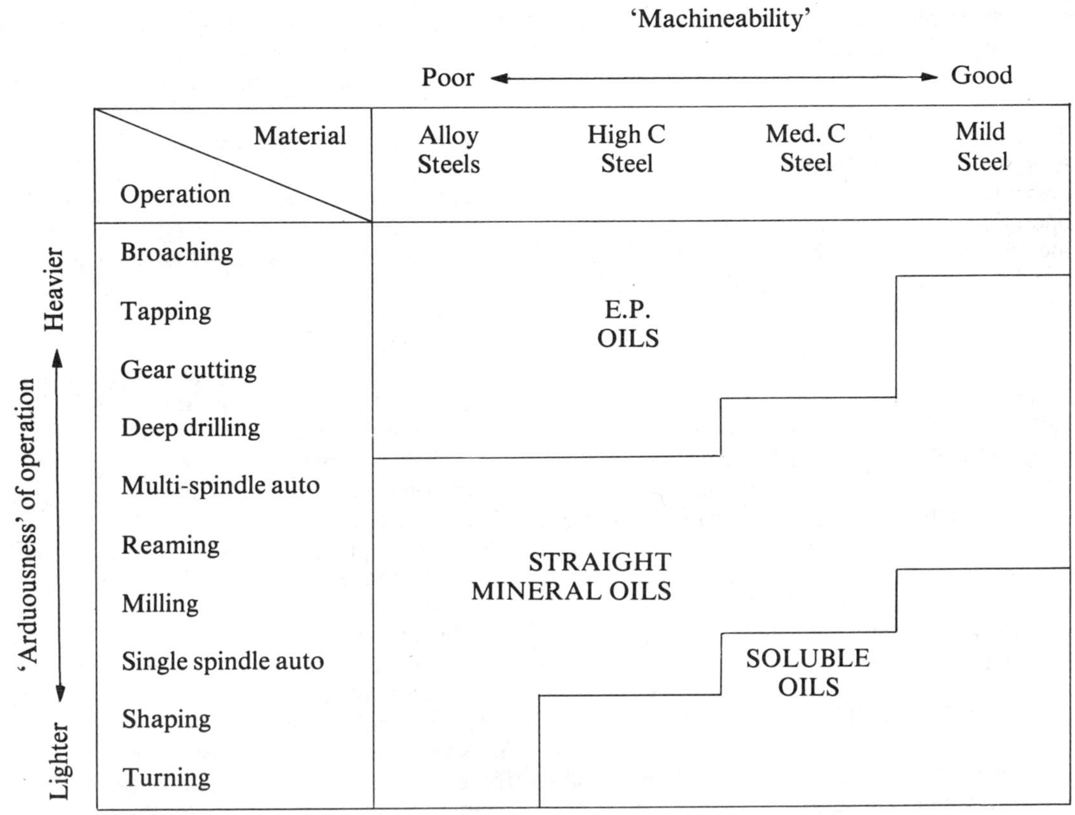

Fig. 5.14. Chart relating machineability of work material to type of operation to give guidance to a choice of cutting fluid

vary from hand application such as use of a brush or oil-can to high pressure application where a prime consideration is the flushing out of swarf. The methods available are as follows:

1. HAND METHODS

Coolant can be applied by brush from a can but it is a method that should only be used where no other is available. It only gives an uncontrolled and intermittent supply of coolant. From a safety point of view it puts the hand into the work area and if a brush is used it is better to be of the type having a wooden handle which can be released if it is picked up in the work. The type with the twisted wire handle is better avoided and should never be used with a finger through the ring at the end. Brushes tend to be used on shaping and milling machines in toolrooms. They cannot supply the flow needed for high production operations.

If coolant is to be applied by hand an oil-can filled with soluble oil used to give a squirt into the cutting zone as required is to be preferred.

2. PUMPED SUPPLY

Most modern machine tools have a sump and a pump to direct a flow of coolant into the cutting zone. For most purposes this is adequate and it must be remembered that in a cutting operation the chip itself tends to prevent the coolant getting where it is most needed – between the chip and the tool face. If a copious flow is used some does tend to get under the chip and at the same time, the more that is used the better will be the cooling effect.

3. HIGH PRESSURE APPLICATION

In some operations coolant is applied at high pressure to flush away the swarf. A particular operation of this type is deep hole drilling where the swarf clogs the flutes of the drill and becomes compressed, eventually seizing it in the hole and spoiling the size and finish, causing overheating due to friction and damaging both the drill and workpiece. Twist drills for this type of work are often produced with straight flutes initially so that holes can be drilled along the body of the drill into the shank where a tapping for the coolant supply can be made. Before heat treating and finish grinding the drills are heated and twisted to give helical flutes. The author has used drills up to 50 mm diameter made in this way. The coolant can then be applied under high pressure to literally force the chips up the flutes and away out of the hole.

CONTAMINATION OF CUTTING FLUIDS

Cutting fluids can become contaminated with bacteria which cause skin diseases, such as dermatitis, which can be passed from one operator to another. This is not so likely on individual machines which are used by one operator all the time but in large machine shops a common sump and drainage system is often used to supply a number of machines. In such cases the oil is filtered, cleaned and may be disinfected but the operator can help himself by following a few simple rules of personal hygiene. These below are taken from the book *Cutting Fluids* published by the Esso Petroleum Company Ltd.

1. Wash hands and forearms *before* starting work and apply a suitable barrier cream. Applying barrier cream over existing oil and dirt does more harm than good.
2. After work – wash thoroughly and change out of oily clothes. Keep to a personal towel.
3. Always wear protective clothing.
4. Use the splash guards on the machine.
5. Keep to your own rags and wipers and dispose of them when dirty.
6. Have proper dressings on cuts, abrasions, boils, etc.
7. Do not spit in the oil or otherwise contaminate it.
8. Do not wear oil soaked overalls longer than necessary.
9. In general, maintain strict personal cleanliness.

QUESTIONS

1. Make a diagram of a single point cutting tool and label the following angles:
   (a) front clearance
   (b) side clearance;
   (c) top rake;
   (d) side rake;
   (e) plan approach;
   (f) plan relief;
   (g) true rake.
2. Make a diagram of a single point cutting tool of the 'straight edged' type and explain the advantages of this type of tool over that sketched for question 1.
3. Name the instrument used for measuring the force acting between tool and work during metal cutting. In practice this force is determined by measuring three components mutually at right angles. Name these components and state:
   (a) which is the power consuming component?
   (b) how are these components affected by an increase in cutting speed?
   (c) how are these components affected by an increase in feed rate?
   (d) what is the effect of an increase in cutting speed on the power consumed.
4. The clearance angle for a single point cutting tool is fixed by the type of operation; other angles can be varied. Assuming that speed, feed and depth of cut remain constant what will be the effect on tool forces and power consumed by:
   (a) increasing the rake angle?
   (b) increasing the plan approach angle?

What limits the value of these angles?

5. Describe with the aid of a diagram two modes of tool failure in metal cutting. In a particular metal cutting operation the relationship between tool life T minutes and cutting speed V metres/min is given by

$$VT^n = C$$

for this particular operation $n = 0.59$
$C = 1320$

(a) Calculate the cutting speed in metres/min which will give a tool life of 100 minutes.
(b) If the speed is fixed by other factors at 50 metres/min, what will the tool life be?

6. List four materials used for metal cutting tools in machining operations and compare their advantages and limitations under the headings: cutting speed, tool mounting, applications and tool life.

7. Explain what is meant by the term 'negative rake', and explain why negative rake tooling is used so often with cemented carbide and ceramic tools. Sketch a clamped tip tool and show how regrinding has been eliminated by the use of such tools for cemented carbide and ceramic tips.

8. List four sources of heat generation in metal cutting and state which can be reduced by lubrication and which cannot.

9. List three characteristics which a cutting fluid should possess and six effects it should not have on the work, machine or operator.

10. List four factors which influence the choice of a cutting fluid and explain how each factor influences the choice of a fluid for a given operation.

11. Name four cutting fluids (other than water) and for each state a typical application and a situation in which each fluid must *not* be used.

12. List nine rules for personal hygiene by which infection due to contaminated cutting fluids can be avoided.

ANSWERS

5. (a) 87.22 metres/min.
(b) 257 min.

6 Machine tools – introduction

General Objective: *The student should appreciate the use and operation of the commonly used machine tools.*

INTRODUCTION

There are certain factors which are common to all machine tools. They are all required to produce components of the required shape and the correct size and in order to achieve these requirements the motion of the cutting tool relative to the work must be controlled in two ways:

(a) the direction and geometry of its motion must be controlled to give the correct shape, and
(b) the position of, or distance moved by the tool must be controlled to produce work of the correct size.

The particular methods of controlling tool motion and position will be discussed in the sections dealing with particular machine tools but those factors common to all machine tools will be discussed here in the introduction.

CONTROL OF SHAPE AND FORM

Specific Objective: *The student should be able to explain, using drawings, how shapes are produced using copying and generating methods, or a combination of both.*

Consider the component shown in fig. 6.1 produced in a lathe by causing the work to rotate and achieving the required shapes by the motions of a single point tool. It can be seen that for each surface produced two motions are required plus a further one to adjust the depth of cut.

These are:

1. rotary motion of the work about a fixed axis, with

either 2. linear motion of the tool parallel to the axis, to generate a cylindrical surface:

or 3. linear motion of the tool at right angles to the axis, to generate a flat surface:

or 4. linear motion of the tool at an intermediate angle to the axis, to generate a conical surface or taper.

The generating motions for a cylinder are the rotary work motion and the parallel tool motion, with right-angled motion of the tool to set the cut. For a flat surface, the right-angled tool motion generates the surface, with parallel motion to set the cut. In generating a taper, the right-angled motion is most frequently used to set the depth of cut. This is typical of a generating process, the points to note being:

1. The shape produced depends entirely upon relative motions between the tool and the work.
2. The tool is designed for efficient cutting and its shape has no affect upon the shape of the work produced.

A similar situation occurs with the shaping machine which may be defined as a machine tool designed to produce flat surfaces by a generating process using a reciprocating single point tool. These surfaces may be horizontal, vertical, or at some angle between the two.

The essential motions for the machine to fulfil these functions are as shown in fig. 6.2 and are as follows:

1. *Reciprocating motion of tool*
This motion supplies the power for cutting and must be adjustable for speed, length of stroke and position of stroke over the workpiece.
2. *Feed motion for horizontal surfaces*
This is the motion of the table across the front

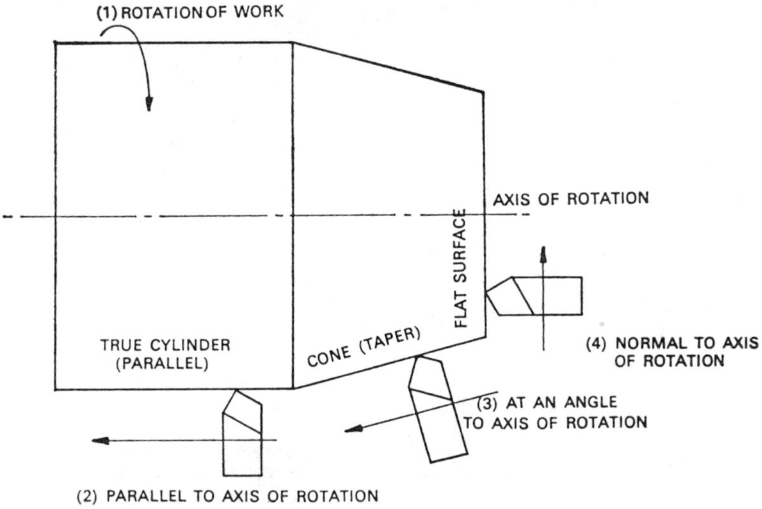

Fig. 6.1. Relative motions of tool and work for generating cylindrical, flat, and conical surfaces on a lathe

of the machine, which may be hand or power fed.

3. *Feed motion for vertical surfaces*

This is normally a hand-fed motion of the toolpost in a head slide mounted on the front of the ram head, but it may be power fed.

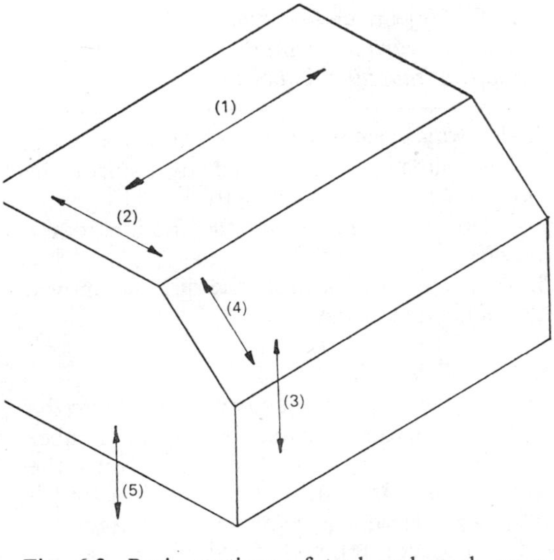

Fig. 6.2. Basic motions of tool and work on a shaping machine

4. *Feed motion for inclined surfaces*

This is a variant of (3) above; the head slide is set over at the angle of inclination required and the tool feeds down at this angle.

5. *Vertical table adjustment*

This is an additional motion not normally used for feed purposes but to adjust the height of the table to accommodate wide variations in the height of the work.

In the case of milling machines which use a multi-toothed rotating cutter to produce a flat surface the situation is slightly different in that, in one direction, the shape of the surface produced is *copied* from the shape of the cutter and in the other direction it is generated by the machine motion. These effects are shown in figs. 6.3(a) and (b), the motions being

1. *Rotation of cutter*

This is the motion through which the power is applied.

2. *Longitudinal table motion*

This is the feed motion which, in conjunction with motion no. 1, produces the cutting action.

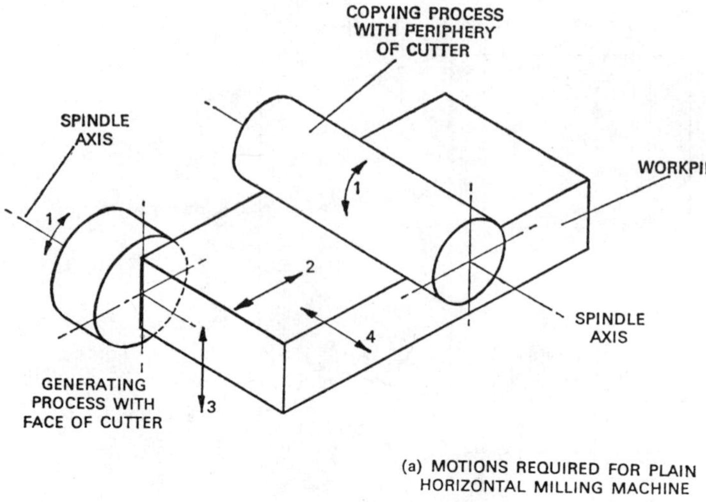

Fig. 6.3(a). Basic motions of tool and work on milling machine

3. *Vertical table motion*
This motion is used to set the depth of cut. On vertical-spindle machines it is also usually possible to feed the cutter axially downwards, an additional motion, no. 5.
4. *Transverse table motion*
This crossward motion of the table is used to position the work under the cutter. It may occasionally be used as a cross-feed.

During cutting, only motions 1 and 2 are normally used, the other motions, i.e. adjustments, being locked.

It might appear at first sight that horizontal machines always utilise a copying process and vertical-spindle machines a generating process, but this is not so. With a face milling cutter mounted in the spindle, or a side-and-face cutter on the arbor, the horizontal machine can be employed for a generating process. Correspondingly, the periphery of an end mill copies a vertical surface when used in a vertical-spindle machine as shown in fig. 6.3(b).

All these cases have one thing in common, *if the geometry of the work is to be correct then the machine motions must be truly linear.* The motions are controlled by the slideway system of machine and it follows that great care must be taken in the design of the slideway system.

SLIDEWAY SYSTEMS FOR MACHINE TOOLS

Specific Objective: *The student should be able to describe various methods used to control the motions of machine tool elements.*

A slideway system consists of three elements:
1. The slideway itself used to control the motion of the moving element
2. A means of imparting the motion to the moving element
3. A means of measuring the distance moved by the moving element.

1. SLIDEWAYS

The slideway itself is usually machined into the machine structure and may be one of a number of different types depending on the requirements. In some machine tools a simple guidance system is acceptable where friction is not too important. Some slideways may be required to resist lifting forces while in others

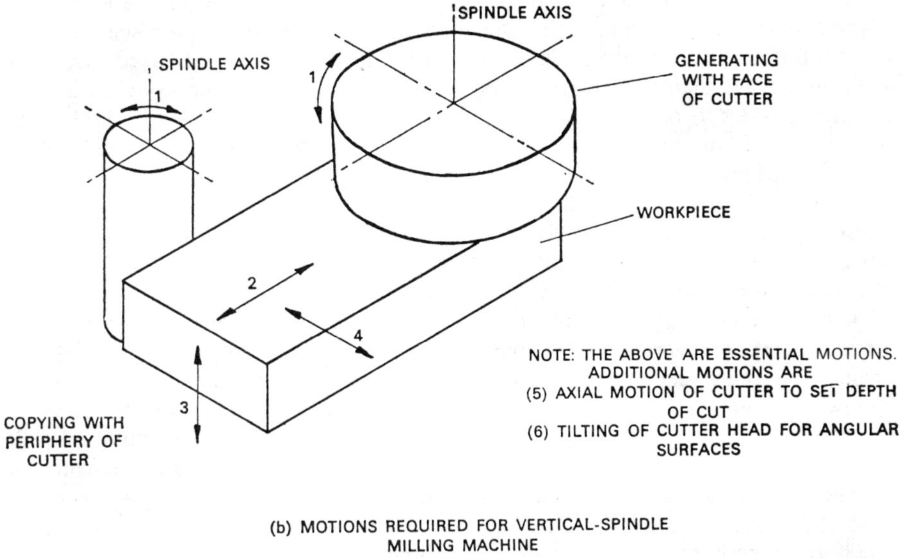

Fig. 6.3(b).

there may be no tendency to lift one element clear of the other. In other cases, particularly with modern numerically controlled machines and high precision machines it is necessary to virtually eliminate friction in order to enable the required high positional accuracy to be achieved.

Dovetail slides

If the forces produced by the cutting action are such that they tend to produce a tilting or lifting effect, dovetail slides are often used. The construction is usually as shown in fig. 6.4 with an adjustable gib key being used to take out any slackness and reduce the tendency to lift. A

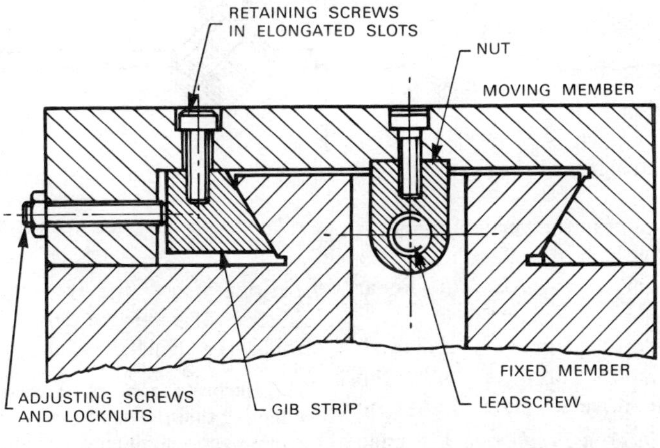

Fig. 6.4. Dovetail slide with adjustable gib strip. This is a cross section of the slide shown in fig. 6.9(b)

point about this type of slide particularly is that it must be designed on a *narrow-guide* principle. Most people are familiar with the dressing table or cupboard drawer which jams when pushed in by a single hand to one side of centre. It can be shown[1] that to avoid jamming the eccentricity of the force must be less than $l_{2\mu}$ where l is the slideway length and μ is the co-efficient of friction. In order to allow for an eccentric force it is therefore necessary to make the slideway length longer in relation to its width.

Dovetail slides are typically used on lathe cross-slides and compound top-slides, shaping machine ram and table slides and milling machine slideways.

Vee slides

This type of slideway would be better described as a vee/flat slide since it is very difficult to achieve the correct geometry with a pair of parallel vees. If such vees are out of parallel in either direction the moving member will ride up or down as it is moved along.

The form of the vee/flat slideway is shown in cross-section in fig. 6.5 and it can be seen that it consists of one vee and one flat. As long as:

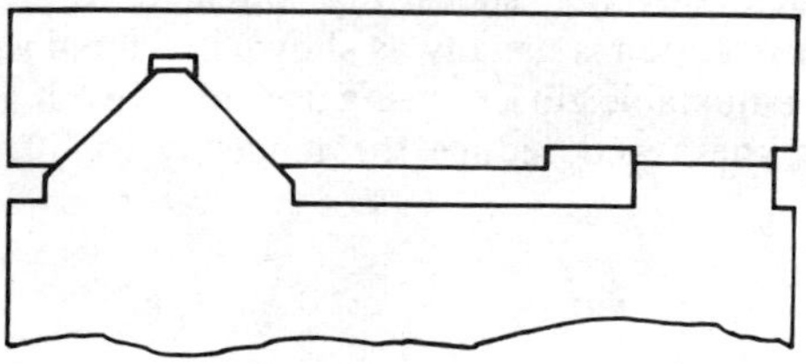

Fig. 6.5. Vee/flat slide

(a) the vee is straight in horizontal and vertical planes,
(b) the flat is straight in the vertical plane,
(c) the vee and flat lie in parallel planes,

the moving member will have true linear motion:

This type of slideway can withstand high loads but the friction force is high. An advantage over the dovetail slide is that no

[1] *Principles of Engineering Production*, Lissaman and Martin. English Universities Press.

adjustment is needed as there can be no side-play when the bearing surfaces wear. Typically this type of slide is used for lathe beds. Examination of most lathe beds reveals two vees and it may be thought that a double vee slide[1] is being used but closer examination reveals two vees and two flats, one vee and flat being used to guide the saddle and the other the tailstock.

Anti-friction slides

In order to achieve the high positional accuracy required on grinding machines, jig-borers, and numerically controlled machine tools, it has been necessary to develop slideways in which friction is reduced to a minimum. This is to enable heavy machine elements such as grinding wheel heads and jig-borer tables to be moved with the minimum effort. At the same time it is necessary that correct geometric principles are adhered to and hence the vee/flat roller slide has been developed. Such a slide by Societé Genevoise for their jig-borers is shown in fig. 6.6. Crossed rollers, whose length is slightly shorter than their diameter, are used in the vee. Thus the table is supported on the

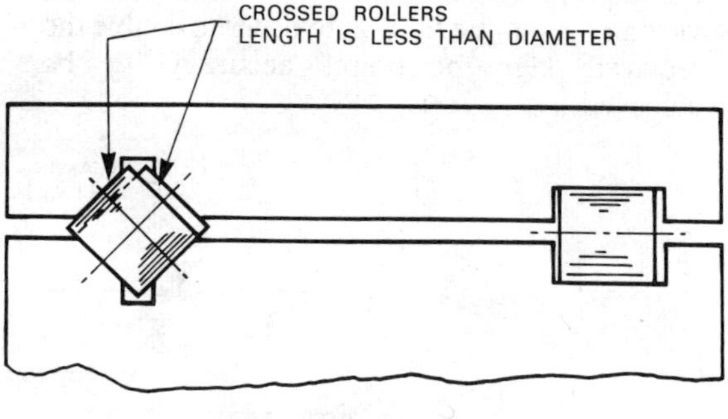

Fig. 6.6. Vee/flat slide with crossed rollers

periphery of the rollers giving minimum friction. If a very heavy load was placed on the table the rollers would indent into the vee

[1] Herbert machines do use two vees, the company claiming that the accuracy of their slideway grinding equipment eliminates problems due to non-parallelism of the vees.

leaving permanent impressions and ruining the slides. Before this happens the roller ends make contact and take the load, thus preventing damage.

Where very light loads are applied ball bearing slideways may be used as long as the track is made very hard (Rockwell C64) in order to prevent permanent indentation, or brinelling as it is called.

A problem with all ball and roller slides is that the movement is limited to twice the movement of the balls. This can be overcome by using a recirculating ball or roller slide in which the balls or rollers are diverted around a groove and carried back to the beginning of the track. This eliminates an unduly long ball track, most of which is empty and gathering dirt. A typical recirculating slideway is shown in fig. 6.7.

Another recent innovation in anti-friction slideway design is the development of *hydrostatic slideways*. Here the moving

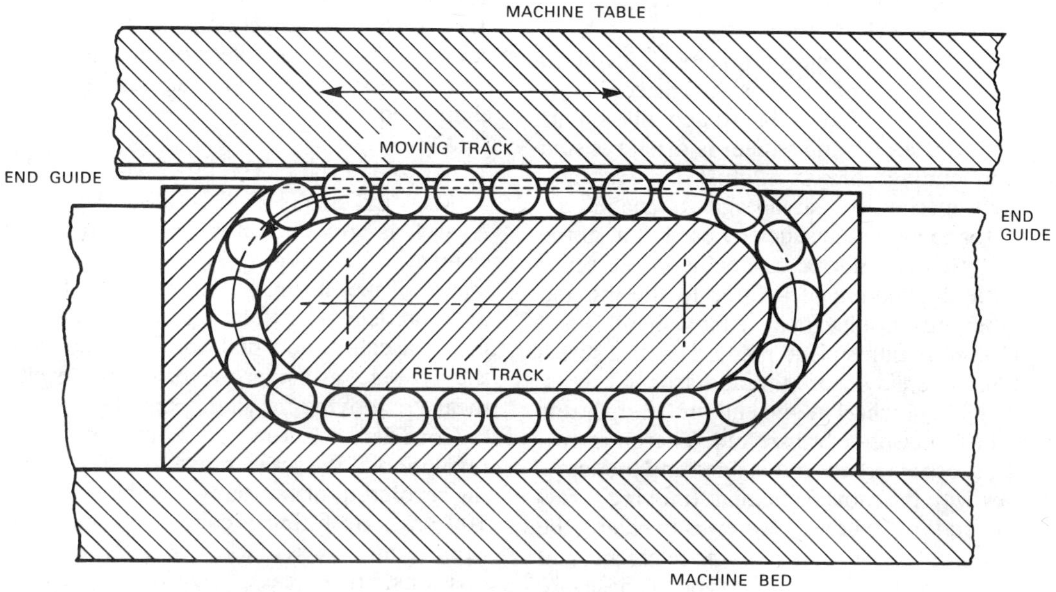

Fig. 6.7. Principle of recirculating ball and roller slideways

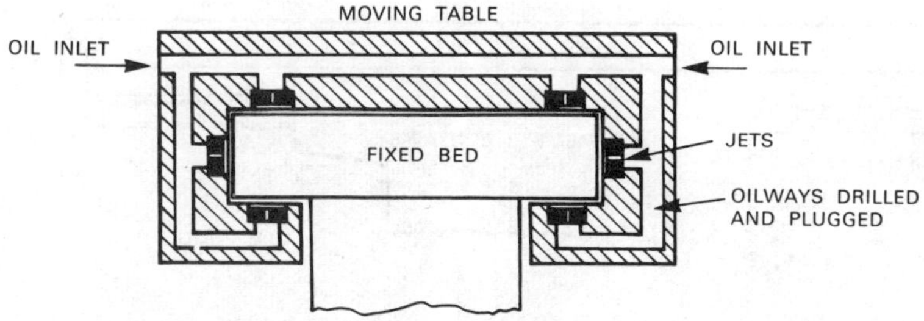

Fig. 6.8. Hydrostatic slide. Note that the jets are recessed. The depth and diameter of recess and the jet diameter are all important in the design of these slideways

member is supported on a series of pads of oil (or air) under pressure. Such a slideway is shown in fig. 6.8. These slideways can be made extremely stiff, i.e. they will carry heavy loads with very little deflection and have the advantage that errors in straightness of the bearing surfaces are averaged out by a factor of about 10:1. Thus to produce a movement where straightness is accurate to within ± 0.001 mm requires the straightness of the fixed member to be of the order of ± 0.01 mm.

2. IMPARTING MOTION TO THE SLIDING MEMBER

It is of little use building a precise slideway to guide the motions of a tool or workpiece if some means is not made available to actually impart the motion. As the movement is often resisted by the cutting forces, it must have a high enough mechanical advantage to enable the operator to move the slideway easily but at the same time must not be reversible in the sense that the application of a force to the moving member must not itself cause movement.

The most convenient method is the screw and nut where a square thread hardened screw having a handwheel at one end for the operator runs in a phosphor-bronze nut. There are two types, those where the nut is fixed and the screw moves with the moving element (moving screw type) and those where the screw is fixed and the nut is attached to the moving element and moves with it (moving nut type). These are illustrated in figs. 6.9(a) and (b). Generally the moving nut type is used where the moving element will not overhang the slideway and the moving screw type where the slideway is shorter than the moving element and overhang occurs. Examples are:

Moving nut – Lathe slideways
Milling machine cross slide
Moving screw – Milling machine table

As with anti-friction slideways, if very low friction is necessary the recirculating ball principle has been applied to the screw and nut. The screw consists of a ground helical track, as does the nut with steel balls in the track which circulate through the body of the unit. Friction is extremely low and backlash is almost eliminated.

3. MEASURING THE DISTANCE MOVED BY THE SLIDING MEMBER

As well as providing movement for the moving element of a slideway there must be some means of measuring the movement. The screw and nut lends itself to this measurement by fitting a calibration dial and virtually making the screw act as a micrometer screw as well as a means of moving the sliding member.

Consider a screw of 5 mm pitch. One revolution of the screw moves the slide 5 mm. If the revolution of the screw can be divided into 100 parts by fitting a calibration dial moving against an index mark, the dial having 100 divisions, then each division represents a

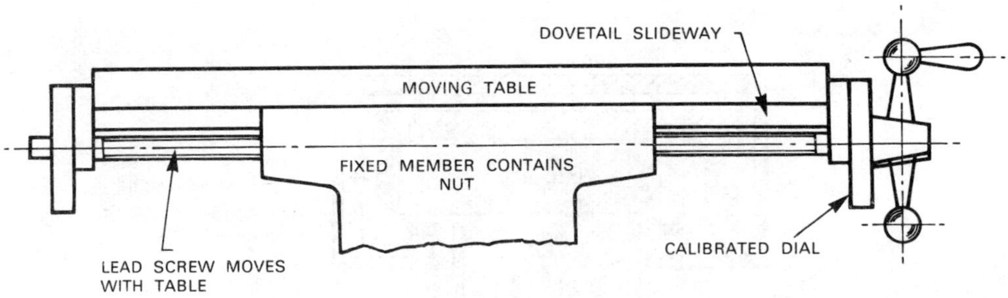

Fig. 6.9(a). Moving screw type slideway. Note calibration dials and provision for mounting handwheel at each end of table to allow for operation from either end

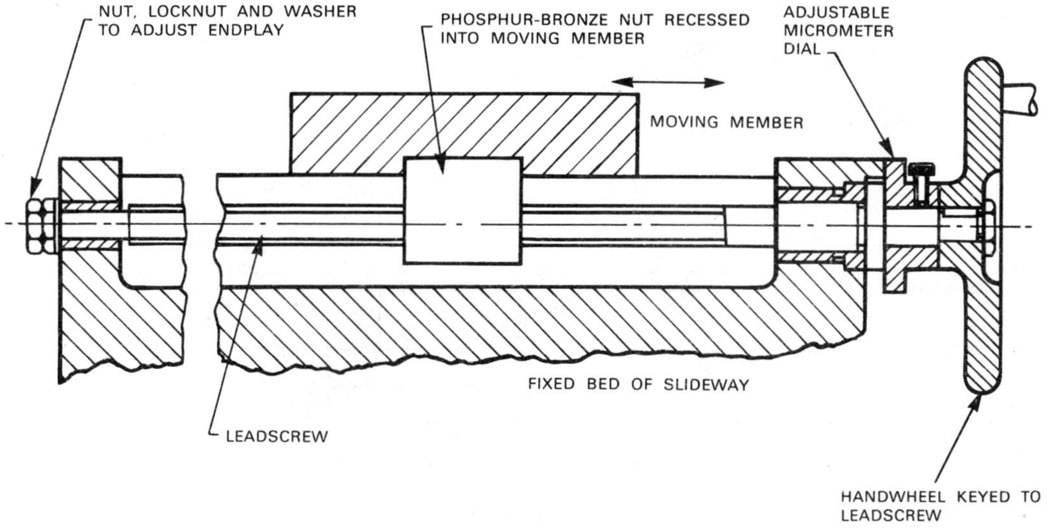

Fig. 6.9(b). Principle of slideway with moving nut. See also fig. 6.4.

movement of 0.05 mm. If each division is to be of an arc length of 5 mm then:

$$\text{Dial circumference} = 5 \text{ mm} \times 100 = 500 \text{ mm}$$

$$\text{Dial diameter} = \frac{500}{\pi} \text{ mm} = 159.13 \text{ mm}$$

As long as the divisions are of equal size the diameter of the dial is not too important and this would probably be rounded up to 160 mm. Such a calibration dial is shown in fig. 6.10. Using such a calibrated dial it would be possible to estimate between the marks to say 1/5th division giving an apparent control over the position of the tool of 0.01 mm. In practice, however, any such estimation is subject to error, not only in estimating the position of the index relative to the marks, but due to pitch errors in the screw. The positional accuracy using such a method would therefore be about ± 0.01 mm.

This accuracy could be improved by using a screw of finer pitch and a calibration dial with more divisions. Thus if a screw of 2.5 mm pitch was used with 250 divisions on the dial then:

$$\text{Movement represented by 1 division} = \frac{2.5 \text{ mm}}{250 \text{ mm}} = 0.01 \text{ mm}$$

With dial division of 5 mm arc length such a dial would be about 400 mm diameter which would be inconveniently large and the divisions would be reduced to say 2.5 mm arc length giving a dial of 200 mm diameter. Such small divisions would make estimation of position between the divisions more difficult and an accuracy of $\pm \frac{1}{2}$ division (0.005 mm) could be consistently achieved using such a dial.

Some machines require higher accuracy than this and accuracies of estimation of say ± 0.001 mm can be attained by either

(a) using finer screws and larger dials

(b) using a dial as in the first example with a vernier to subdivide the divisions into 0.001 mm.

(c) gearing the dial down so that it rotates at one tenth the speed of the screw.

In general all such systems calling for very high precision suffer from trying to make the screw do two things at once, both move and measure the movement of the slideway. On

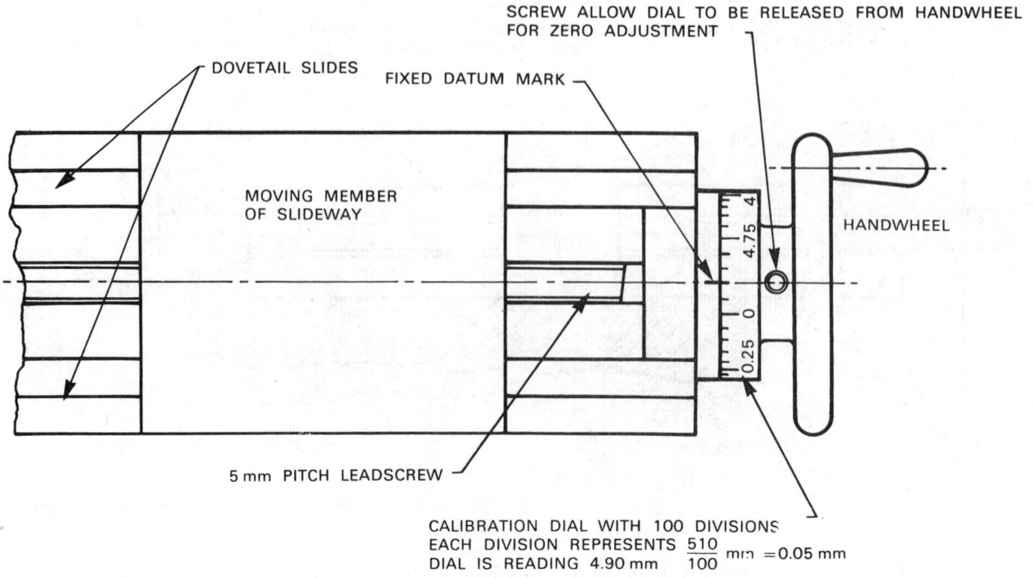

Fig. 6.10. Plan view of dovetail slideway showing calibration dial to indicate distance moved by slideway

many modern high precision machines the functions are separated by using the screw only to provide the movement of the slide and providing a separate, usually optical, measuring system to measure its movement.

This chapter has dealt with basic principles applicable to all machine tools. The following chapters will deal with the machine tools themselves, their capabilities and the way in which work can be carried out to give consistently good results.

QUESTIONS

1. Considering a screw cutting operation using a form tool on a centre lathe, state which part of the operation is copying and which part is generating. List the characteristics of the copying and generating processes.
2. Describe, with the aid of diagrams, two types of machine slideways which do not use anti-friction devices.
3. Describe with the aid of diagrams three types of anti-friction devices used in machine slides and discuss the advantages and disadvantages of each.
4. Explain the difference between 'moving nut' and 'moving screw' slideways, naming a particular application of each.
5. What function does a leadscrew often have as well as imparting motion to the moving part of the slideway?

7 Lathe work

PART I. THE CENTRE LATHE

General Objective: *The student should be able to describe the function and uses of a centre lathe.*

INTRODUCTION

Specific Objective: *The student should be able to identify the component parts of a centre lathe.*

A centre lathe is basically a machine tool used to produce cylindrical, flat, and conical surfaces by a generating process and the use of a singlepoint tool. From a previous discussion on cutting tools (Chapter 5) it will be recalled that the shape of surface produced by a generating process depends entirely on the relative motion between tool and work, not on the shape of the tool. In generating a surface, two motions are required, plus a further one to adjust the depth of cut, this adjustment remaining fixed while each cut is in progress.

For the application of this principle to the lathe, the motions shown in fig. 7.1 must be provided. These are:

1. rotary motion of the work about a fixed axis, with

either 2. linear motion of the tool parallel to the axis, to generate a cylindrical surface;

or 3. linear motion of the tool at right angles to the axis, to generate a flat surface;

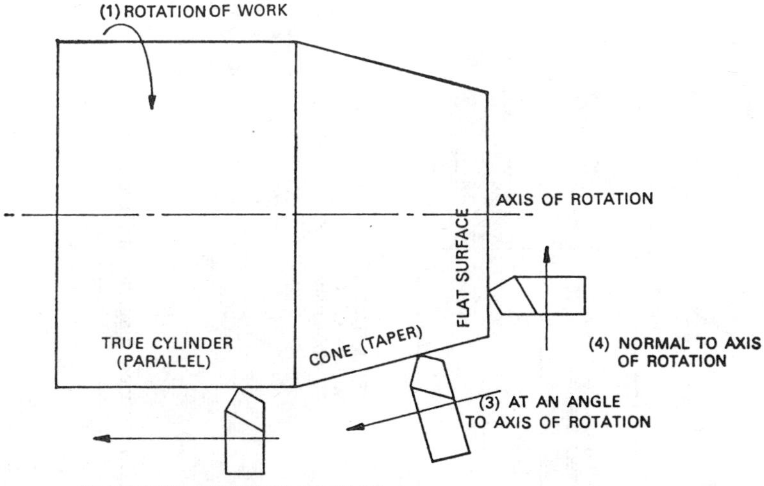

Fig. 7.1. Relative motions of tool and work for generating cylindrical, flat, and conical surfaces on a lathe

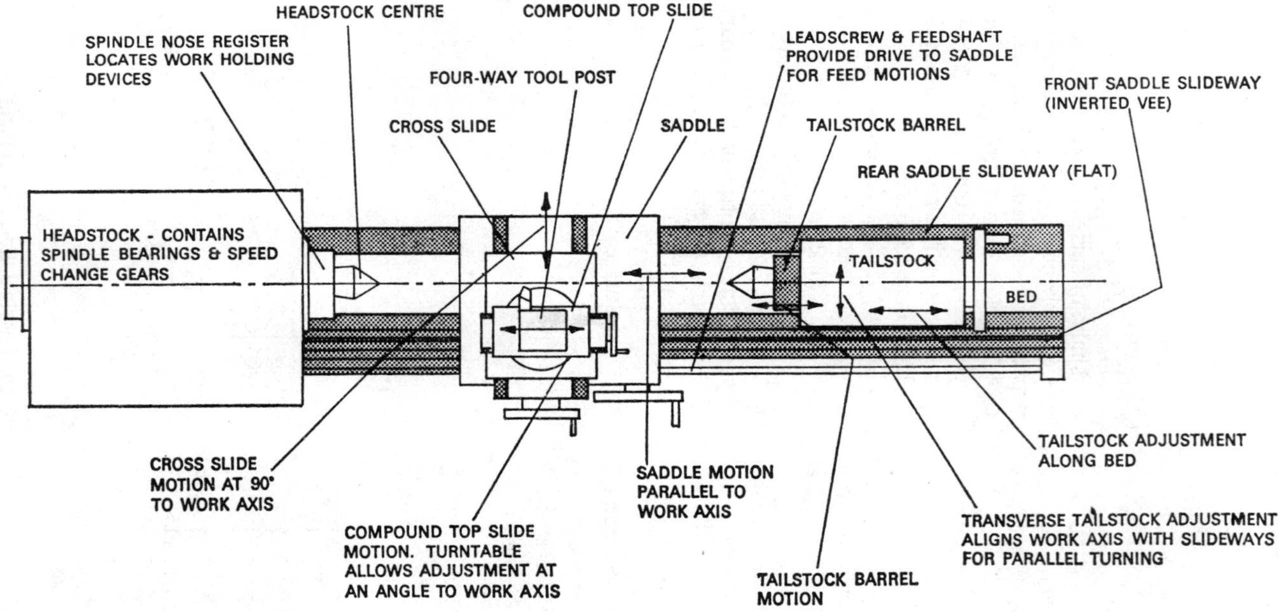

Fig. 7.2. Plan view diagram of lathe showing how motions shown in fig. 7.1 are obtained. Shaded parts are slideways used to guide moving parts

or 4. linear motion of the tool at an intermediate angle to the axis, to generate a conical surface or taper.

The generating motions for a cylinder are the rotary work motion and the parallel tool motion, with right-angled motion of the tool to set the cut. For a flat surface, the right-angled tool motion generates the surface, with parallel motion to set the cut. In generating a taper, the right-angled motion is most frequently used to set the depth of cut. Thus if work is to be produced to a high order of accuracy and surface finish, the following physical requirements must be met:

1. Provision for work holding must be so made that the work rotates but is not moved in any other way by the applied cutting forces. This implies support to prevent the work from deflecting during cutting.
2. The feed motions must have a constant speed to produce a uniform and good surface finish. This demands power feed for all feed motions, if possible, and certainly for the longitudinal and transverse feeds.
3. The depth of cut must be transferable to a high degree of accuracy.

These basic alignments and physical requirements were realised many years ago, and an early lathe built a century ago is basically little different from the modern machine except in power and robustness. In those days, rates of metal removal, using plain carbon-steel tools, were much lower than today and the resulting loads on the machine structure were less. However, a turner who finished his apprenticeship at the end of the last century would have no difficulty in adapting himself to a modern lathe. The techniques are very similar and most of the knobs and handles are in the same places – a few new ones have been added, but the basic outline is still as shown in fig. 7.2.

WORK-HOLDING METHODS

Specific Objective: *The student should be able to select a suitable work-holding method for a given workpiece.*

It is a general requirement of cylindrical work that all diameters should be concentric and all faced surfaces should be square to the cylinder axis. (This is what we mean when we talk of work 'running true'.) The best way to meet these requirements is to perform as many operations as possible at one setting, that is, without removing the work from its setting in the machine. This is not always practicable, so a number of work-holding methods have been devised to suit various types of job and to enable the accuracy requirements to be met.

1. SELF-CENTRING CHUCK

Undoubtedly the most convenient, and therefore most used, method of holding work is the self-centring three-jaw chuck. This takes a wide range of work diameters, and with the provision of a second set of 'outside' jaws its range can be increased, as shown in fig. 7.3. The chuck consists of a cylindrical body, across the face of which the three jaws move in equally-spaced radial slots, driven by a scroll plate inside the body. This plate has a spiral groove, or 'thread', cut across its face in the form of a scroll, which mates with similar 'threads' cut on the back of each jaw, so that when the scroll plate is rotated the scroll moves all three jaws in or out. The jaws and slots are numbered and the jaws must be assembled in sequence in their individual slots or they will not centralise correctly.

The back plate of the scroll plate forms a large bevel gear which can be driven by any of three pinions which are let into the rim of the chuck. Each pinion has a square socket at its outer end into which the chuck key fits. Rotation of any pinion thus turns the scroll plate and moves the jaws.

The chuck body is mounted on the back plate, which may be bored and threaded internally to screw on the nose of the lathe spindle. A better form of attachment has the back plate recessed to engage a register on the spindle nose, and positive holding is then provided by four bolts. This type cannot come loose if the machine stops suddenly.

If the chuck is treated with care, it will

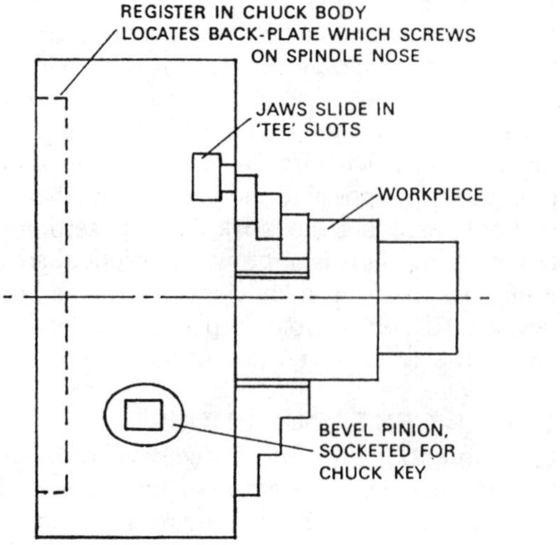

(a) OUTSIDE VIEW OF CHUCK WITH STANDARD JAWS

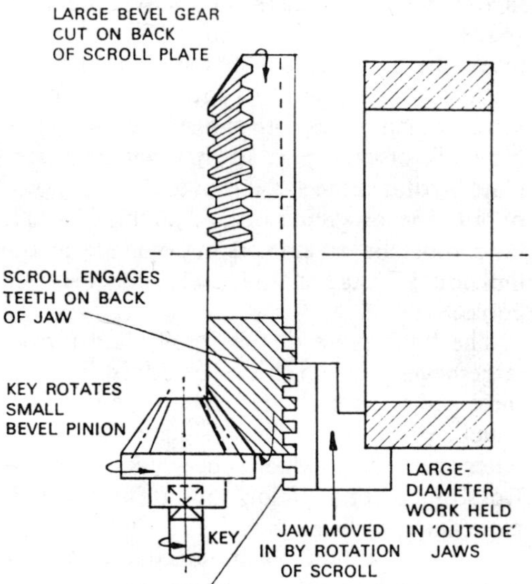

(b) DIAGRAM OF CHUCK ACTION

Fig. 7.3. Self-centring three-jaw chuck

centralise work-pieces fairly accurately for a long time, but the scroll and jaws do become worn and the chuck becomes inaccurate. A slight inaccuracy is often immaterial, e.g. when parts are to be machined all over at one setting from bar stock held in the chuck. Because of the possibility of inaccurate setting all operations on a workpiece should be completed without removing it from the chuck.

Where accurate setting or concentricity with an existing diameter is required a self-centring chuck should *not* be used.

2. THE INDEPENDENT FOUR-JAW CHUCK

This chuck has four jaws, each working independently of the others in its own slot in the chuck body and actuated by its own separate square-thread screw. By suitable adjustment of the jaws, a workpiece can be set to run either true or out of truth, as required. If it is removed for any reason, it can be replaced and reset to run concentrically by checking with a dial gauge and adjusting each jaw as necessary. The check on the workpiece should be carried out near the chuck and repeated as far from it as the workpiece permits, to ensure that the work is not held in the chuck at an angle to the axis of rotation.

The independent adjustment also provides the facility of deliberately setting work off-centre to produce an eccentric workpiece, as shown in fig. 7.4. The concentric diameters are turned first and the chuck is then adjusted for eccentricity by a dial guage which, for accuracy, must register at a radial position. Note that the dial guage readings must vary by *twice* the amount of eccentricity required.

3. THE COLLET CHUCK

A collet is a chuck which is tightened through being drawn into a tapered socket in the spindle nose by a draw tube which bears against the back end of the spindle. Each collet is made for a single diameter and a set of collets is provided for standard diameters of work. Apart from this limitation, collets give a high repeatability of position and work can be removed and replaced

Fig. 7.4. Setting of independent four-jaw chuck for turning an eccentric crank pin

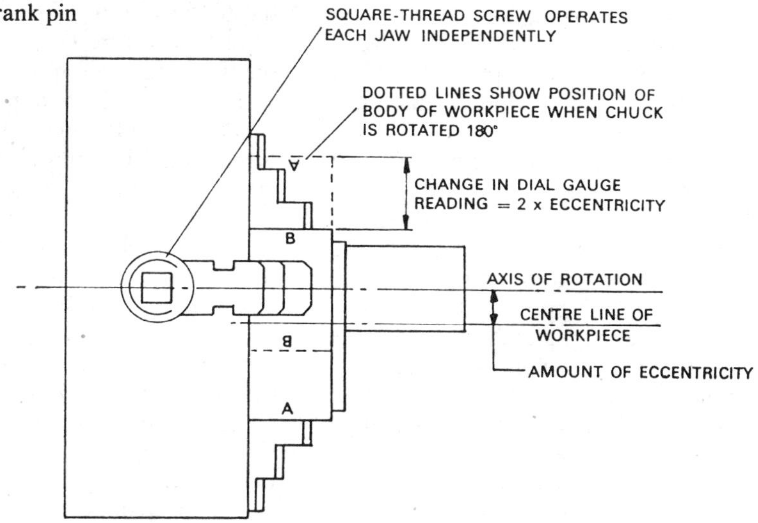

in them with the knowledge that any resulting eccentricity will be very slight. Furthermore, work of very small diameter can be held and, since it is gripped round virtually its full circumference, no damage is done to its surface.

A collet chuck with a typical workpiece is shown in fig. 7.5; the workpiece has a small diameter and requires machining at both ends, and the ends are to be concentric. It is too small to grip in a four-jaw chuck and would be difficult to turn between centres, so that it is ideal for collet work, particularly if a number of

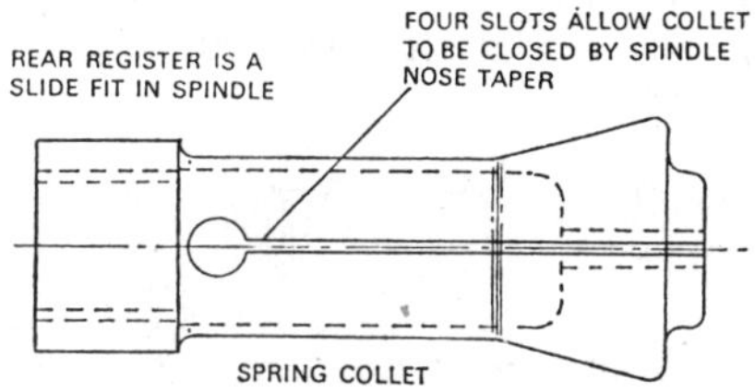

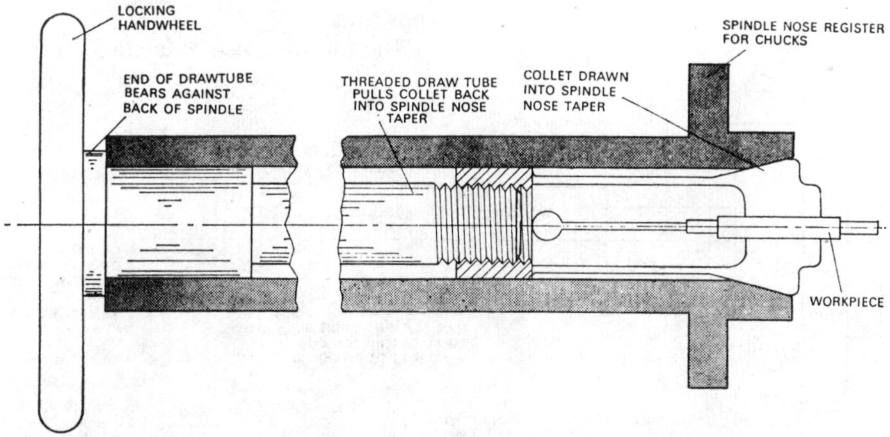

Fig. 7.5. Collet chuck. The rear register and taper are ground concentric with the gripping chuck bore to ensure true running of workpiece and concentricity of machining

workpieces is required. A further point to note is that the absence of rotating jaws, as found on three- and four-jaw chucks, makes it possible to cut close to the collet with minimum length of work protruding, deflections due to tool loads thus being minimised.

The collet is made from medium-carbon steel and is hardened and spring-tempered so that when unlocked it will spring open and release the workpiece. To enable it to close on the workpiece three or four narrow slots are cut from the body to the gripping end. The outer taper of the gripping end matches the taper bore of the machine spindle nose and as the collet is drawn into the spindle its gripping end is closed on the workpiece. The draw tube and the back end of the collet are hollow to permit bars to be fed through the spindle for repetition turning and parting-off of workpieces.

After the collet has been hardened and tempered its rear register in the spindle, its grip taper, and its gripping bore are fine-ground to size so that they are concentric, to ensure true running of workpieces and concentricity of machining.

4. HOLDING WORK BETWEEN CENTRES

If a workpiece is solid, i.e. not bored, and can have a centre hole in each end, it can be supported between a running centre in the headstock spindle and a fixed centre in the tailstock spindle or barrel. Provided the headstock centre (live centre) is running true, the work can be removed and replaced at will, and the repeatability of position will be virtually exact. Thus one end can be machined, the work turned round in position and the other end machined with matching concentricity.

It is, of course, necessary to drive the work, so a 'catch' plate with a driving pin is fitted on the spindle nose and a work carrier or 'driving dog' is attached to the work as shown in fig. 7.6.

Parallelism in turning can be achieved only if the axis of rotation, i.e. the line between centres, is parallel with the tool motion. The transverse position of the tailstock across the bed can be adjusted to correct any error which may exist in this alignment.

The headstock centre must 'run true' for concentricity to be realised, and the centres must be clean and free from damage.

Since the headstock centre does not rotate relative to the work it need not be hard and, if the taper of the spindle nose is for any reason out of truth, a soft centre can be fitted and turned in position to the correct point angle. It will then be concentric and remain so as long as it is not removed. However, soft centres are prone to damage and should not be used unless necessary. It is better to maintain the spindle in good condition so that a normal hard centre runs true.

The tailstock centre is stationary and does

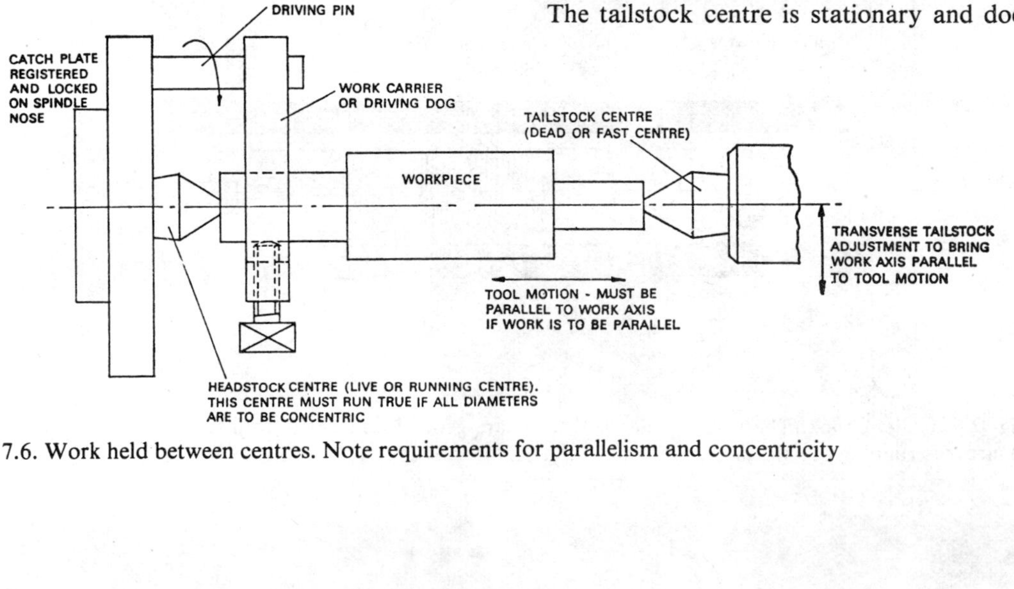

Fig. 7.6. Work held between centres. Note requirements for parallelism and concentricity

not affect the concentricity, but it must be hard to resist wear and kept lubricated to reduce overheating by friction. The tailstock, i.e. the position of the tailstock centre, must be adjusted so that the work is free to rotate but has no end play. After some running the work heats up and expands, so the tailstock barrel requires periodic adjustment.

If the work has been bored or reamed to a standard diameter it can be mounted on a mandrel as shown in fig. 7.7(a) and the outside diameters then turned. As long as the mandrel is straight and runs true on its centres the work will be concentric. Standard plain mandrels are made to standard sizes and have a very slight taper, of 1 in 200 on diameter, so that workpieces can be driven tightly on them. If a taper mandrel is used, the feed force of the tool should normally be towards the large end, otherwise the work may be pushed along the mandrel and freed.

Fig. 7.7(b) shows a parallel mandrel in which the work is gripped between a shoulder at one end of the mandrel and a nut on a screw thread at the other. The concentricity of machining will depend upon the closeness of fit between the mandrel and the bore in the work. However, the grip is more positive than with a taper mandrel, and the part is always located in the same place along the axis of the machine.

5. THE FACEPLATE

The faceplate is simply a circular worktable, mounted on the spindle nose, to which large or awkwardly shaped work is bolted or clamped by the slots provided. Work of awkward shape is then usually out of balance and it is generally necessary to fit a balance weight at an appropriate point on the faceplate to stop vibration.

SAFETY NOTE

The forces which cause vibration due to an out-of-balance weight *are not removed* **by balancing – they are simply balanced by an equal and opposite force. Thus the stresses due to the out-of-balance weight are still present and since the force, and hence the stress, increases with the** *square of the speed,* **the speed must be kept low.**

The method of setting the work on the faceplate depends largely on the accuracy required. If we consider the flanged tee-piece shown in fig. 7.8(a), in which dimension *Y* is important but dimension *X* is not critical, we can examine two methods of setting the work-piece, to suit these conditions.

As dimension *Y* is important it would be as

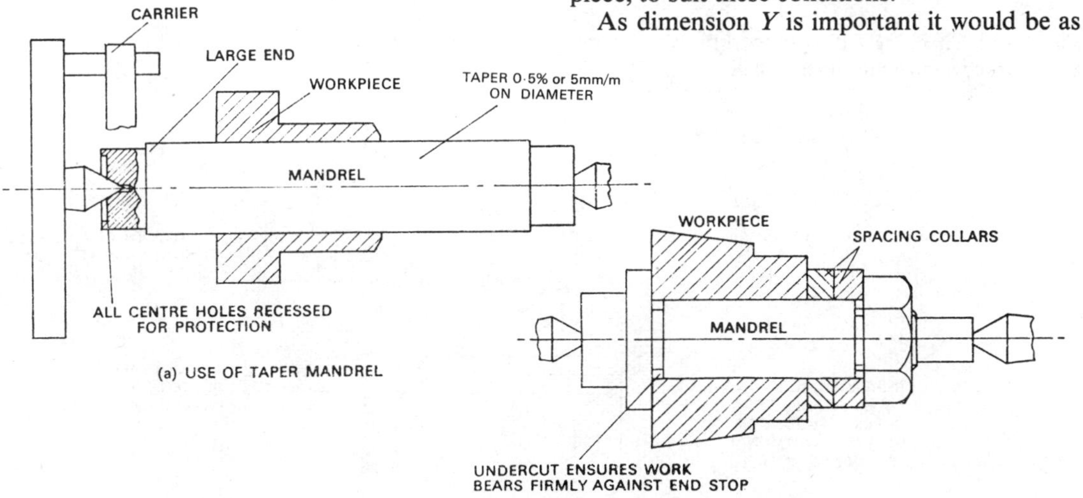

Fig. 7.7. Workpieces mounted on mandrels between centres

(b) USE OF PARALLEL MANDREL WITH LOCKING NUT

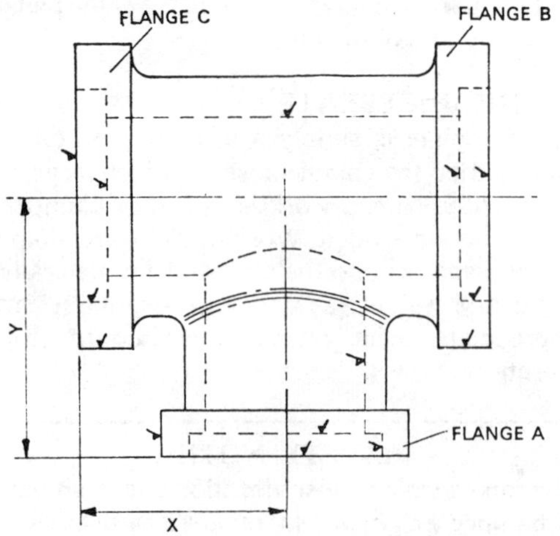

Fig. 7.8(a). Workpiece suitable for machining on faceplate

well to machine flange A and its associated bore first to give us a datum from which we can work. The casting is first marked-out, establishing the centres of the cored holes as accurately as is practicable and ensuring that there is enough metal on each face to allow for machining, i.e. that all faces will clean up.

The hole in flange A is plugged with a piece of wood or lead and its centre marked in the plug. An angle plate is lightly bolted to the faceplate in the approximate position and the tee-piece is set up on it as shown in fig. 7.8(b). The setting is adjusted until the marked centre is stationary when the faceplate is rotated slowly. Checking with the point of a surface gauge against the marked centre is accurate enough for this dimension. The bolts holding the angle plate and the bolts or cramps holding the tee-piece to the angle plate are tightened and the setting given a final check. The lathe spindle is put in a free position so that the balance of the faceplate can be tested as a suitable balance weight is fixed in the appropriate position on it. The balance weight is secured firmly. The hole plug is removed and flange A and its associated bore are then machined, with the facing taken to its marked line so that the distance Y is approximately correct.

To set for flange B, a test mandrel is placed between the lathe centres and the distance between the angle plate and the mandrel is set with the aid of workshop-grade gauge blocks as shown in fig. 7.8(c). The height h of the gauge blocks is given by:

$$h = \text{Dimension } Y - \tfrac{1}{2} \text{ diameter of mandrel}$$

Whatever other setting may be necessary, dimension Y is now fixed accurately. The

Fig. 7.8(b). Checking setting of workpiece to a marked centre for dimension X

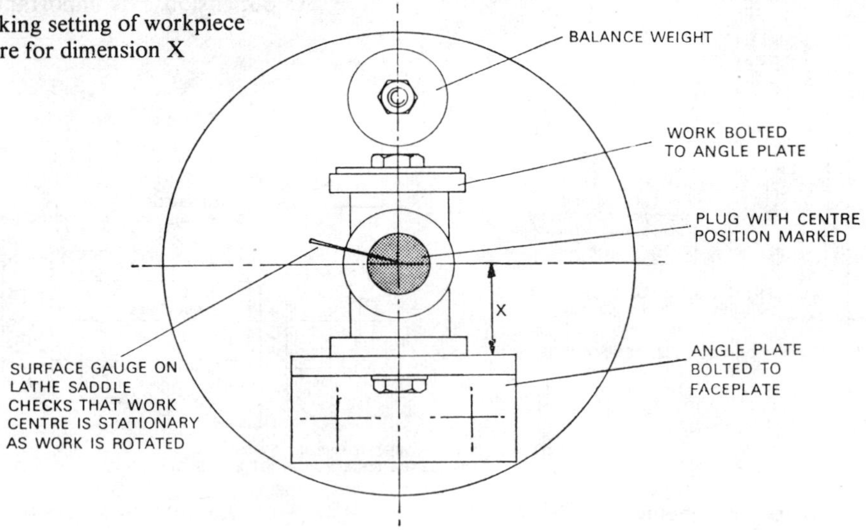

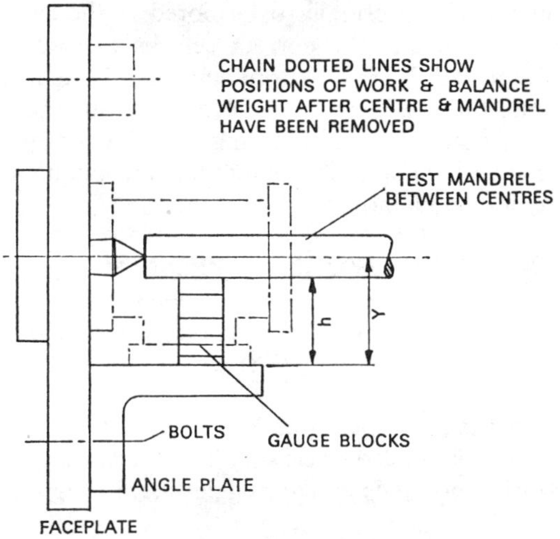

Fig. 7.8(c). Pre-setting angleplate to give precise dimension Y

workpiece is clamped to the angle plate on the machined datum face of flange A and adjusted sideways until a surface gauge indicates that the outside diameters of flanges B and C are running approximately true. Note that setting an unmachined surface of a casting with a surface gauge is adequate; a dial gauge should never be used on the rough surface of a casting.

A suitable balance weight is mounted in the appropriate position and the tightness of all bolts and cramps is checked. Flange B and its associated bore are then machined, the tee piece is removed, and the angle plate and balance weight are dismantled. Flange B can now be clamped to the face plate and set true by checking the machined bore with a dial gauge. Flange C and its recess can then be machined. Alternatively, flange B can be held in a four-jaw chuck for the machining of flange C.

The accuracy of work such as this depends very largely on the care taken in setting up. This is unavoidably time-consuming but not nearly so expensive as working in a slapdash way and scrapping an expensive casting.

THE USE OF TOOLMAKERS' BUTTONS

Specific Objective: *The student should know how to accurately position work on the faceplate.*

If a workpiece is to be bored on the faceplate and its hole centre is to be positioned in both directions, e.g. from two edges, to a high degree of accuracy, toolmakers' buttons will enable this positional accuracy to be obtained. Consider the part shown in fig. 7.9(a), where the hole centre is to be a positional tolerance of 0.008 mm diameter from the true position (TP). Assuming the part to be machined accurately all over, apart from the hole, then the procedure would be as follows:

1. mark-out the hole centre position with a height gauge:
2. at the marked centre position, drill and tap a no. 4 BA hole:
3. fit a toolmakers' button over the hole but do not fully tighten the screw.

A toolmakers' button is a small, hardened and ground cylinder of steel, usually 10 mm diameter, whose end is accurately square to its cylindrical axis. The hole in the button is well in excess of 5 mm diameter to give ample clearance around the no. 4 BA screw which holds the button in place, so that the button can be adjusted into the correct position even though the drilled hole and screw are slightly out of correct position, as is likely.

4. Adjust the button position by piles of gauge blocks equal to the TP dimensions minus 5 mm (button radius) as shown in fig. 7.9(b).
5. When the button position is correct, lock the screw and check that the button position has not changed.
6. Mount the work on the faceplate (or in a four-jaw chuck) and adjust the position of the work, not the button, until the button runs true as shown by a 0.002 mm dial gauge.

As the button has been positioned accurately on the axis of the required hole the work is now rotating on this axis.

7. Remove the button and drill out the tapped

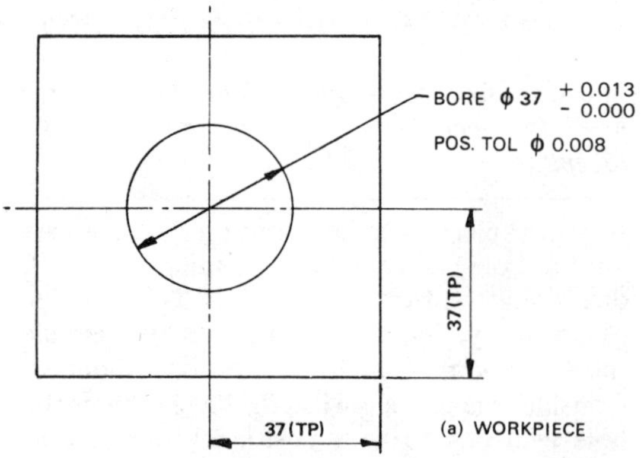

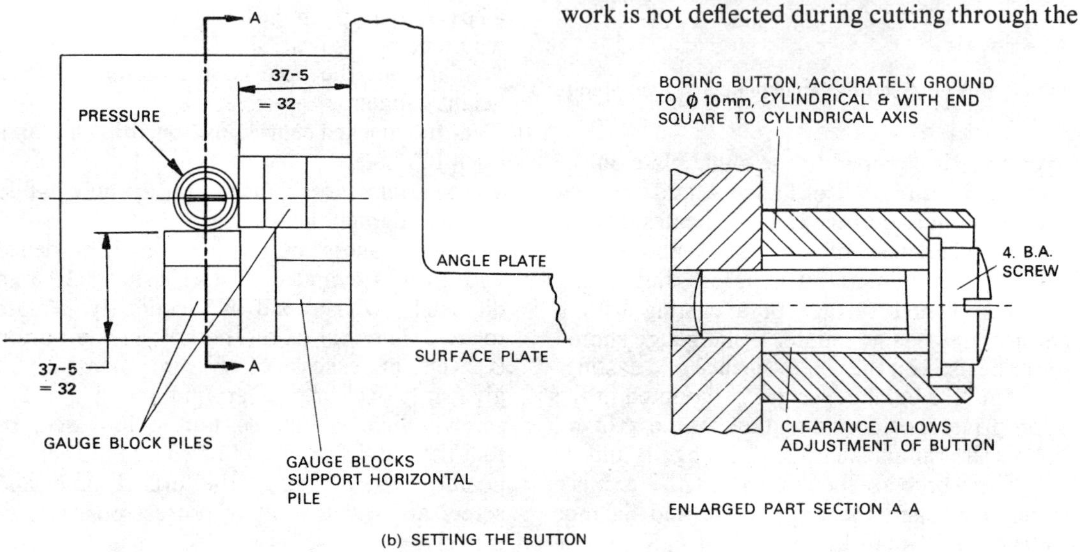

Fig. 7.9. Use of toolmakers' buttons

hole, which will probably be eccentric, to a convenient size for boring.

8. Bore the hole to size.

In practice, a faceplate is better for this work than a four-jaw chuck, since its face can usually be relied upon to be square to the axis of rotation, so that the problem of setting the workface square to the axis of rotation does not arise.

This technique of button boring is not, of course, limited to one hole per workpiece. A job may require a number of holes whose centre distances are critical to be bored in the same workpiece and the correct centres can all be preset with boring buttons and bored one at a time. Such a job could be carried out much more quickly on a jig boring machine, or even on a vertical-spindle milling machine, but it can be done on a lathe by this method.

EXTERNAL WORK SUPPORTS

Specific Objective: *The student should be able to select a suitable work-steady when required.*

Apart from gripping and driving the work in a chuck, it is also necessary to ensure that the work is not deflected during cutting through the cutting forces. Work with a large overhang is usually supported by the tailstock centre but if it is long and slender it requires additional support close to the tool to prevent it from bending. In this case a *travelling steady* is used, as shown in fig. 7.10, bolted to the cross-slide on the saddle and travelling with the tool as it feeds along the work. Note that the steady rests are positioned to withstand the two forces which would bend the work, i.e. the tangential cutting force F_T and the radial force F_R set up by the cutting action.

Fig. 7.10. Travelling steady: used with long slender work to prevent deflection by cutting forces

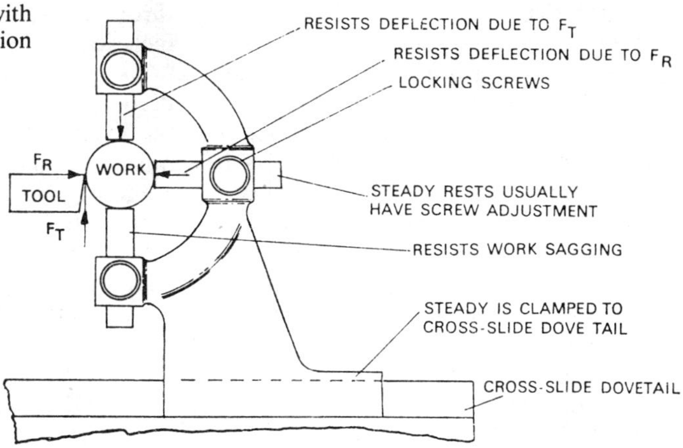

Consider now a tube 2 m long in which one end has to be bored to size. Obviously the tailstock cannot be used as an end support, so a *fixed steady* is used, bolted to the bed near the end of the tube. The tube must be set to run true at the chuck end and the fixed steady must be adjusted to support the tube parallel to the lathe bed and parallel to the tool motion, or tapered work will result. These adjustments are carried out with the aid of a dial gauge or a surface gauge, first to ensure that the headstock end is running true and then to check that the gauge reading remains constant in both planes along the length of the tube. A fixed steady is illustrated in fig. 7.11.

The pads or rests of both types of steady act as bearings and are usually made of brass. Like any other bearing they require lubricating from time to time to avoid excessive wear and overheating.

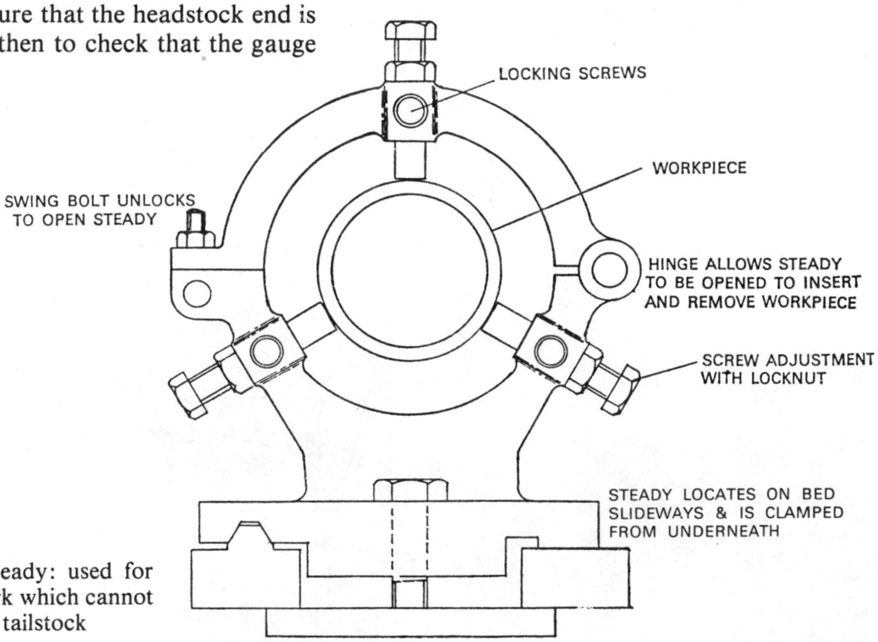

Fig. 7.11. Fixed steady: used for supporting long work which cannot be supported by the tailstock

TAPER TURNING

Specific Objective: *The student should be able to describe three methods of taper turning.*

In order to produce a taper, the tool must be caused to move at an angle to the work axis in the horizontal plane. One method is to set the tailstock off-centre, where a fine taper is required and the work is mounted between centres. However, the main purpose of the tailstock adjustment is to enable the tailstock to be set to give truly parallel work, and to have to reset it after adjustment to produce a taper is often more trouble than it is worth. The two other methods of taper turning both cause the tool direction to be adjusted, not the axis of rotation.

1. TAPER TURNING WITH THE COMPOUND TOP SLIDE

The top slide on a lathe saddle is normally set parallel to the axis of the work and is used for setting the depth of cut when facing. However, it is mounted on a rotary base, graduated in degrees, and can be swung to produce a taper as shown in fig. 7.12. Note that the slide is swung through the semi-angle of taper, not the included angle.

This method has the advantage that it can be used for any angle and direction of taper, but the length of travel is limited and hand feed must be used. The angular scale is of comparatively small diameter and the unit graduations of degrees are therefore small, so the accuracy of setting by the angular scale is somewhat limited. If a test bar or specimen component of the correct taper is first mounted in the lathe, however, the top slide can easily be set to it without reference to the angular scale and will then reproduce the same taper.

2. THE TAPER-TURNING ATTACHMENT

Many modern lathes have a taper bar fitted at the back of the bed, which can be adjusted through a range of angles to the spindle axis. The bar carries a sliding block which, during taper turning, is attached by a link to the back of the cross-slide, as shown in fig. 7.13. The lead screw of the cross slide is released so that it no longer controls the depth of cut setting, and the slide is thus freed. When the saddle is moved along the bed the cross-slide now follows the

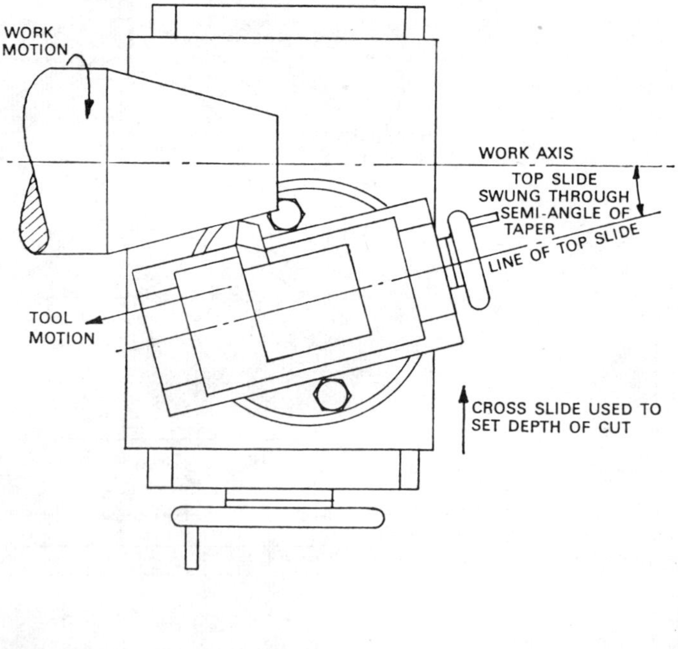

Fig. 7.12. Generating a taper with the compound top slide

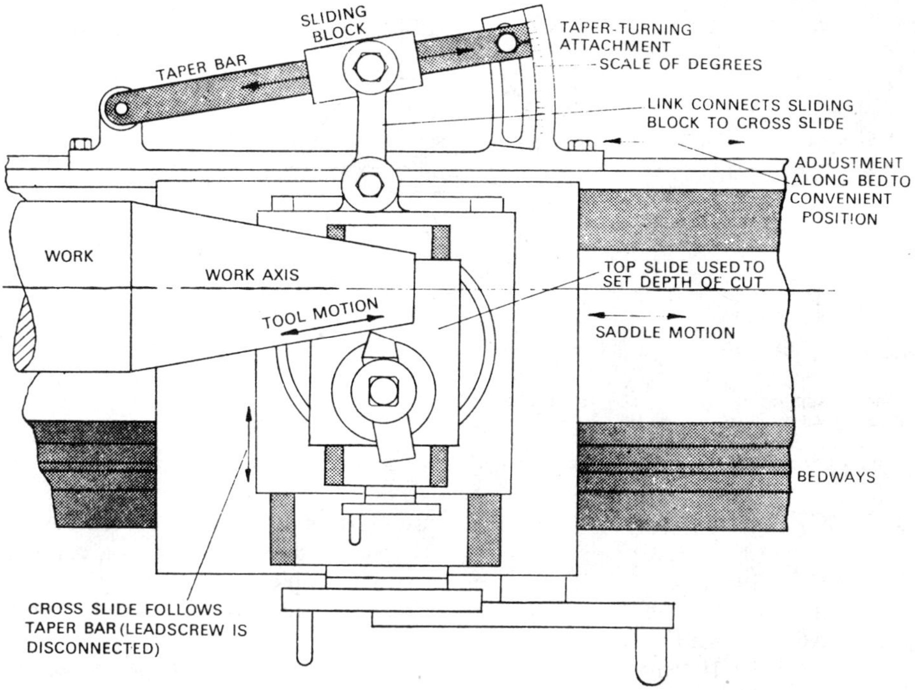

Fig. 7.13. Generating a taper with a taper-turning attachment

taper bar, so that the tool moves parallel to the bar and a taper is produced. The top slide is swung through 90° to lie at right angles to the work so that it can be used to apply the cut.

The length of the taper bar allows large degree divisions on its degree scale, which enable accurate settings to be carried out, an angle vernier often being incorporated. The taper is produced by the movement of the saddle under power feed, giving improved and controllable surface finish, and a long taper is possible. It is, however, limited to semi-angles of taper of about 15° (30° included angle).

THE EFFECT OF INCORRECT TOOL SETTING ON TAPER TURNING

For cutting a taper it is important that the tool height should be set correctly to the spindle axis. If the tool is set low, two faults will occur:

1. the angle of taper will be incorrect;
2. the taper will not be straight but will be a hollow curve of complex mathematical shape.

Both effects can be demonstrated by deliberately setting a tool too low. Let us assume that the tool has been set 2.0 mm low, and the machine is set to produce a taper whose radius increases by 5 mm for each 10 mm traversed (taper of 0.5 mm per mm on radius, or 1 mm per mm on diameter).

However high or low the tool has been set, it will still move away from the machine centre line by 5 mm for each 10 mm traversed. Thus in figs. 7.14(a) and (b) the distance AB is 5 mm for each 10 mm traversed. In fig. 7.14(a) the outside radius $R = r + \text{AB}$, but in fig. 7.14(b) this statement is not true. Let $r = 20$ mm, and the tool be set 2.0 mm low. The machine is traversed 10 mm, so the tool withdraws a distance $\text{AB} = 5$ mm.

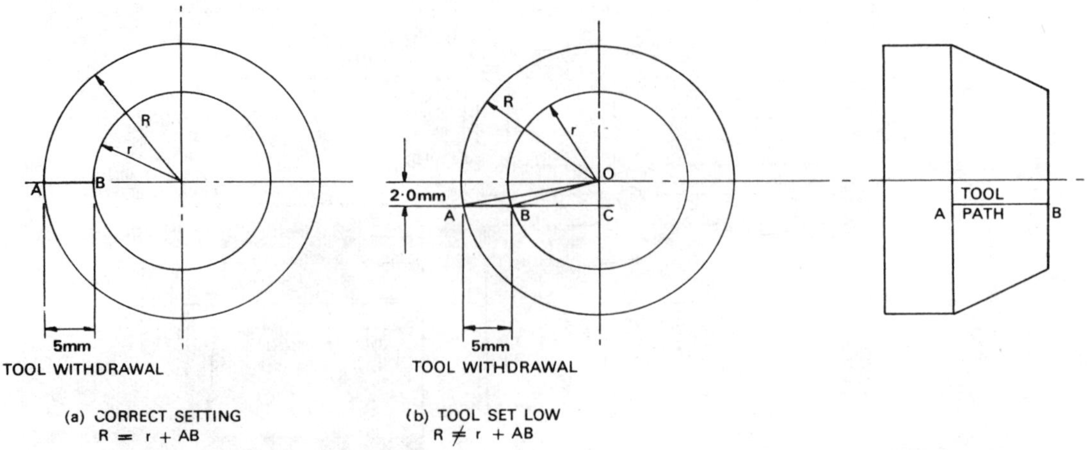

Fig. 7.14. Effect of incorrect tool setting on taper turning

$$\begin{aligned}
\text{In } \Delta\ \text{OBC}\ (BC)^2 &= r^2 - (OC)^2 \\
&= (20)^2 - (2.0)^2 \\
(BC)^2 &= 396\ \text{mm}^2 \\
BC &= 19.9\ \text{mm} \\
AC &= AB + BC \\
&= 5 + 19.9\ \text{mm} \\
&= 24.9\ \text{mm} \\
R^2 &= (AC)^2 + (OC)^2 \\
&= (24.9)^2 + (2.0)^2 \\
&= 620.01 + 4.0 \\
&= 624.01\ \text{mm}^2 \\
R &= \sqrt{624.01} = 24.98\ \text{mm}
\end{aligned}$$

But for a correct taper R should be $(r + 5.0)$ mm or 25 mm. Thus there is an error of 0.02 mm per mm length on radius, or 0.04 mm per mm length on diameter simply due to setting the tool low.

The straightness of the taper can easily be checked by sighting it against a straight edge, and if the tool has been set too low the resulting curve is readily seen.

SCREWCUTTING ON THE LATHE

Specific Objective: *The student should be able to describe a method of screwcutting on the centre lathe using a single point tool.*

Generating processes have already been discussed and copying processes mentioned. The screwcutting process is a combination of the two in which:

1. the thread form is copied from the tool shape;
2. the thread helix is generated from the machine motions.

SCREWCUTTING TOOLS

The form of thread is copied from the tool, which must therefore be ground to the correct shape, i.e. to the correct thread angle and nose radius. It must also be set correctly, square to the work and at the correct height, and must not have top rake, or the thread form will be incorrect.

TOOL GRINDING

The tool shape can be ground by hand to give cutting clearance and the correct angles can be obtained by comparing them with a screw-cutting plate gauge. The same gauge is used for setting the tool square to the work surface, as shown in fig. 7.15. It is important to note that a hand-ground single-point cutting tool must not have a rake angle or the form will not be reproduced correctly.

Fig. 7.16 shows, greatly exaggerated, the effect of grinding top rake on a screwcutting tool.

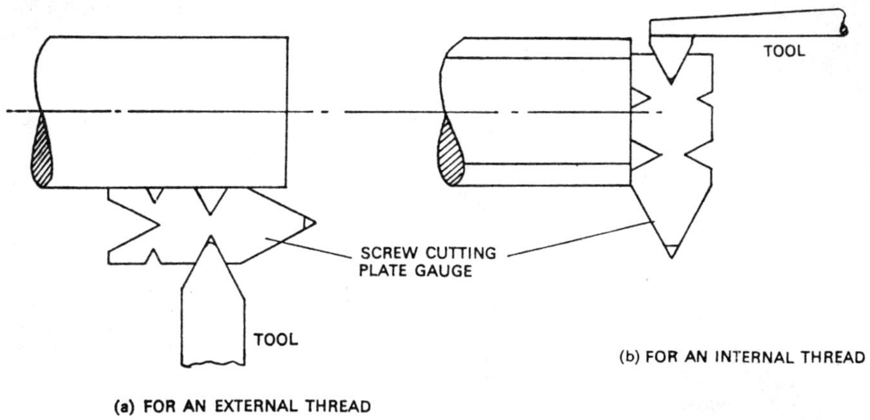

Fig. 7.15. Setting a screwcutting tool in the lathe

The tool has been fed in an amount AB, giving a depth of cut of *D* in. But due to the top rake angle the tool actually contacts the work surface at C, not at B. Projecting these points down on to a plan view of the tool we see that the tool flanks contact the work surface at C and C, giving a width of form *W*. This, however, is produced at a distance *D* backwards from the point, and the angle produced, which is the angle actually cut, is shallower than that of the tool itself.

If a screwcutting tool is to be given top rake it must be 'form ground', that is, the tool form must be corrected for the rake angle. Circular

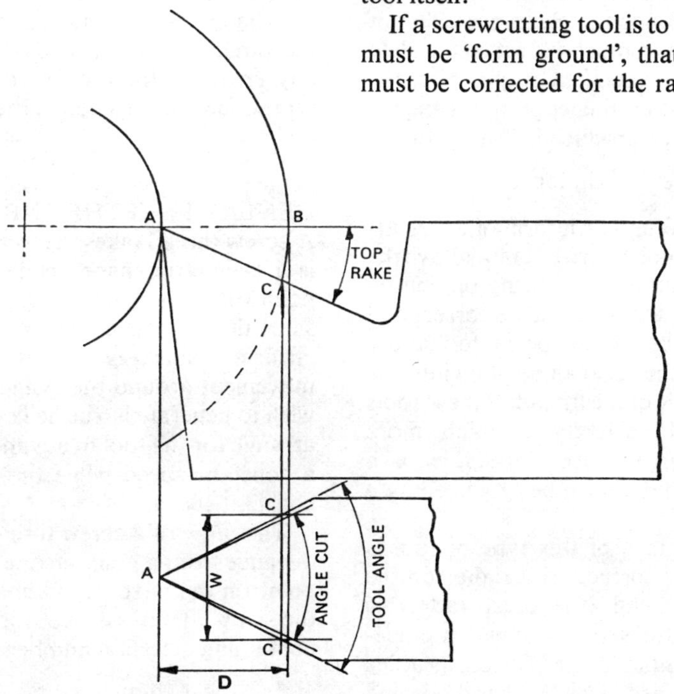

Fig. 7.16. Effect of top rake on thread form

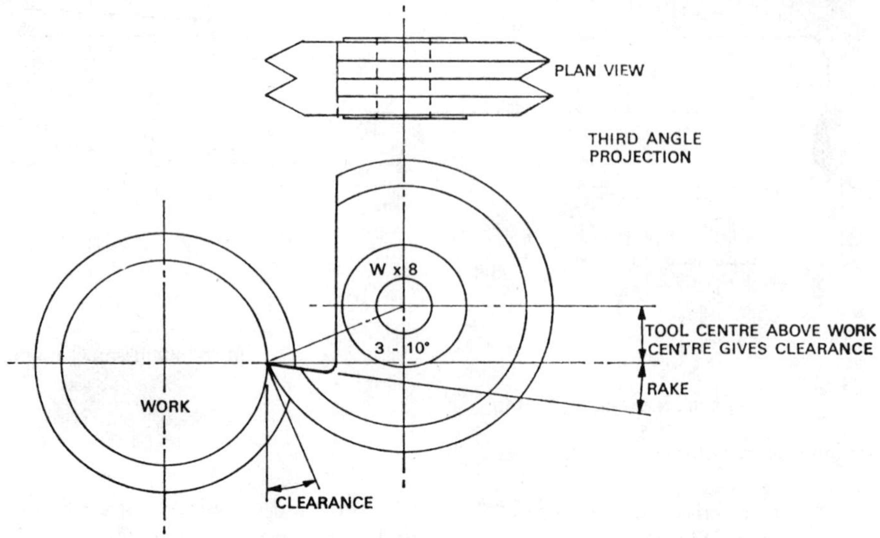

Fig. 7.17. Circular form-tool for screwcutting

form-tools of the type shown in fig. 7.17 are made by Johansson Ltd, and are corrected for rake angle, and also for the clearance effect of a circular tool. They must be set accurately with the tool centre a specified height above the work centre, and the cutting edge on centre, for the correct form to be produced. That shown in fig. 7.17 bears the inscription $\frac{W \times 8}{3-10^\circ}$. This means that it is of Whitworth form of 8 threads per inch and if the tool centre is 3 mm above the work centre, the cutting edge being on centre, the clearance angle and form will be correct and the rake angle will be 10°. A special toolholder is required so that the tool can be tilted into the line of the helix angle of the thread. If these tools are set and used correctly a much more accurate screw thread form results, with a better surface finish than can be produced by a single-point tool.

A further advantage of this type of tool is that it produces the correct crest radius on the Whitworth-form thread (the crest radius is optional on the Metric series) and with a single-point tool this is impracticable. The crest radius has then to be finished either by hand chasing or, if the screw is a standard size, by running a solid die-nut down the thread as a finishing operation.

Whatever the method used, the production of the thread form is a copying process, so that any errors in the tool or in its setting will be reproduced or reflected in the thread form.

GENERATING THE THREAD HELIX

A screw thread takes the form of a *helix*, which is a geometric shape defined as the path of a point which moves around a cylinder and at the same time moves along the cylinder so that the linear movement is proportional to the angular movement around the cylinder. If therefore we wish to generate a true helix on a lathe we must arrange for the tool to advance along the bed at a constant speed while the work rotates at a constant speed.

The *pitch* of a screw thread is defined as the distance from a point on one thread to a similar point on the next thread, measured parallel to the axis. Thus, if we measure the length containing a certain number of threads:

$$\text{Pitch} = \frac{\text{Length}}{\text{No. of threads}} \text{ in units of mm per thread}$$

Note: Metric threads usually only quote pitch in mm.

In screwcutting, the saddle is driven by the leadscrew of the lathe and this in turn is driven from the machine spindle through a gear train, shown in diagrammatic form in fig. 7.18. Note that the idler has no effect on the gear ratio, but is provided to give the DRIVEN gear on the leadscrew the same direction of rotation as the DRIVER gear on the spindle. The gear ratio between the DRIVER gear and the DRIVEN gear clearly affects the speed of rotation of the leadscrew in relation to that of the spindle. It thus affects the traverse of the saddle and, in fact, controls the pitch and the helix of the thread being cut. To determine the pitch, therefore, we are concerned about the numbers of teeth on the driver and driven gears.

Consider a lathe whose leadscrew is of 5 mm pitch cutting a thread of 1 mm pitch, and let the tool move along the work 50 mm. For these conditions the work must make 50 revolutions and the leadscrew 10. Thus:

$$\frac{\text{Revs of work}}{\text{Revs of leadscrew}} = \frac{50}{10}$$

As the spindle is rotating faster than the leadscrew it will have the smaller gear, and therefore

$$\frac{\text{No. of teeth on DRIVER gear}}{\text{No. of teeth on DRIVEN gear}} = \frac{10}{50} = \frac{1}{5}$$

or

$$\frac{\text{Driver}}{\text{Driven}} = \frac{\text{Pitch of work}}{\text{Pitch of leadscrew}}$$

Considering the 1 mm pitch thread and the 5 mm pitch leadscrew we find that:

$$\frac{\text{Driver}}{\text{Driven}} = \frac{1}{5}$$

It is not practicable to use a gear with less than 20 teeth, and a gear larger than 120 teeth becomes unwieldy, so most machines are provided with gears ranging from 20 teeth to 120 teeth in steps of 5 teeth, the 20-tooth gear being duplicated. We must therefore find a pair of gears which give us the ratio $\frac{1}{5}$ out of those available.

In calculations of this nature it is important to remember that the fraction must remain unchanged and that whatever is done to the numerator of the equation must also be done to the denominator.

$$\text{Thus:} \quad \frac{\text{DRIVER}}{\text{DRIVEN}} = \frac{1}{5} \times \frac{20}{20} = \frac{20}{100}$$

so we can use a 20-tooth driver and a 100-tooth driven gear set up as shown in fig. 7.19. This is a *simple train* in which the idler gives the leadscrew the same direction of rotation as the machine spindle, so that a right-hand thread, like that of the leadscrew, is cut.

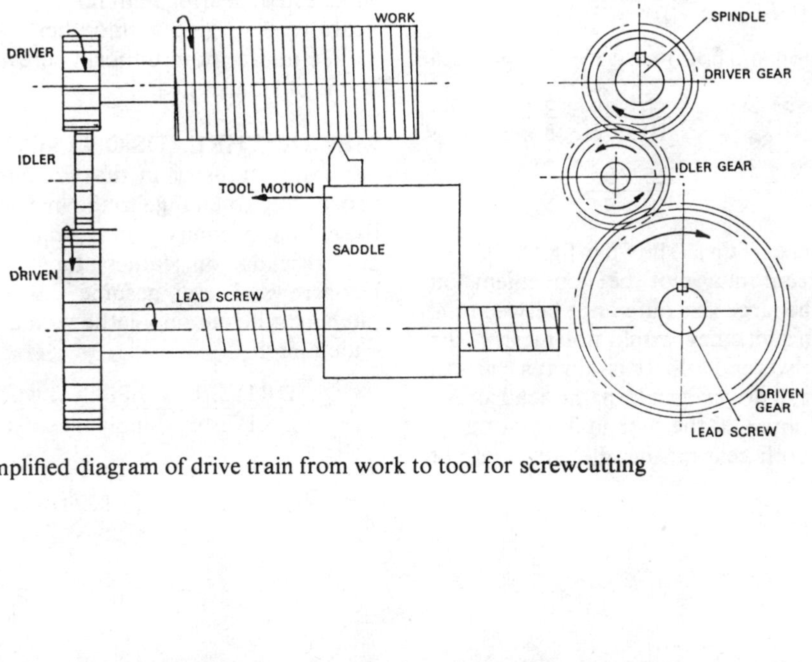

Fig. 7.18. Simplified diagram of drive train from work to tool for screwcutting

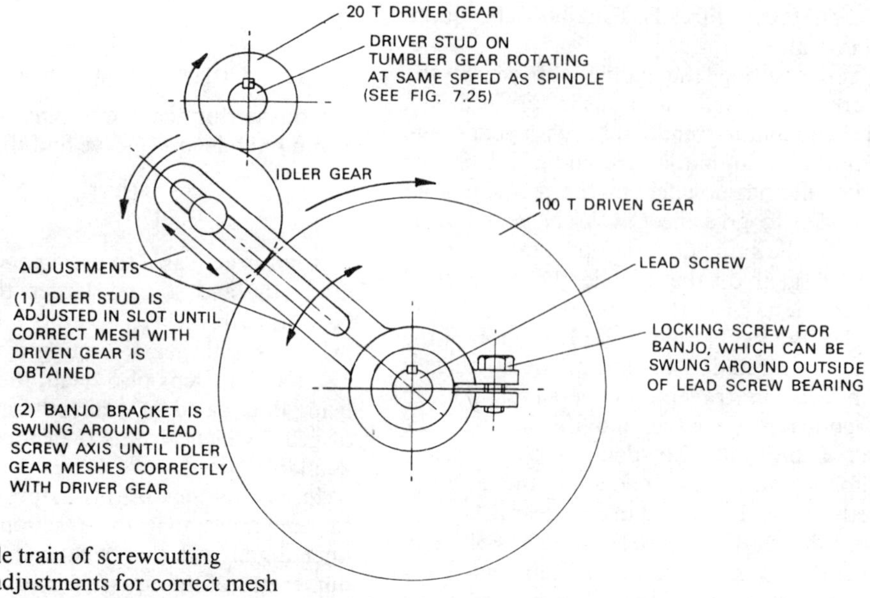

Fig. 7.19. Simple train of screwcutting gears showing adjustments for correct mesh

Now consider a thread of 0.7 mm pitch to be cut on a lathe with a 3 mm leadscrew.

$$\frac{\text{DRIVER}}{\text{DRIVEN}} = \frac{0.7}{3} \times \frac{50}{50} = \frac{35 \text{ teeth}}{150 \text{ teeth}}$$

We do not have a 150-tooth gear, so a *compound train* with two pairs of gears is required. The gears can be found by factorising the fraction into:

$$\frac{\text{DRIVER}}{\text{DRIVEN}} = \frac{35}{150} = \frac{5 \times 7}{10 \times 15}$$

Multiplying top and bottom by 5×5 we get

$$= \frac{5 \times 7}{10 \times 15} \times \frac{5 \times 5}{5 \times 5}$$

$$= \frac{25 \times 35}{50 \times 75}$$

These gears are set up as shown in fig. 7.20.

Such a train may not be convenient on account of the large size difference between the gears, and in fact, they would not mesh if the arrangement were $\frac{25}{75} \times \frac{35}{50}$. It is always wise to put the largest driven gear on the leadscrew. Note that in fig. 7.20 the second 35-tooth gear and the 50-tooth gear on the idler stud are not idlers but are keyed together and part of the gear ratio.

If we wish to change these gears, we can do anything to suit our convenience so long as we do not change their overall ratio, and any of the following trains will give the same result.

$$\frac{25}{100} \times \frac{70}{75};\ \frac{20}{60} \times \frac{70}{100};\ \frac{25}{60} \times \frac{70}{125} \text{ etc.}$$

From these, the most convenient train should be selected, bearing in mind that trains of larger gears tend to give a smoother traverse to the saddle and a more uniform screw thread than trains of smaller gears.

METRIC THREADS ON INCH LATHES

Although industry in the United Kingdom is proceeding to change to the metric system, it is likely that for many years it will be necessary to cut threads on lathes having inch pitch leadscrews. Let us assume a thread of 2 mm pitch is to be cut on a lathe with a leadscrew of $\frac{1}{4}$ inch pitch.

$$\frac{\text{DRIVER}}{\text{DRIVEN}} = \frac{\text{Pitch of work}}{\text{Pitch of leadscrew}}$$

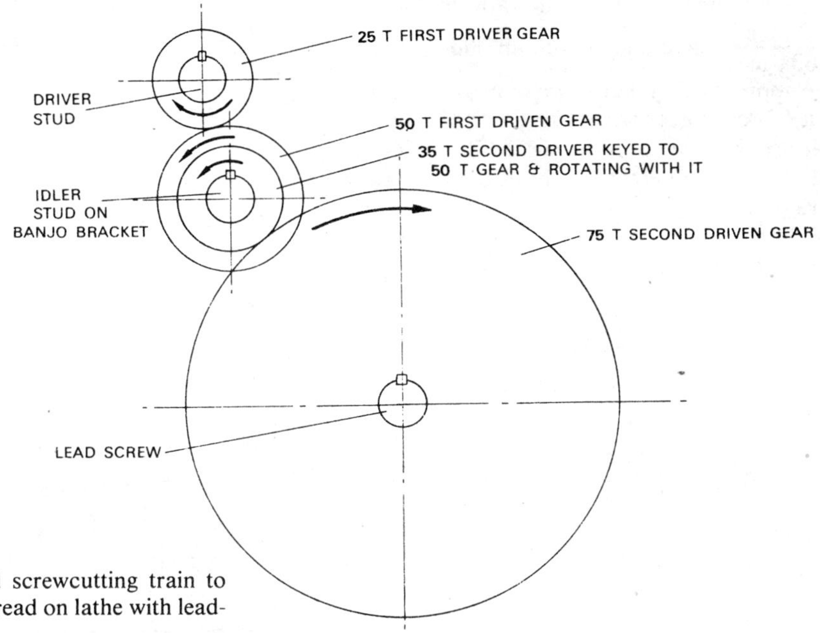

Fig. 7.20. Compound screwcutting train to cut a 0·7 mm pitch thread on lathe with lead-screw of 3 mm pitch

$$=\frac{2\text{ mm}}{\frac{1}{4}\text{ in}}$$

We cannot divide millimetres by inches so we must convert the 2 mm to inches.

Now $1\text{ in}=25.4\text{ mm}$ or $1\text{ mm}=\frac{1}{25.4\text{ in}}$

Therefore $2\text{ mm}=\frac{2\text{ in}}{25.4}$

$$\frac{\text{DRIVER}}{\text{DRIVEN}}=\frac{\frac{2\text{ in}}{25.4}}{\frac{1}{4}\text{ in}}=\frac{8}{25.4}\times\frac{5}{5}$$

$$=\frac{40\text{ teeth}}{127\text{ teeth}}$$

This ratio is set up in fig. 7.21.

The number 127 is a prime number, i.e. its only factors are 1 and 127, and it cannot be subdivided. Thus to cut an accurate metric thread on a lathe which has a leadscrew whose pitch is based on inches we require 127-tooth gear in the gear train.

Any gear larger than 120 teeth is rather

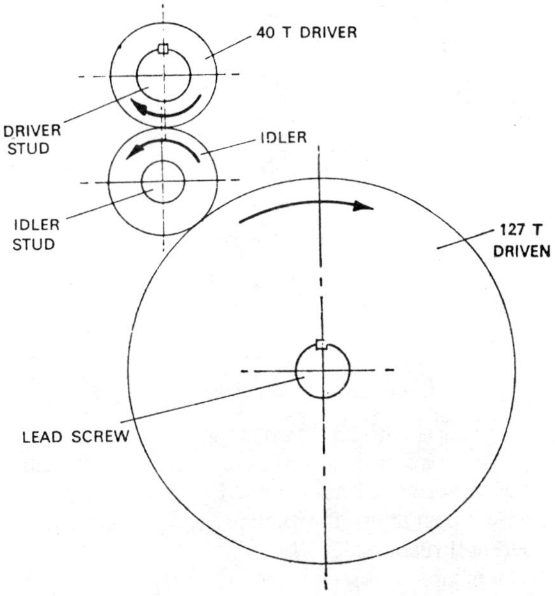

Fig. 7.21. Simple train for cutting metric thread of 2 mm pitch on lathe with leadscrew of $\frac{1}{4}$ in pitch (4 T.P.I.)

unwieldy, but if we divide the ratio by two we get $\frac{20 \text{ teeth}}{63\frac{1}{2} \text{ teeth}}$, and a gear with an odd halftooth is an apprentice's joke! If we approximate and use a 63-tooth gear we get an error of 1 in 127 or approximately 8 parts per 1000. Thus the pitch would be in error by about 8 mm in each metre of length.

If, however, the 63-tooth gear is used as a driver in the example worked out above, we get:

$$\frac{\text{DRIVER}}{\text{DRIVEN}} = \frac{8}{25.4}$$

Substituting 63 for the driver we get:

$$\frac{63}{\text{DRIVEN}} = \frac{8}{25.4}$$

$$\therefore \quad \text{DRIVEN} = \frac{25.4 \times 63}{8}$$

$$= \frac{1600.2}{8}$$

Dropping the fraction 0.2 gives us an error of only 1 in 8000 and therefore:

$$\text{DRIVEN} = \frac{1600}{8} = 200$$

$$\therefore \quad \frac{\text{DRIVER}}{\text{DRIVEN}} = \frac{63}{200}$$

$$= \frac{9 \times 7}{10 \times 20}$$

$$= \frac{4.5 \times 3.5}{5 \times 10} \times \frac{10 \times 10}{10 \times 10}$$

$$\therefore \quad \frac{\text{DRIVER}}{\text{DRIVEN}} = \frac{45 \times 35}{50 \times 100}$$

This ratio is shown set up in fig. 7.22.

Thus, we now do not need the 63-tooth gear and the approximation used gives an error of only 1 part in 8000, which is negligible for most screw threads.

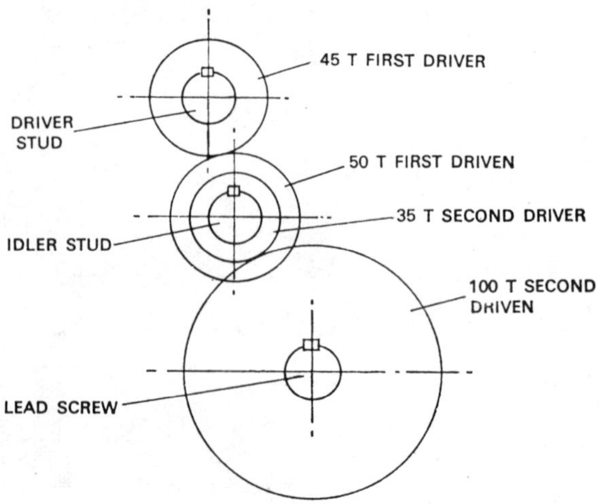

Fig. 7.22. Approximate gear train for metric thread of 2 mm pitch giving a pitch error of 1 part in 8000. This is based on a 63T driver, which is not in fact used *(see text)*

SAFETY NOTE

One of the most efficient devices for crushing fingers is a gear train driven by a powerful motor. *Never* start up a lathe after changing the screwcutting gears without replacing the guard.

MODERN SCREWCUTTING LATHES

THE NORTON GEARBOX

Most modern lathes do not require a new gear train to be fitted for each thread pitch to be cut. They are fitted with a quick-change type of gearbox, often known as a Norton gearbox, which covers all the ratios likely to be wanted.

As shown in fig. 7.23, it has an input shaft driven from the headstock which drives a long roll gear R. Through the tumbler gears T_1 and T_2, shown in the end view, gear R can be engaged with any one of the gears numbered 1 to 9 in the 'cone cluster' of gears. A sliding cluster of three gears *a*, *b* and *c* is engaged with gears 10, 11 and 9 respectively on the cone cluster shaft to give three alternative output ratios.

Since this three-speed output gearbox is

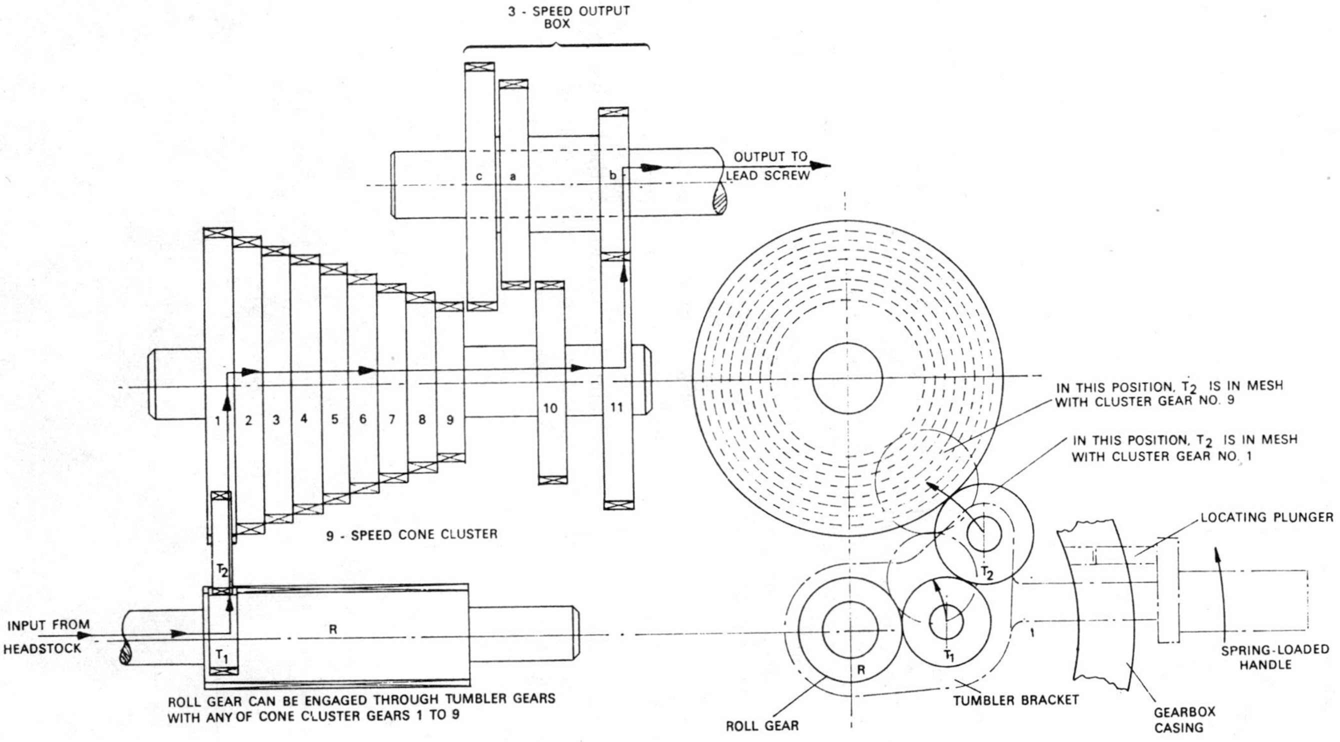

Fig. 7.23. Layout of Norton-type screwcutting gearbox

compounded with the nine speeds from the cone cluster, the gearbox provides a total of 27 different gear ratios which cover a very wide range of screw pitches and satisfy most needs.

In the engagements shown in fig. 7.23, the drive is from R to cluster gear no. 1, and from gear no. 11 to gear *b* and the overall gear ratio is:

$$\frac{\text{Roll gear R}}{\text{Gear 1}} \times \frac{\text{Gear 11}}{\text{Gear } b}$$

ENGAGEMENT OF LEADSCREW

The saddle feed for normal turning is produced by a feed shaft which is geared to the leadscrew and drives through a worm and wheel arrangement at the saddle. The revolutions of the feed shaft are thus in a constant ratio with those of the leadscrew, and the sliding feeds, which are expressed in cuts per inch or inches per revolution of work, are in a fixed ratio to the number of threads per inch obtained when the same gear train is used for screwcutting rather than sliding.

In screwcutting, the leadscrew is engaged by a split-nut arrangement at the saddle. The phosphor-bronze nut which engages the leadscrew is split and the two halves slide vertically in dovetail slides. The half nuts are moved to the engaged or disengaged position by a cam plate, shown in fig. 7.24, which is turned by the operating lever on the outside of the saddle apron.

LEADSCREW REVERSAL

The tumbler gear shown in fig. 7.23 is used to allow gears of different diameters to be engaged with another gear at a fixed centre distance. A similar arrangement of tumbler gears can be used to reverse the leadscrew, to enable left-hand threads to be cut. In practice, the first driver gear of a screwcutting train is not fitted directly to the lathe spindle but is mounted on a driver stud which rotates at the same speed as the spindle. The driver gear which rotates with and drives the driver stud is driven from the spindle driving gear and since they have the same speed they must be of the same size. To reverse this driven gear we must be able to insert an extra idler at will and to do this we interpose two tumbler gears, mounted on a bracket which can swing around the axis of the driven gear as shown in fig. 7.25. Tumbler gear A is always in mesh with the driven gear and in mesh with its fellow tumbler gear B. In fig. 7.25 the drive is direct through tumbler gear A and tumbler gear B is idle. If the tumbler bracket is moved upwards, tumbler gear A rolls around the driven gear until it is out of mesh with the driver gear, and tumbler gear B moves into

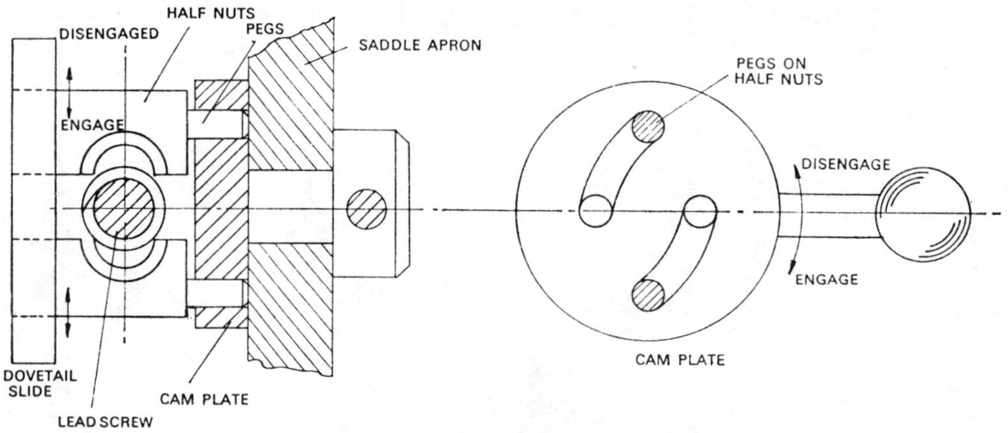

Fig. 7.24. Split-nut method of engaging leadscrew, shown disengaged

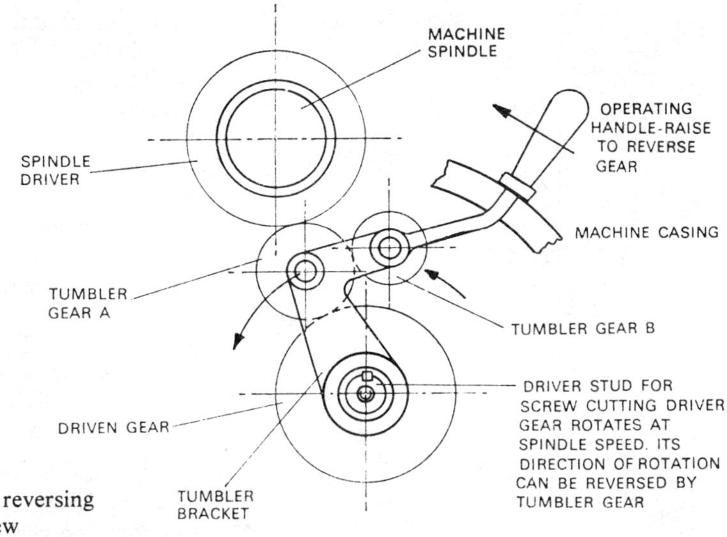

Fig. 7.25. Tumbler gearbox for reversing direction of rotation of leadscrew

mesh with the driver, reversing the direction of the driven gear. Thus the two trains available are:

B not engaged

Forward: Driver → A → Driven
Reverse: Driver → A → B → Driven

THE CHASING DIAL

When a screw thread is being cut in the lathe, the traverse of the saddle is caused by the engagement of the leadscrew by the split-nut device described on page 144 and shown in fig. 7.24.

Once the first cut has been taken and a helix produced on the work, the leadscrew must subsequently be engaged in such a way that the tool always picks up the same helix on successive cuts until the thread is completed. The *chasing dial*, fitted usually at the left-hand end of the saddle, enables this to be done correctly. It consists of a disc which rotates when the saddle is stationary but stops rotating when the saddle is traversed by the leadscrew. The disc has four equi-spaced marks numbered 1 to 4 with intermediate marks, giving eight in all.

With leadscrews of 5 mm pitch, it is driven by a 16-tooth worm wheel which engages the leadscrew and makes $\frac{1}{16}$ of a turn for each turn of the leadscrew. The leadscrew can thus be engaged as each mark passes the index on the saddle and also halfway between each mark, giving sixteen positions on the dial where the leadscrew can be engaged. The plan view and drive to the dial are shown in fig. 7.26. As each successive cut is taken, the tool must be dropped in, i.e. the leadscrew must be engaged, only when the work thread and the leadscrew thread coincide. Since it takes 16 revolutions of the leadscrew to give one revolution of the chasing dial, one revolution of the dial represents an 80 mm length of leadscrew moving past a fixed point or index mark. If we lay off this 80 mm length of leadscrew, with its 16 threads to scale, against a length of 2.50 mm pitch work thread as shown in fig. 7.27, we see that each leadscrew thread, representing a position on the dial at which the leadscrew can be engaged, coincides with a work thread and that this is true for all work threads which divide exactly into 5. This means that for such threads we can engage the leadscrew at any point.

If we now compare the leadscrew scale with a thread whose pitch will divide exactly into 10 mm (two marks), we find that only every other

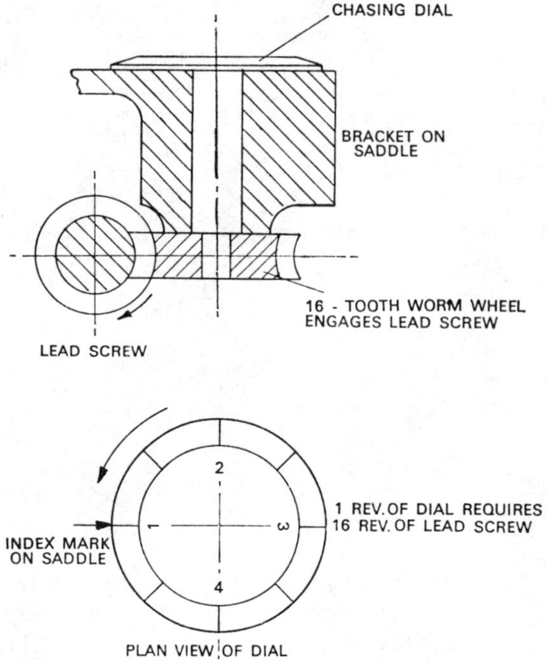

Fig. 7.26. Chasing dial for screwcutting

point coincides, so we can only drop in at every other position, i.e. at every mark on the dial but not at intermediate positions.

Comparing a scale representing a thread pitch which will only divide exactly into 20 mm (four marks), we find coincidence at each fourth position, i.e. every second mark on the chasing dial, so we can drop in at every numbered mark but not at intermediate marks or unmarked places on the dial.

If the pitch will only divide exactly into 40 mm, e.g. 8 mm, we find that we can engage at only two positions, i.e. every alternate number on the dial. If we start cutting on number 2 we can only engage on number 2 or number 4, and if we start on 1 we can continue on 1 or 3. A thread which will only divide exactly into 80 mm, e.g. 16 mm pitch, allows only one engaging position on the dial, since it does not coincide anywhere on our scales except at each end. Other threads will not coincide anywhere on the scales except at the single position at which they were engaged. This means that if the leadscrew is disengaged it will not 'pick up' the work thread again. Therefore such threads should be cut without disengaging the leadscrew; the tool is withdrawn with the leadscrew still engaged and the machine reversed to traverse the tool back to the beginning of the thread.

All these scales are shown in fig. 7.27 which refers to a 5 mm pitch leadscrew and a 16 tooth wormwheel with an eight-mark (16-position) dial.

A little thought will show that a rule for finding the divisions of a dial suitable for 'dropping in' is:

The pitch to be cut must exactly divide into the length of leadscrew represented by the fraction of a revolution of the dial at which it may be dropped in.

In fact the suitable fractions are $1/n$; $\frac{1}{8}$; $\frac{1}{4}$; $\frac{1}{2}$; and 1/1 revolutions of the dial, where n is the number of teeth on the wormwheel and hence the number of possible positions for engagement.

This rule can be written as a mathematical expression:

$$\frac{\text{Length of leadscrew per rev. of dial}}{\text{Pitch to be cut}} \times \frac{1}{N} = I$$

where I must be the smallest whole number (integer) obtainable if $1/N$ is one of the above fractions. E.g. if a 16 tooth, 8 mark dial, is used with a 5 mm pitch leadscrew find the positions on the dial at which the leadscrew may be engaged when cutting a thread having a pitch of 4 mm.

Length of leadscrew per rev. of dial $= 5$ mm $\times 16 = 80$ mm

$$\frac{80}{4} \times \frac{1}{N} = I$$

The value of $1/N$ which will make I the smallest whole number is $\frac{1}{4}$. Thus the leadscrew may be engaged every $\frac{1}{4}$ turn or, from fig. 7.27, every numbered mark.

Working from this principle, the pitches that can be cut by engaging the leadscrew at various

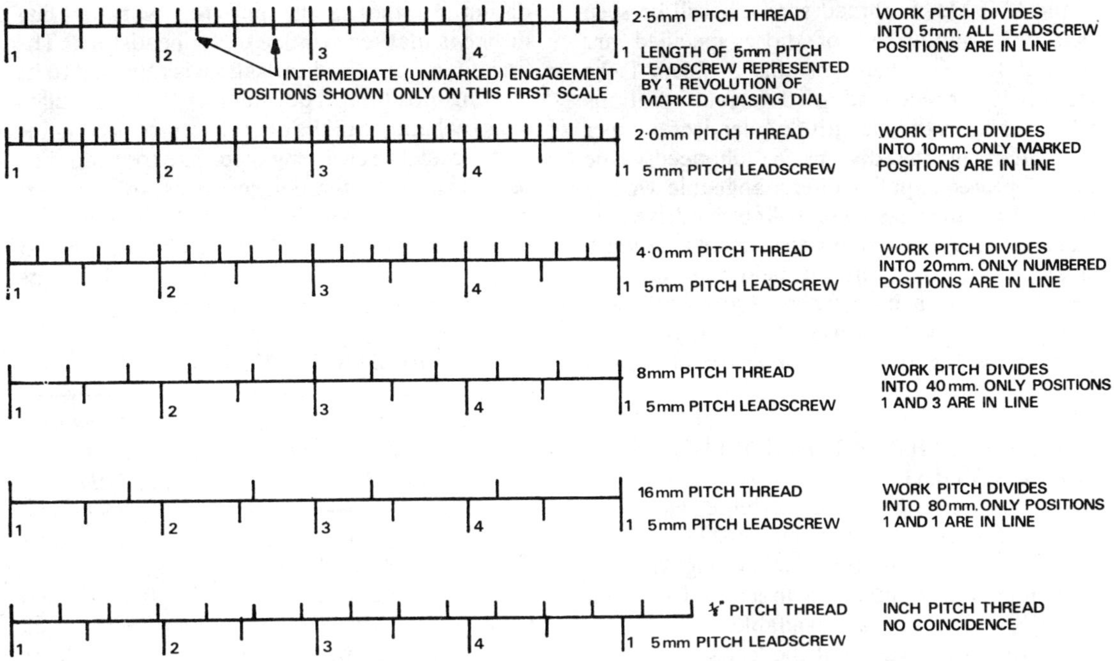

Fig. 7.27. Scales showing coincidence between a leadscrew thread of 5 mm pitch and work threads of various pitches as indicated by marked positions on chasing dial

positions of this chasing dial-leadscrew system are:

Any position (1/16 rev.): 0.25 mm; 0.50 mm; 1.00 mm; 1.25 mm; 2.50 mm; 5.00 mm
Alternate positions ($\frac{1}{8}$ rev.): 2.00 mm; 10.00 mm
Numbered marks ($\frac{1}{4}$ rev.): 4.00 mm; 20.00 mm
Opposite marks ($\frac{1}{2}$ rev.): 8.00 mm
Same mark always (1/1 rev.): 16.00 mm

For other threads the leadscrew must not be disengaged.

Another combination of leadscrew and chasing dial in use is a 6 mm pitch leadscrew used with a 20 tooth wheel and a 20 mark dial. Again consider a thread of 4 mm pitch to be cut.

Length of thread per rev. of dial $= 20 \times 6$ mm $= 120$ mm

$$\frac{120}{4} \times \frac{1}{N} = I$$

The suitable fraction $1/N$ which will make I the smallest whole number is $\frac{1}{2}$, and therefore the leadscrew can only be engaged with the chasing dial on opposite marks.

The points of engagement for this system are:

Any position (1/20 rev.): 0.50 mm; 0.60 mm; 0.75 mm; 1.00 mm; 1.20 mm; 1.50 mm; 2.00 mm; 3.00 mm; 6.00 mm
Numbered positions ($\frac{1}{4}$ rev.): 1.25 mm; 2.50 mm; 5.00 mm; 10.00 mm
Opposite marks ($\frac{1}{2}$ rev.): 4.00 mm
Same mark each time (1/1 rev.): 8.00 mm

For other pitches the leadscrew must not be disengaged.

If the pitches available in either of these systems are compared with the pitches specified in the ISO Metric thread series, it will be seen that only about half of those specified are catered for. To overcome this problem lathe manufacturers have adopted various solutions which allow the length of leadscrew per revolution of the dial to be changed. One manufacturer supplies interchangeable wormwheels to fit the chasing dial. Another drives the dial from the leadscrew and wormwheel via a small gearbox built into the saddle. It may well be that this is a case where adoption of the metric system will increase rather than decrease complexity due to the small number of factors of 10.

SCREWCUTTING WITHOUT THE CHASING DIAL

If a screw is to be cut on a lathe without a chasing dial some other means must be adopted to ensure that the tool always engages the thread correctly after the first cut has been taken. Two methods are available.

1. At the end of each cut the leadscrew is not disengaged but the tool is withdrawn and the lathe is reversed until the tool is back at the tailstock end of the bed. The tool is then advanced to the correct depth for the next cut and the machine restarted forwards.

SAFETY NOTE

Always traverse back well beyond the start of the cut so that, when the tool starts to cut, any backlash in the leadscrew has been taken up.

---

2. Marking the machine. The tool will only re-engage the thread correctly if:
   (a) the position of the saddle,
   (b) the angular (rotational) position of the work and
   (c) the angular position of the leadscrew are all three exactly the same as they were for the start of the original cut.

These requirements can be met if, before the first cut is taken, the leadscrew is engaged and a mark is made or a stop is fitted on the bed so that the saddle can always be returned to its original starting position. At the same time the leadscrew is marked against a convenient datum on the saddle and the work-holding device is marked relative to the headstock. The first cut is taken and the saddle is returned to its starting position. The machine is now pulled over by hand until the other two marks coincide and the leadscrew is engaged. The tool can then be advanced to the correct depth for the next cut and the machine started in the sure knowledge that the tool will re-engage the first cut, cut correctly on this and, with the same preliminaries, on successive cuts.

HOLE PRODUCTION

Specific Objective: *The student should be able to describe the machining of holes to a given specification (B.S. 4500) on a centre lathe.*

TAILSTOCK OPERATIONS

The hole in the front end of the tailstock barrel is bored to a Morse taper, so that the tailstock can be used in performing all the operations that can be carried out on a drilling machine. The tailstock centre is simply replaced by a drill chuck, drill, spot-facing tool or counterbore, etc., while the workpiece is rotated at the headstock.

The commonest operation is drilling, and the barrel is usually marked with a scale so that the depth of the drilled hole can be controlled by noting the scale reading when the drill touches the work and continuing to drill until the required depth is reached. A drill tends to wander and cannot always be relied on to produce a straight concentric hole, and if the geometry of the hole is important, the hole should be bored out to ensure concentricity and straightness.

Reaming can be carried out from the tailstock but a reamer will only bring a hole to size and will not necessarily correct its geometry. Thus, where possible, the hole should be bored to correct its alignment before reaming. If the hole has not been bored and is out of true, the side loads on the reamer may result in an oversize hole or even break the reamer. The reamer should therefore be supported in a holder which allows it to 'float', or by the tailstock centre, so that it can follow the hole. In the latter case the work is kept stationary and the reamer is rotated by a wrench.

When a hole is to be tapped, the tap can also be supported by the tailstock centre to ensure that it is kept square to the work. The tap is turned by a tap wrench and followed up by the centre to ensure continued support to the tap.

BORING OPERATIONS

To ensure that a hole is positioned correctly relative to other features, or is 'running true', it is necessary to bore it, since a drill will tend to wander and subsequent tailstock operations will tend to follow the original drilled hole. To prevent the boring bar being deflected by the varying loads caused by a non-true drilled hole it should be as stiff as possible, that is as large in diameter and as short as possible to do the particular job.

Precise positioning of a bored hole in a non-cylindrical workpiece is described on pages 131–34 as it involves precisely positioning the workpiece relative to the axis of rotation and is a workholding problem.

The correct depth of a bored hole can be achieved simply, if not to a high degree of accuracy by marking the slideways with a soft pencil (not a scriber). More precise depth of holes can be achieved by using a micrometer stop fitted to the slideways so that its spindle is in line with a machined face on the saddle. For shallow holes the micrometer can be used directly within its range of 25 mm but its range can be extended by the use of grade 2 gauge blocks. Assume that a hole of 142.5 mm depth is required. Gauge blocks could be made to 125 mm and the micrometer set to the difference.

$$142.5 - 125\ \text{mm} = 17.5\ \text{mm}$$

The tool is now positioned to be just touching the edge of the hole and, with the micrometer stop set to 17.5 mm it is adjusted to trap the gauge blocks between the micrometer spindle and the saddle. When the blocks are removed the saddle will have to be advanced 142.5 mm to contact the spindle. Note that the feed should be disengaged before the stop is reached and the remaining distance fed by hand. The use of the micrometer stop is illustrated in fig. 7.28.

The student is referred to the tables in B.S. 4500 for the accuracies which can be achieved by different types of operations. They are given as tolerance grades in B.S. 4500 data sheets, those for lathe work being as follows:

High quality turning and boring	Tolerance grade 7
Boring, turning and reaming	Tolerance grade 8
Drilled holes	Tolerance grade 11

SUMMARY

This section of the chapter has made no attempt to explain in detail how to operate a lathe but has tried rather to cover the principles involved. The only way to become proficient in the operation of any machine tool is to understand its underlying principles and then use it. For the student, this practical experience is normally provided during off-the-job training at work, and the principles are explained and demonstrated in college workshops.

The principles involved in the centre lathe will be found to apply in other machines used for large work, such as a vertical boring machine, which is basically a lathe standing on its headstock, and on high-production machines such as capstan and turret lathes, single-spindle and multi-spindle automatic lathes, chucking automatics, etc. None of these machines will produce work of correct geometry unless the basic alignments are correct. Parallel work is produced only when the tool motion is parallel to the axis of rotation. Flat facing occurs only when the tool moves at right angles to the axis of rotation, and a true taper is produced only when the tool is set correctly.

SAFETY NOTE

The workpiece on a lathe rotates, usually at high speed and with a considerable amount of power behind it. All rotating and moving parts

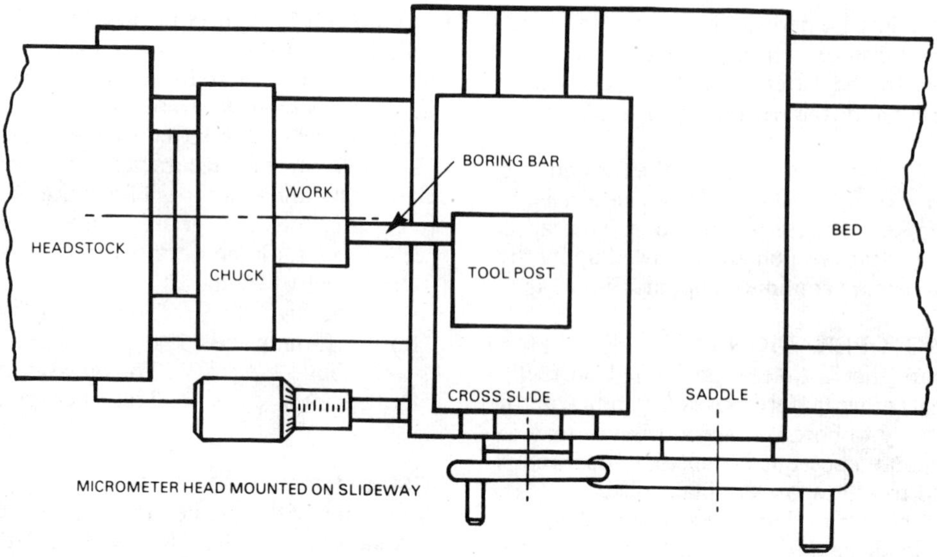

Fig. 7.28. Micrometer method of obtaining correct depth of a bored hole

represent a hazard, and the hands should be kept clear. Loose clothing can be picked up and drawn in, and a necktie in a rotating mechanism is as efficient as, if slower than, a hangman's noose. Do not be tempted to touch swarf, stationary or moving, with the bare hands or a deep cut can result; stop the machine and use a swarf-hook. Coolant should be fed to the job through pipes, not brushed on, and the machine should be allowed to stop itself – hands are too delicate and precious to use as brake bands on the chuck. Eyes need protection; swarf is hot and sharp, and if it gets into the eye can cause damage. A man with a lot of luck and quick reflexes catches the swarf particle between his eyelids where it sticks and blisters them, which is most painful. ***Wear safety glasses.***

QUESTIONS

1. Name four different methods of workholding on a centre lathe and for each suggest one example of a type of work for which it is particularly suited.
2. If a lathe is to produce parallel parts by turning and flat parts by facing, the tool and work must have certain relative motions. State what these motions are and name the features of the lathe which control them.

If each of these motions contains an error show with the aid of a diagram the shape of workpiece which will be produced.

3. Explain in some detail *two* methods setting up a centre lathe to enable a hole to be bored accurately in a workpiece relative to another feature or datum surface.
4. Lathe steadies may be bolted directly to the bed of the lathe (fixed steady) or to the saddle (travelling steady). Sketch each type, showing the adjustments available, and explain the type of work each is used for giving a typical example of each type of work.
5. Name three methods of producing a taper on a centre lathe, one to be used for each of the following tapers.
   (a) 35° angle of taper over a length of 55 mm of workpiece.
   (b) 5° included angle of taper over a length of workpiece of 250 mm.
   (c) A mandrel having a taper of 1 mm per metre length.

In each case make a line diagram of the set-up

and explain how the necessary adjustments are made.

6. During any turning operation it is important that the tool is set at the correct height. It is particularly essential that it is set correctly for taper turning. Explain why this is so.

7. (a) Using simple diagrams, explain how large angles of taper can be turned using the settings available on the compound slide and toolpost.
   (b) Accurate circular scale settings are not normally available on the compound slide. Explain any way in which an accurate male taper (i.e. one that will fit the other member accurately) of say 60° included angle can be turned. Give details of any measurements or tests required.

8. A thread having a pitch of 1.6 mm is required and is to be cut on a lathe having a leadscrew of 5 mm pitch.
   (a) Calculate the gears required if they are to be set up by the operator.
   (b) Explain how the tool is to be set.
   (c) How would the chasing dial be used in this case.

9. Second-operation turning is often done by mounting the work on a temporary mandrel made by turning a peg or spigot.
   (a) Give an example of a typical component which could be machined in this way, and show a sequence of operations.
   (b) A high degree of concentricity can be achieved by this method. Explain the reason for this.

10. The component shown in fig. 7.29 is to be produced so that the 12 mm H8 hole is accurately concentric with the 30 mm outside diameter.
   (a) Using standard tables of limits and fits state the limits on the hole.
   (b) Name a method of producing the hole to these limits.
   (c) Explain briefly how concentricity could be achieved whichever method of machining the hole was selected in (b).

11. In a certain rough turning operation a 50 mm diameter mild steel bar has to be reduced to 30 mm diameter for a length of 75 mm. This roughing could be done by taking one cut at a suitable rate of feed or by taking several lighter cuts at higher feed rate.
   (a) Compare the advantages and disadvantages of these two methods of procedure.
   (b) State which of the methods you consider best, and why.
   (c) Give the following information:
      (i) cutting speed to be used in m/mm,
      (ii) spindle speed in rev/min,
      (iii) depth of cut and rate of feed,
      (iv) tool shape and cutting angles,
      (v) material of which the tool is made.

12. (a) The workpiece in fig. 7.30 has to have a 40 degree taper produced upon it by turning the lathe. Using simple

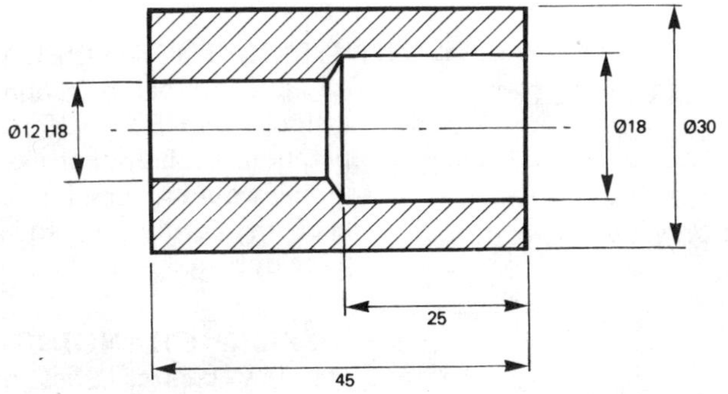

Fig. 7.29. Turned component—Question 11

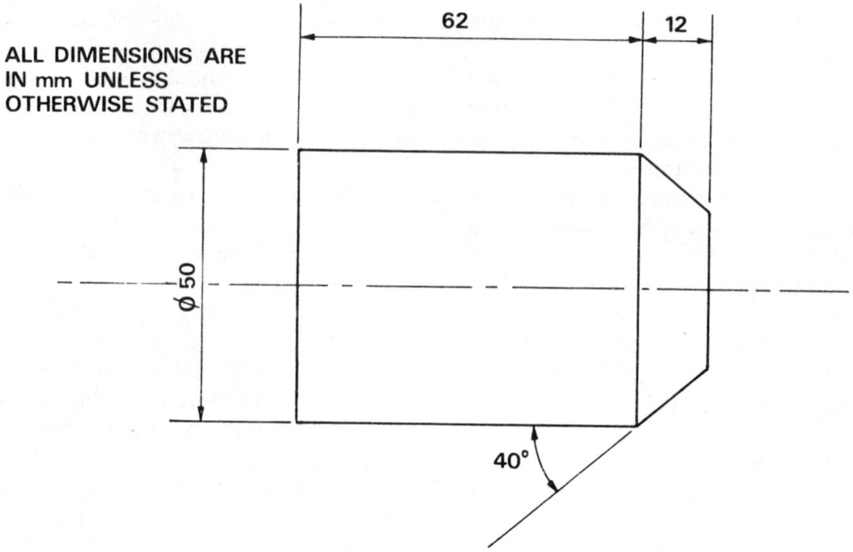

Fig. 7.30

diagrams of tool shapes and tool movements, explain clearly the difference between forming and generating as basic methods of machining this taper.

(b) With reference to one of the above methods, outline the practical machining process, including the tool shape and feed, the work-holding arrangement and the means of controlling the dimensions of the taper.

ANSWERS

8. (a) $\frac{30}{50} \times \frac{80}{100}$.

PART 2. CAPSTAN AND TURRET LATHES

General Objective: *The student should be able to compile simple block diagrams of capstan and turret lathes.*

INTRODUCTION

For the performance of repetitive turning, drilling and other operations, the centre lathe has certain limitations. These include the following:

(a) MANUAL SKILL
Each dimension is arrived at by the skill of the operator in measuring the workpiece and manipulating the machine controls to give the desired size.

(b) TOOL CHANGING
The number of tools mounted in the toolpost is severely limited and to perform a large number of operations involves virtually resetting the machine each time the tools are changed.

(c) TAILSTOCK OPERATIONS
Only one tool can be mounted at a time in the tailstock and to perform a sequence of operations such as centre-drilling, reaming and tapping would require four changes of tool, and the depth of drilling would have to be controlled by the operator.

(d) SPEED CHANGING
Very few centre lathes have provision for changing quickly the spindle speed or reversing its directions of rotation. The machine must be

stopped if gnashing of gear teeth and possible damage are to be avoided.

If a lathe is required for repetitive work, certain modifications of it must be made to eliminate the above deficiencies. These modifications, which resulted in the development of capstan and turret lathes for repetitive production, are as follows:

(a) SIZING METHODS
If the manual skill is to be eliminated, the size required must be produced by pre-setting the machine. All slideways are equipped with stops and the sliding member is fed in until the stop is reached.

(b) TOOLING SET-UPS
The cross-slide is equipped with a multi-tool post at the front which can be rapidly indexed into position. A single rear toolpost is also fitted.

(c) TAILSTOCK TOOLING
To facilitate the mounting of a larger number of tools which can be fed along the axis of the work, the tailstock is replaced by a hexagonal turret which has six tooling stations. Each station has its own stop so that the depth of feed can be controlled, and power feed is available.

(d) SPEED CHANGING
The headstock is fitted with a constant-mesh gear box, i.e. the gears are always in mesh and the required speed is selected, forward or reverse, by engaging clutches which can be thrown in or out of mesh without stopping the machine. A later development has been the introduction of 'plugboard control', in which the speeds are changed by an electro-pneumatic system. The required speed for each turret position is selected on a plugboard and as the turret is indexed a set of micro-switches brings in the appropriate circuit to engage the pre-selected clutch and so give the desired speed.

Thus the lathe has now taken on a new configuration, as shown in fig. 7.31, and is known as a *capstan* or *turret lathe*. In a capstan lathe the turret is mounted on a subslide which can be adjusted and locked in position along the bed, but the length of travel of the turret is limited by the length of the sub-slide. The turret of the turret lathe is mounted directly upon the lathe bed and its length of travel is therefore limited only by the length of the main slideways. Also shown in fig. 7.31 are the arrangements of the length stops on the saddle, cross-slide, and turret.

CAPSTAN LATHE TOOLING

Specific Objective: *The student should be able to describe typical machining operations on capstan and turret lathes.*

As the capstan lathe evolved, specialised tooling set-ups were developed to enable its full capabilities to be used. Most of the tools used in the cross-slide toolpost are of the standard type of turning tool, although knife tools with zero plan-approach angle are frequently used to save a second operation to square up and clean out corners. The main development in tooling for the capstan lathe has been in turret-mounted equipment, so that the turret has not only replaced the tailstock for co-axial machining operations such as drilling and reaming, but is also used for turning and boring operations.

CROSS-SLIDE OPERATIONS
Most of the cross-slide work is straightforward turning, facing and parting off, a normal set of turning tools being used. Where a complex form which it is not practicable to generate is required, a form tool can be used.

Two common types of form tool used are the *radial* type, shown in fig. 7.32(a), and the *tangential* type, fig. 7.32(b). In both cases it is essential that the tool is re-ground and set correctly or an error of form will result. To give precise height adjustment and rigidity, the radial tool is mounted on a serrated wedge-type toolpost and the tangential tool is usually gripped in a dovetail. Both tools are sharpened by grinding the rake face; obviously it would be too costly to re-grind the form. The extra length

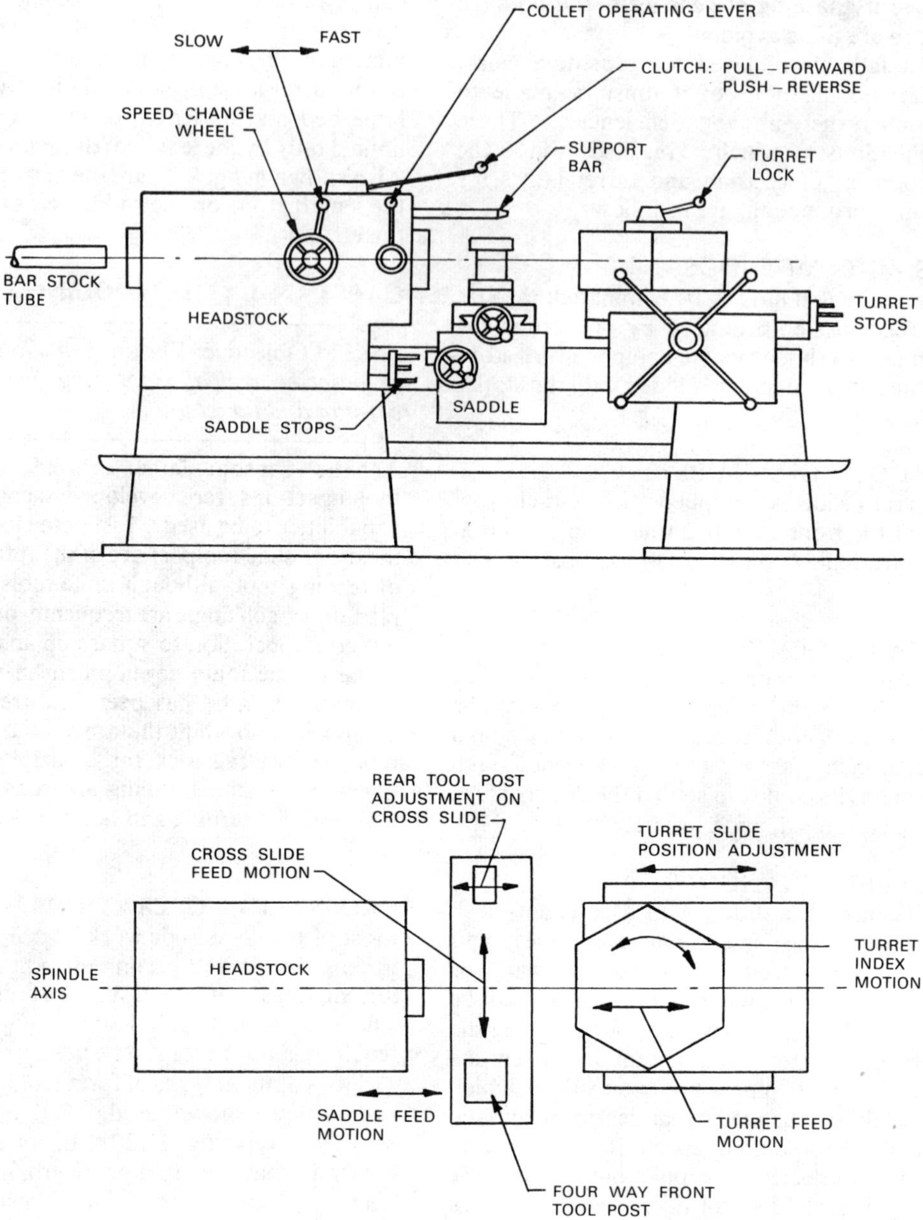

Fig. 7.31. Block diagram of capstan lathe showing main controls, motions and adjustments

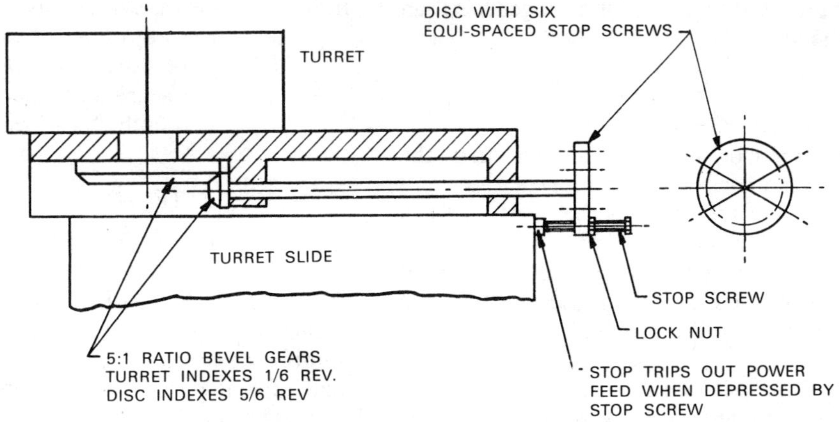

ARRANGEMENT OF TURRET SLIDE STOPS

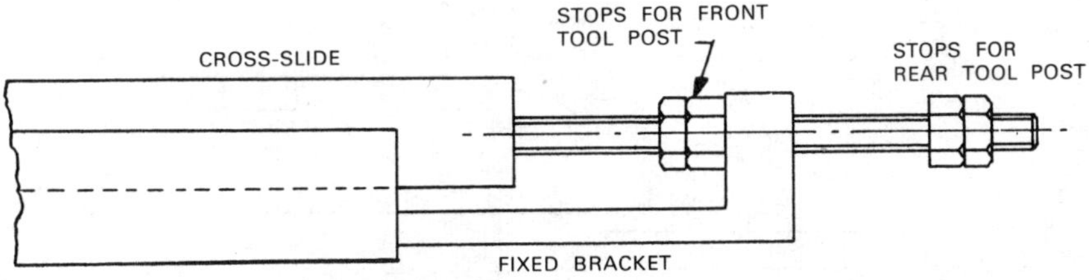

ARRANGEMENT OF CROSS-SLIDE STOPS

of the tangential tool gives it a longer life than the radial tool.

FORM TOOL GEOMETRY

It can be seen from fig. 7.33 that the form on the tool is ground parallel to the clearance face and its depth is not the same as the depth of form which it produces on the work. The depth of form on the tool depends on the depth of form on the work, the clearance angle and the rake angle, and it must be calculated.

(*a*) *Flat form tool with zero rake angle*
This is the simplest case, shown in fig. 7.33(a)

If R = outside radius of work
r = root radius of work
$(R-r)$ = depth of form on work
θ = clearance angle
x = depth of form on tool

$$\text{Then } \cos\theta = \frac{x}{(R-r)}$$

and $$x = (R-r)\cos\theta$$

(*b*) *Flat form tool with positive rake angle*
In this case, shown in fig. 7.33(b), the geometry is more complex as correction must also be made for the rake angle. The solution is in three stages.

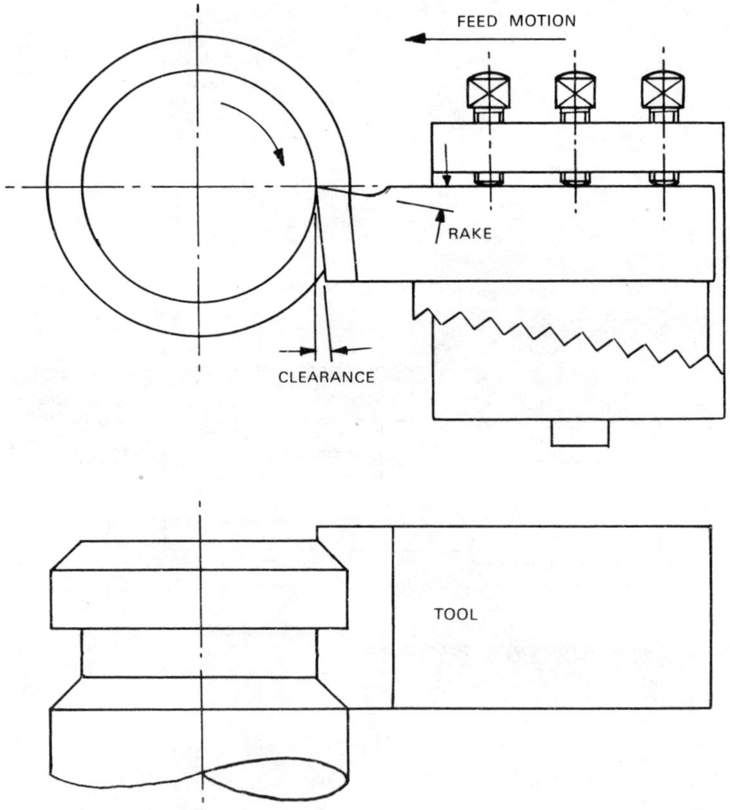

Fig. 7.32(a). Radial type form tool and holder. Note serrated wedge for height adjustment.

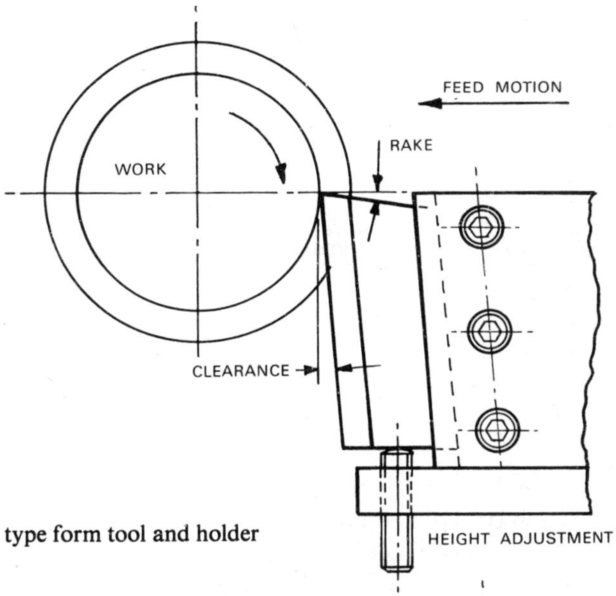

Fig. 7.32(b). Tangential type form tool and holder

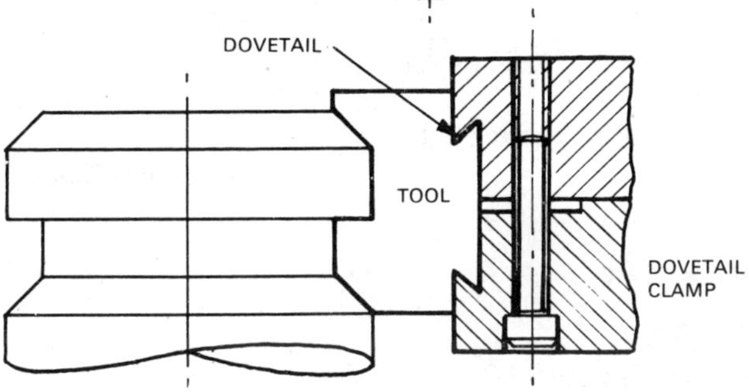

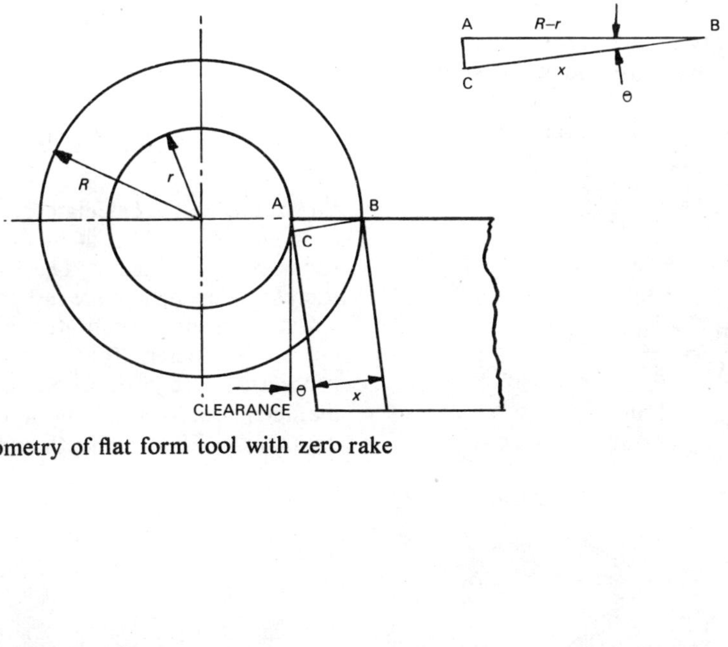

Fig. 7.33(a). Geometry of flat form tool with zero rake

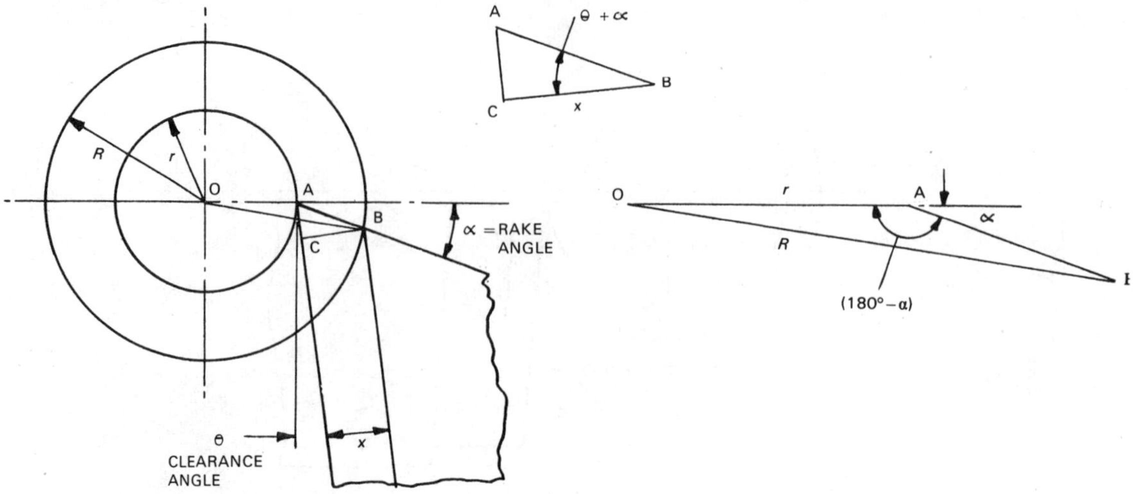

Fig. 7.33(b). Geometry of flat form tool with positive rake

(i) In triangle ABC

AB = length of rake face in contact with the work

Angle ABC = $(\alpha + \theta)$, where α = rake angle

$$\frac{x}{AB} = \cos(\alpha + \theta)$$

$$x = AB\cos(\alpha + \theta) \qquad \ldots(1)$$

To find AB, use triangle AOB

Angle OAB = $(180° - \alpha)$

OA = r

OB = R

The remaining two stages are:

(ii) Use the sine rule to find all the angles in the triangle.

(iii) Knowing all angles, again use the sine rule to find AB.

(ii) $$\frac{R}{\sin A} = \frac{r}{\sin B}$$

$$\sin B = \frac{R}{r\sin A} \qquad \ldots(2)$$

(iii) $$\frac{AB}{\sin O} = \frac{R}{\sin A}$$

$$AB = \frac{R\sin O}{\sin A}$$

But O = $(\alpha - B)$

A = $(180° - \alpha)$

$$\therefore \quad AB = \frac{R\sin(\alpha - B)}{\sin(180° - \alpha)} \qquad \ldots(3)$$

in which B is found from stage (ii)

α = rake angle

R = outside radius of work

θ = clearance angle

Note also that $\sin(180° - \alpha) = \sin\alpha$

Summarising stages (i), (ii), and (iii) we have

(i) $$x = AB\,.\,\cos(\alpha + \theta) \qquad \ldots(1)$$

(ii) $$\sin B = \frac{R}{r\sin\alpha} \qquad \ldots(2)$$

(iii) $$AB = \frac{R\sin(\alpha - B)}{\sin\alpha} \qquad \ldots(3)$$

Substitute in (i) to calculate x.

It will be noted that in these solutions the terms R and r occur. The shape of the tool therefore depends on the radius of the work and will only produce accurate forms on bar of the correct diameter. If a tool is designed to produce a vee groove on a bar of 20 mm diameter but is used on a bar of 100 mm diameter, neither the depth nor the angle of the groove will be correct. However, as most form tools are designed for one particular component this problem does not often arise.

Similar methods must be applied to form correction for circular form tools. In each case it is advisable to determine first the length AB of the rake face in contact with the work.

TURRET OPERATIONS

The major proportion of the work on a capstan lathe is carried out with turret-mounted tooling, the tools being mounted in sequence and the turret automatically indexing as it is withdrawn by the operator.

TURNING FROM THE TURRET

The machining of external diameters falls into two categories, depending on the method of work holding.

(a) Chuck-held work

Work which is held in a chuck or turning fixture is usually fairly rigid and needs no additional support while cutting takes place. Such turning is usually carried out with a *knee tool holder*, as shown in figs. 7.34(a) and (b). Note that the rigidity of this type of tool holder can be increased by using the overhead support bar protruding from the headstock. This engages the guide bush shown on the tool holder and maintains the alignment of the turret *and* the spindle axis, preventing deflection of the turret by the cutting force applied.

A boring bar can also be fitted so that a hole

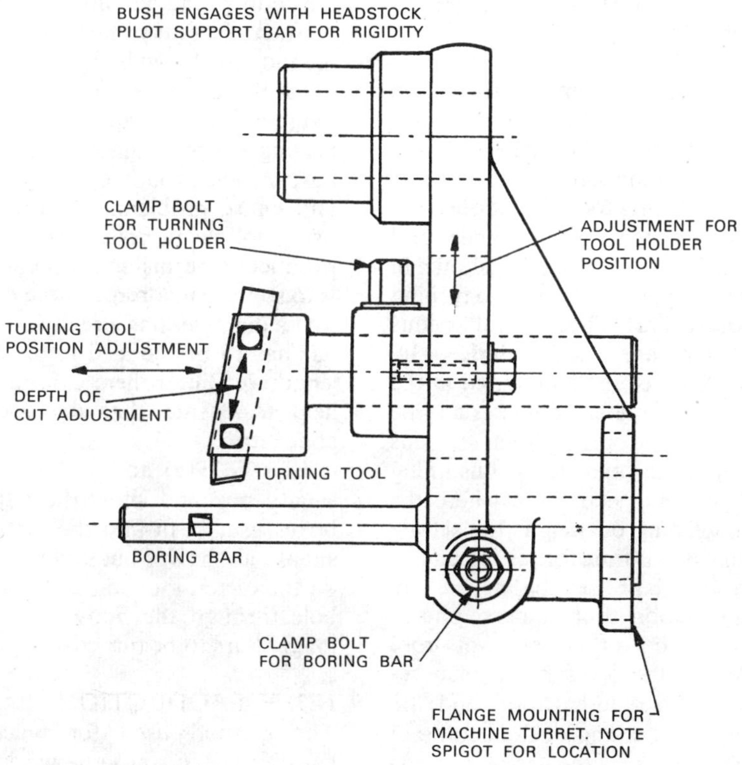

Fig. 7.34(a). Elevation of knee tool holder.

Fig. 7.34(b) Knee-turning tool holder – note provision for fitting with boring bar. (Reproduced by kind permission of H. W. Ward & Co.)

can be machined at the same time as the external surface. The chuck or fixture should be fitted with a guide bush to pilot the end of the boring bar and again maintain the geometry of the workpiece by reducing tool deflections.

A further development of the knee tool holder is the *combination tool holder,* shown in fig. 7.35(a), which can be fitted with two turning tools and a boring bar. Thus, simultaneous machining operations are possible. Indeed, by increasing the number of tools mounted in the boring bar, multi-diameter bores can be machined at the same time. Such set-ups involve more complex and time-consuming toolsetting and are only economic where the time saved in machining outweighs the setting time. The machine down-time for setting can be reduced by the use of pre-set tooling, i.e. by having two combination tool holders, one in the machine and the other out for tool maintenance. When the tools have been re-ground they are pre-set in position and the complete tool holder is changed. Pre-setting of boring tools is facilitated by the use of a boring bar cutter micrometer, as shown in fig. 7.36.

Comparing the knee-type tool holder with the combination-type tool holder, although the knee type only has provision for one turning tool it has a greater range of adjustment, for the tool post can be moved up the body of the holder. The adjustment in the combination-type tool holder is merely that of the tool overhang in the toolpost.

(*b*) *Bar Work*

When work is produced from bar stock it is usually not as firmly supported by the chuck and is itself rather less rigid than the castings and forgings which comprise much of chuck and fixture-held work. Thus some form of steady is required to prevent deflection of the work through the cutting load. For such work, therefore, a *roller steady box* is generally used on which is mounted, as well as the tool, a pair of rollers which bear against the work and resist the cutting load, as shown in fig. 7.37.

The rollers may be either leading or following the tool, as shown in fig. 7.38. Rollers following the tool are used when the operation is the first reduction in diameter on a new bar. Obviously leading rollers cannot be used on non-round bar, and even round bar may have irregularities and cause problems. A further advantage of using following rollers is that their pressure produces a burnishing effect and the roller finish is to some extent reproduced on the turned bar.

If a diameter has already been turned and the bar has to be reduced in diameter for a short length, leading rollers can be used. They then help to ensure that the two diameters are concentric.

Figs. 7.39(a) and 7.39(b) illustrate a roller steady box and show the adjustments. Small boxes usually fit into the turret by means of a shank, as shown, but larger boxes bolt directly on the turret, locating in a register, and have a hole through the body of the box to enable longer bars to be turned.

HOLE PRODUCTION

The method used for machining holes in capstan and turret lathe work, as in any other machining process, depends on the accuracy

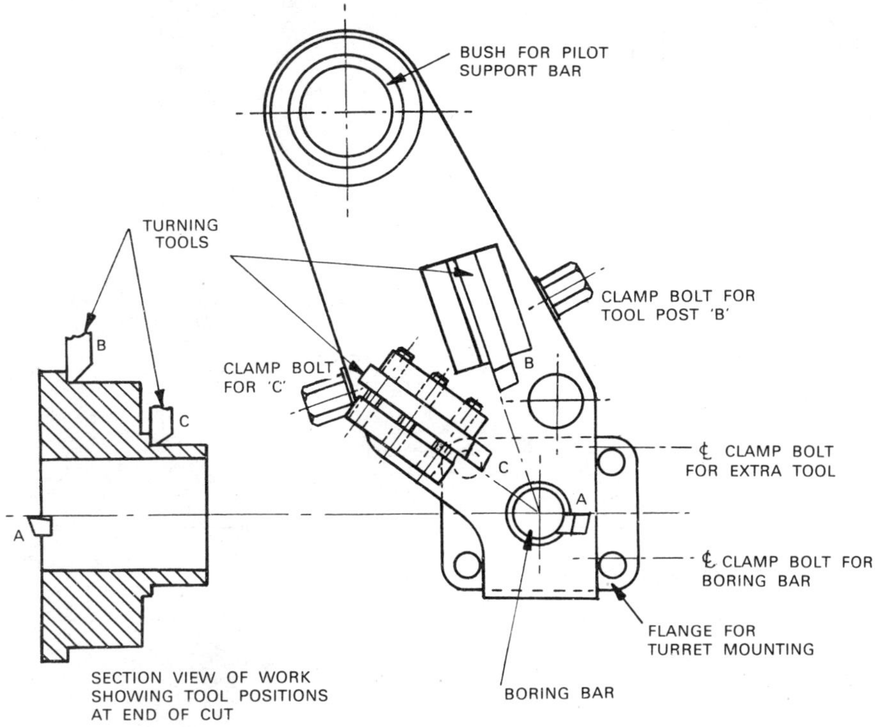

Fig. 7.35(a). Combination boring and turning tool holder. Front view showing relative positions of tools, machine axis and pilot support bar axis

Fig. 7.35(b). (Reproduced by kind permission of H. W. Ward & Co.)

required. Thus holes with large tolerances are simply centred and drilled. Drilling, however, is not only prone to errors of size but the drill can wander and give rise to a non-straight hole. Where accuracy of diameter is required the hole can be reamed, but a reamer is a sizing tool and will not correct all geometrical errors and rather tends to follow the drilled hole. Thus, for the highest degree of accuracy, the hole must be centred, drilled, bored and finally reamed.

Centring

Centre holes for starting a drill can be produced with a centre drill. These tools are, however, notoriously delicate and easily broken. A better method is to use a short flat drill as shown in fig. 7.40. This is extremely stiff and has less tendency to wander or break.

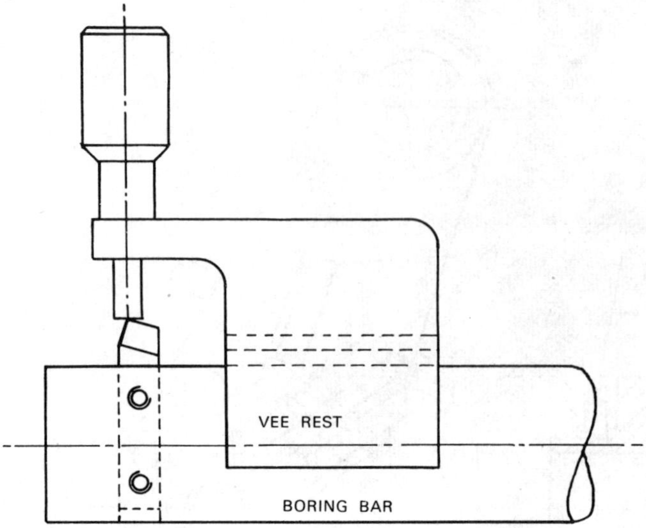

Fig. 7.36. Boring bar setting micrometer

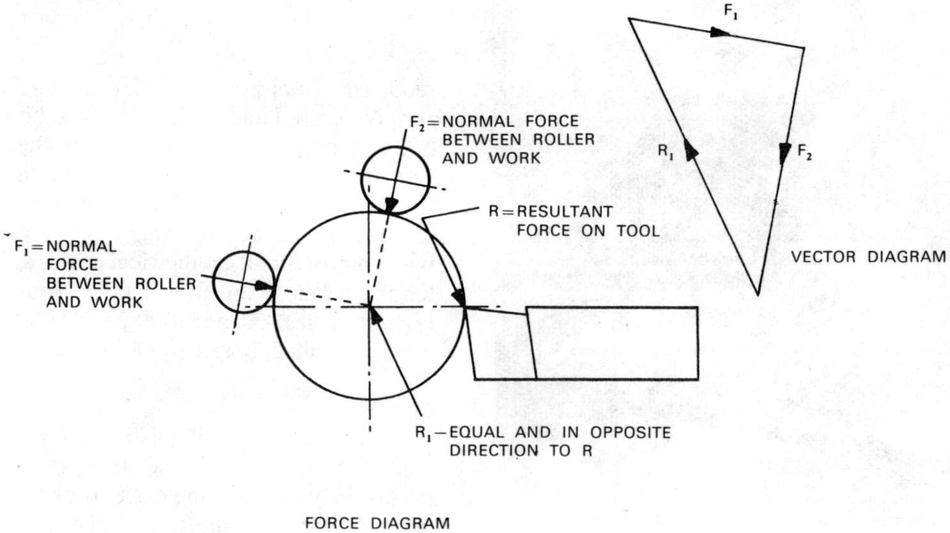

Fig. 7.37. Geometry of roller box showing forces on tools and rollers

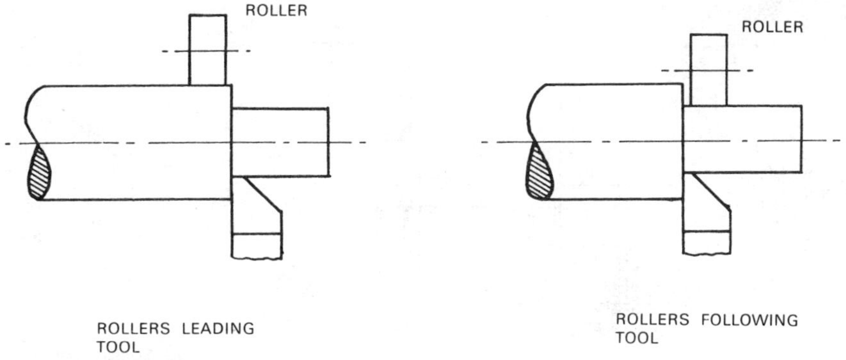

Fig. 7.38. Alternative positions of rollers in relation to tool

Drilling

Normal twist drills are used for capstan lathe work. They can, even with the smaller sizes, be obtained with a taper shank to fit the hole in the drill holder mounted on the turret. Where straight shank drills are used an adaptor with a clamping bush, as shown in fig. 7.41, is used. This method of holding can also be used for taps and reamers.

Deep holes present a problem in getting the swarf out of the hole and the cutting fluid into the cutting area. For this reason, larger drills are formed initially with straight flutes so that holes can be drilled up the solid part of the drill. They are then twisted hot to the required helix angle before being hardened and finish ground. In operation, cutting fluid is forced down the holes under pressure to cool the cutting edge and flush the swarf out.

Reaming

If a hole has been bored prior to reaming, the reamer can be held rigidly in a taper adaptor or a drill holder as shown in fig. 7.41. If a drilled hole is to be reamed, however, and the hole is not straight, or is eccentric, a rigidly held reamer might be broken. In these circumstances the reamer must be mounted in a floating reamer holder, fig. 7.42, which allows enough free movement for the reamer to follow the hole.

Recessing

If a blind hole is to be tapped or a stepped hole is to have a squared-out shoulder, a recessing operation may be necessary. This can be accomplished with the recessing attachment shown in fig. 7.43. The tool is mounted on a dovetail slide so that it can be moved across the face of the turret by the lever shown. The operator enters the tool into the work with the handle drawn towards him, the slide thus being moved away, advances the turret to the stop

Fig. 7.39(a). (Reproduced by kind permission of H. W. Ward & Co.)

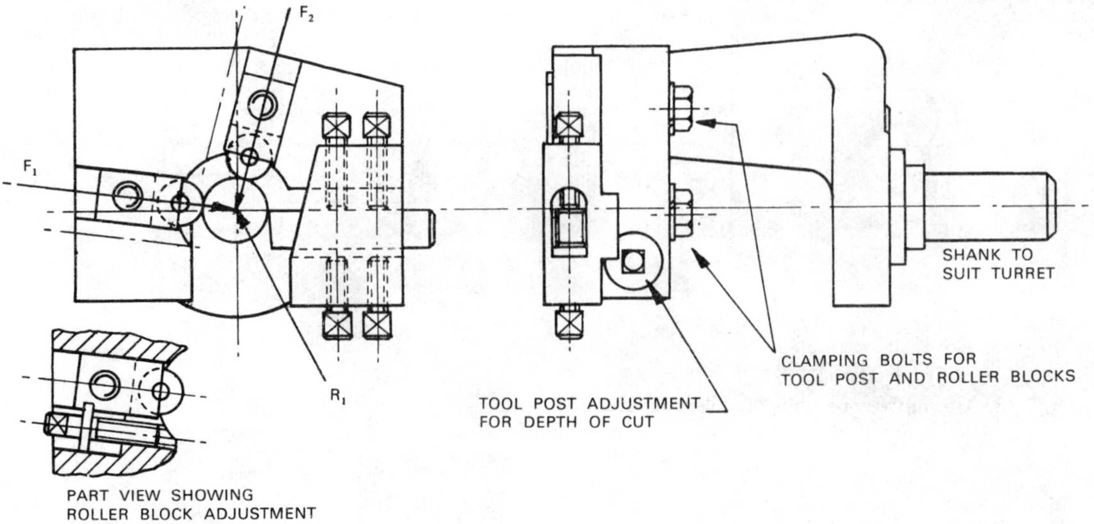

Fig. 7.39(b). Roller steady tool holder showing adjustments. Forces F_1, F_2 and R_1 refer to fig. 7.37.

and pushes the handle away from him to form the recess. The handle is then pulled back and the turret withdrawn.

TAPER AND PROFILE BORING

Form tools cannot very well be used for machining bores. If a straight taper is required the machine may be fitted with a taper-turning attachment similar to that used on a normal centre lathe. If this is not available, an attachment similar to the recessing attachment is used, with the slide connected by a link and roller to a cam plate mounted on the cross-slide, as shown in fig. 7.44. The depth of cut is set by the cross-slide and when the operation is completed the turret is withdrawn so that the roller clears the slot. The turret can then be indexed in the normal way.

This device can also be used for external turning to avoid a second operation. A case in point is the machining from copper of spot-welding tips as shown in fig. 7.45. Originally these were straight-turned, threaded, and parted off, the machining of the taper being a second operation. Forming the taper by a form tool was impossible since, due to the weak nature of the work material, the component broke off before forming was completed. By using the cam plate the taper was turned by a series of fine cuts and the component was finally parted off when completed.

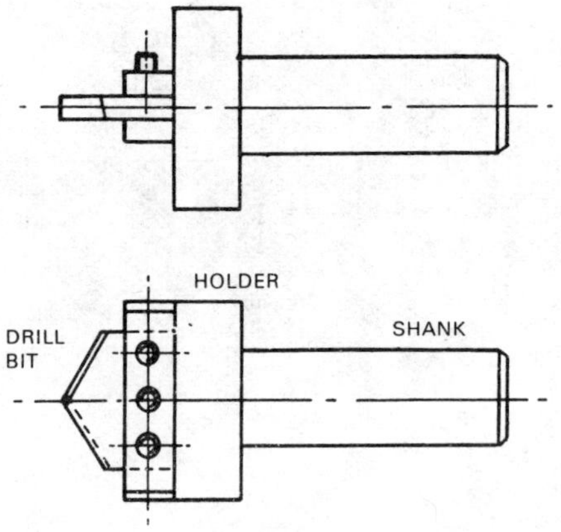

Fig. 7.40. Flat or spade drill for starting holes

STOCK LENGTH STOP

When parts are being machined from bar stock it is necessary to control the stock length fed from the chuck. This can be done by means of

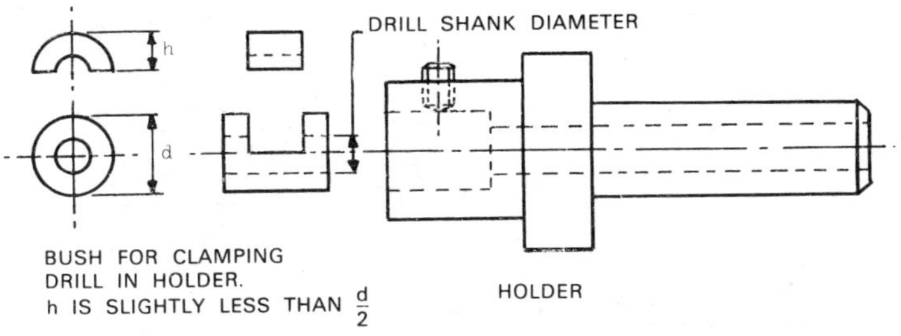

Fig. 7.41. Adaptor for straight shank drills. The same clamping arrangement may be used in tapping attachments and reamer holders

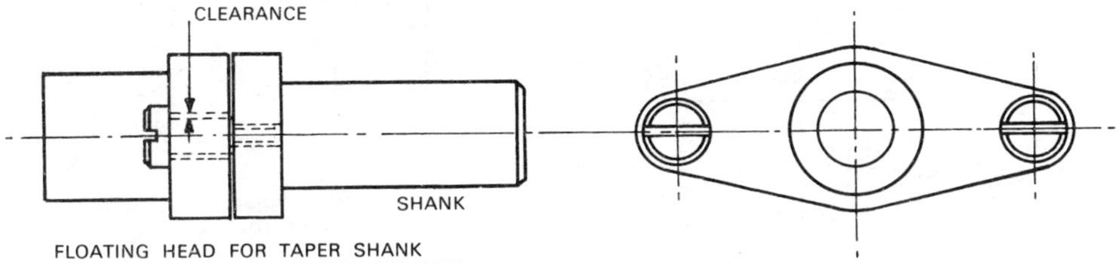

Fig. 7.42. Simple floating reamer holder. Clearance between the shoulder screws and holes in the front flange allows the reamer to float

a simple piece of bar held in the turret and adjusted by the turret stop. A better method is to use a stop bar with a screw and locknut, as shown in fig. 7.46. This makes for easier fine adjustment and control of the length fed from the collet.

An alternative which saves the use of an extra turret station and indexing movement is to combine the length stop with a centre drill holder on a swinging arm. With the stop in position the stock is fed to length and the collet locked. The turret is now withdrawn to allow the centre drill to be brought into position and then advanced for centre drilling. This attachment is shown in fig. 7.47.

SCREW THREAD MANUFACTURE

Much of the work carried out on capstan and turret lathes involves a threading operation, either externally or internally.

EXTERNAL THREADS

These can be produced by a screwcutting process like that used with centre lathes but with the following difference. The capstan lathe is usually fitted with a simple gear box with ratios of 1 : 1, 2 : 1 and 4 : 1 with the spindle, from which interchangeable leader screws are driven. Thus to cut a thread of 2 mm pitch a leader screw and nut of 8 mm pitch would be fitted and driven using the 4 : 1 ratio. Since the gearbox only has these ratios, the problem of picking up the thread using a chasing dial does not arise. This method is generally used only on large diameter work. For threads up to about 40 mm, *self-opening dieheads* or *thread rolling boxes* are used.

If one considers cutting an external thread with a button die in a special holder, one must realise that after the die has been screwed on it must also be screwed off by reversing the machine. Since this work is carried out at low speed it would be much more efficient if the threading device could be withdrawn directly along the thread without reversing the machine. The self-opening diehead enables this to be done

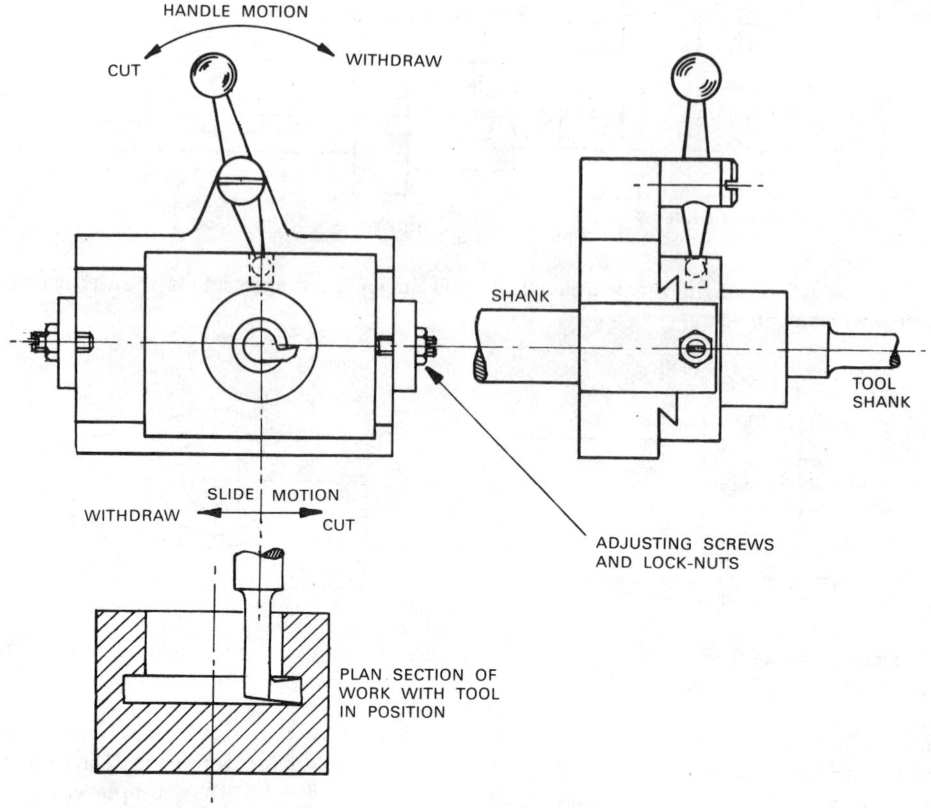

Fig. 7.43. Recessing tool holder

by cutting the thread by four equally-spaced chasers, as shown in fig. 7.48, which can be withdrawn clear of the thread crests when the end of the thread is reached. The diehead is then withdrawn.

The chasers fit into slots in the body of the diehead and are surrounded by a cam plate. As the cam plate is rotated in a clockwise direction (from the back of the machine), i.e. the handle is pushed away from the operator, the cams close the chasers to the correct depth of cut, the cam plate being held in position by a spring-loaded pin or detent pin. The diehead is fed to the work by the operator who then allows it to feed itself along the work and follows up with the turret. The turret stop is set slightly short and when the axial movement of the turret is arrested the front portion of the diehead continues to feed forward, under the self-feeding action of the chaser, until it is pulled clear of the detent pin. The diehead is then opened by springs (not shown in the diagram), the chasers being positively withdrawn by opening cams which engage in slots in the backs of the chasers.

It is important to note that the chasers are in fact sections of a thread and allowance must be made for the helix angle of the thread. They are therefore made in sets, each having its thread form offset by a quarter of one pitch. It is essential that they are fitted into the diehead in the correct order and to enable this to be maintained they are numbered as shown in fig. 7.48.

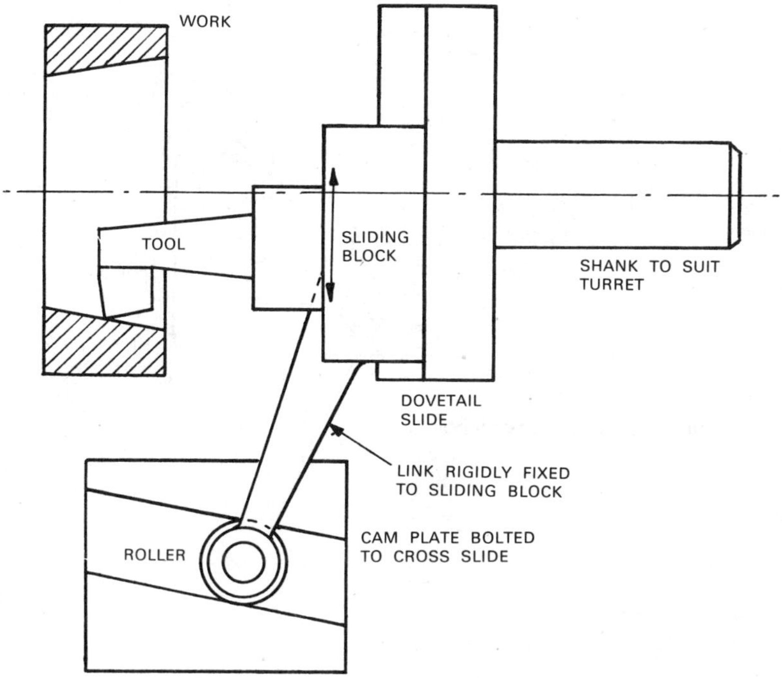

Fig. 7.44. Taper turning and boring attachment

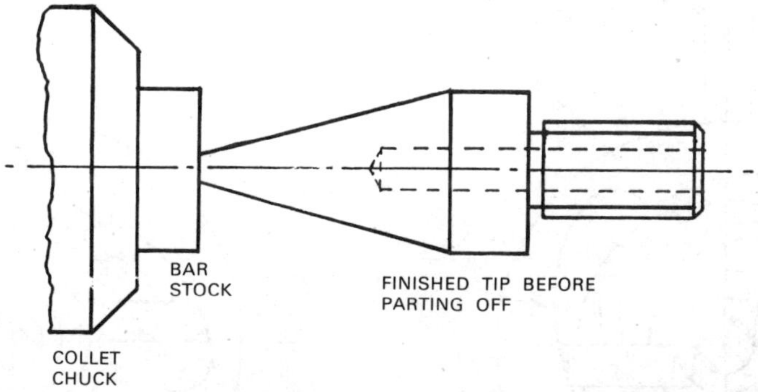

Fig. 7.45. Spotwelding tip. Use of a taper-turning attachment saved a second operation to produce the taper

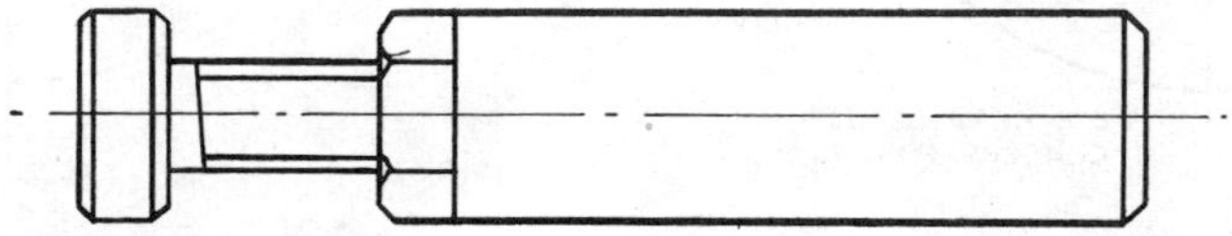

Fig. 7.46. Adjustable stop for bar work

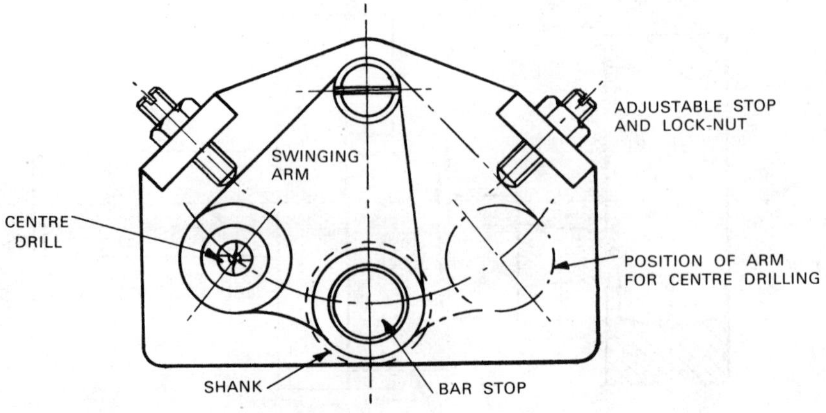

Fig. 7.47. Combined bar stop and centre drill holder

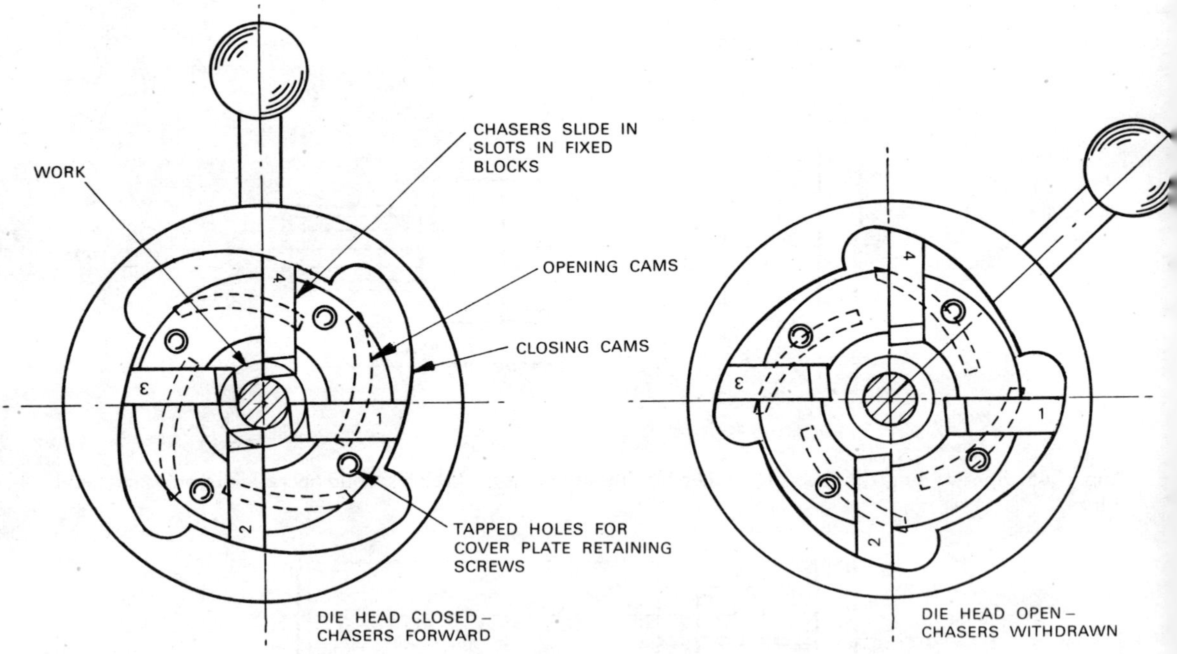

Fig. 7.48. Self-opening diehead. The cover plate is removed to show the cams

The chasers cut only on the front or cutting section shown in fig. 7.49. The remainder of the chaser, known as the self-feeding section, engages the thread already cut and provides the feed motion. Note that the chasers are radial.

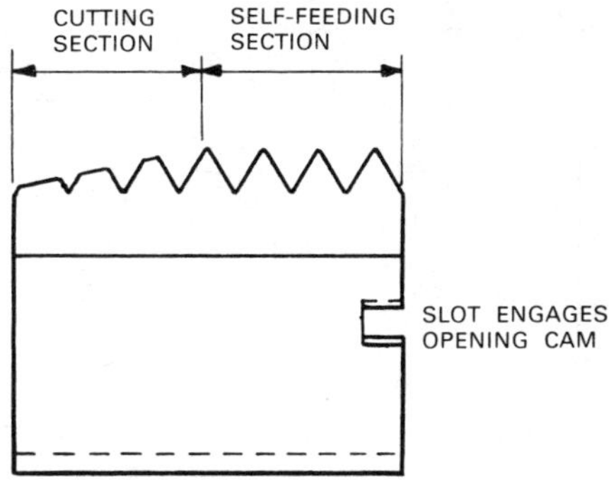

Fig. 7.49. Radial-type chaser

They are not suited to heavy loads which occur with coarser threads and may not be ground on the rake face. They are ground on the cutting portion of the clearance face in a jig, to ensure that they all contact the work at the same time and take an equal cutting load.

A general view of this type of diehead is shown in fig. 7.50

Another form of diehead uses tangential chasers. The principle is similar but the chasers are presented to the work in a tangential manner as shown in fig. 7.51. Note that although these chasers are fed axially along the work they are essentially form tools and can be compared with those shown in figs. (a) and (b). In the tangential chaser the cutting force is directed into the body of the chaser, rather than on to an overhanging section as with a radial chaser. Tangential chasers are therefore capable of withstanding heavier loads and are used for larger threads.

With both types of diehead it is unlikely that manufacturing precision would allow the chasers to be assembled and cut a thread of the correct diameter without adjustment. The knurled screw shown in fig. 7.50 adjusts the angular position of the cam plate when the diehead is closed, which adjusts the depth of cut. When the adjustment is correct the adjusting screw is locked by the nut shown. Another feature of both types of diehead is the ability to cut threads in two passes, a roughing cut and a finishing cut. This is achieved by making the end of the detent pin into a form of cam. When the lever on the end of the pin is towards the operator the diehead is opened slightly for roughing, and when it is turned through 180° the diehead is closed to the finishing position.

Cutting threads with dieheads is essentially a slow-speed operation; cutting speeds range from 2 m/min for tough steel to 15 m/min for free-cutting steel, higher speeds being used for brass and aluminium.

A later technique combines the principle of the diehead with that of thread rolling in the *self-opening thread-rolling head.* The self-opening principles are similar to those of the dieheads described above but the chasers are replaced by free-running accurately machined rollers carrying the thread form. These are set at the helix angle of the thread and feed themselves along the bar, which is made slightly larger than the mean diameter of the thread. The rolling action forces excess material out of the thread valley to form the crest as shown in fig. 7.52, which shows two rollers operating. In practice, three rollers are used which are offset relative to each other by $\frac{1}{3}$ of the pitch.

A little thought will show that as the centroid of the crest area is at a greater radius than the centroid of the root area there is a greater volume of material above the mean line than below it. This extra material is provided by making the blank diameter slightly larger than the mean diameter of the thread. The actual blank diameter can be calculated by application of the theorem of Guldinus and Pappus, but as some trial and error is inevitable a good guide is to make the blank diameter equal to the mean thread diameter plus 0.1 mm. The blank requires a 30° chamfer to less than the thread root diameter and the speed used should be in excess of 1000 rev/min. As the rolling head therefore moves along the bar very rapidly, the

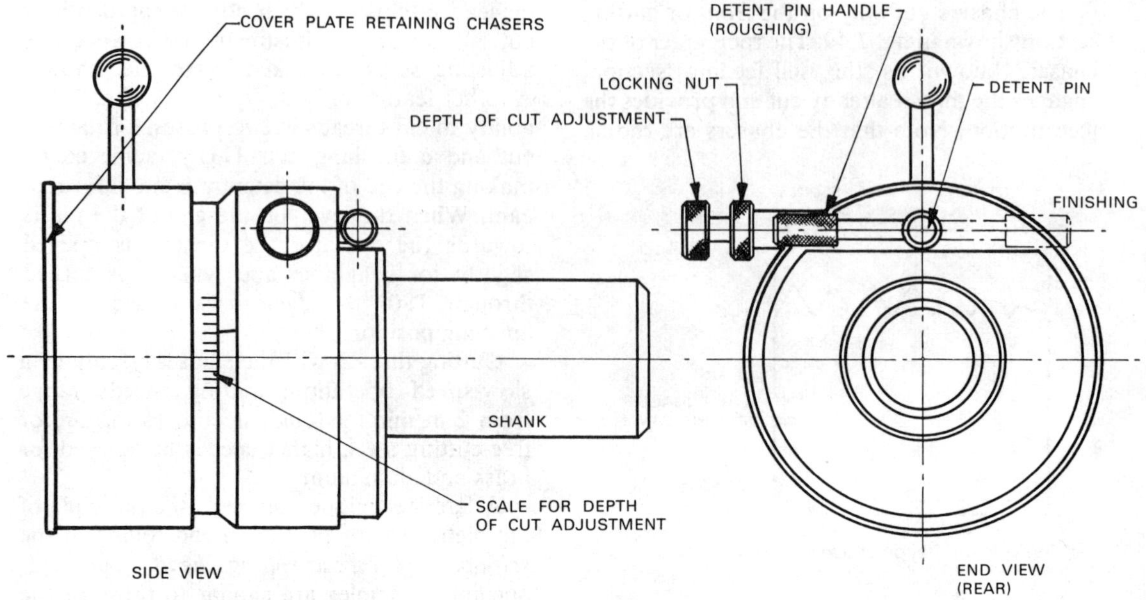

Fig. 7.50. External view of self-opening diehead

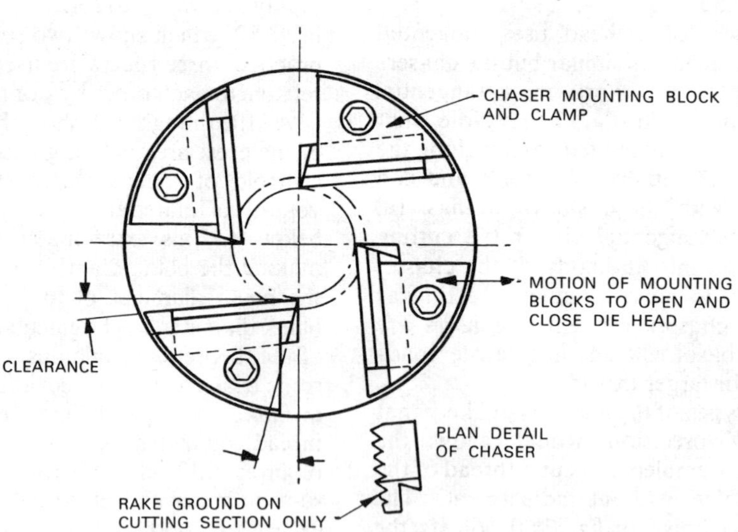

Fig. 7.51. Front view of tangential type of self-opening diehead

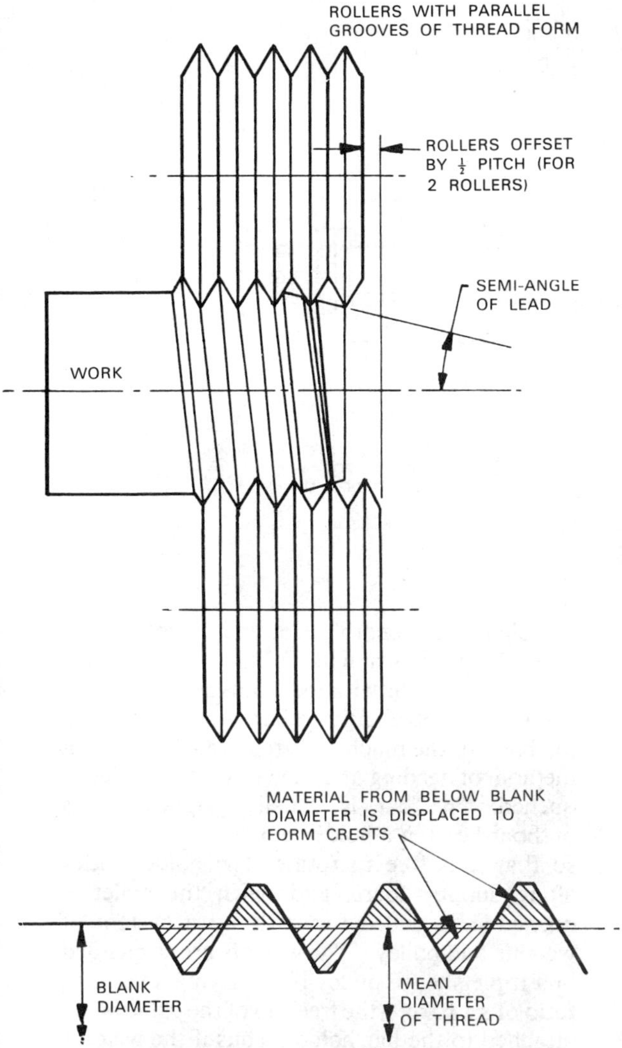

Fig. 7.52. Principle of thread rolling. In plan view the rolls are inclined to coincide with the helix angle of the thread

turret head must be fed at the required speed up to the blank. This looks almost brutal but the process produces accurate threads of high finish very rapidly. A further advantage is that the grain flow of the material follows the profile of the thread, giving a rather stronger thread than is produced by cutting, where the lines of grain flow are cut by the thread form.

INTERNAL THREADS

These are produced by tapping and it is unfortunate that, in sizes below about 25 mm, a tap having been screwed in must be screwed out by reversing the machine. As reversal is not instantaneous some arrangement is required to enable the tap to rotate freely, and thus feed no further forward, until the operator can reverse the machine, when the tap is again prevented from rotating and screws itself out. Note that in both directions the tap is self-feeding, i.e. the full-form threads along the body of the tap provide the feed effect by engaging with the threads already cut by the cutting portion on the nose of the tap. In both forward and reverse directions the operator merely causes the turret to follow the natural motion of the tap.

The simplest device for tapping consists of a tap holder whose body is free to rotate. It is restrained from doing so by the operator gripping a knurled portion of the body and then releasing it, to allow free rotation, when the full depth is reached. He then reverses the machine and again grasps the tap holder to prevent rotation so that the tap screws itself out. This device can only be used for small-diameter threads of fine pitch where the tapping torque can easily be provided by the grip of the operator.

A more common device is that shown in fig. 7.53, which consists of a tap holder mounted on the moving portion of a spring-loaded dog clutch. As with a diehead, the turret stop is set short so that the turret reaches the limit of its travel before the tap reaches the full depth of the hole. The tap continues to feed forwards and draws the clutch dogs out of engagement thus allowing the tap to rotate freely until the machine is reversed. The clutch on the tapping head shown has angled engaging faces and is spring loaded into engagement. If the tap meets an obstruction and the torque increases, the clutch is forced out of engagement, against the spring, before the tap breaks.

WORK-HOLDING METHODS

The method of holding the work in a capstan lathe depends largely on the type of work being

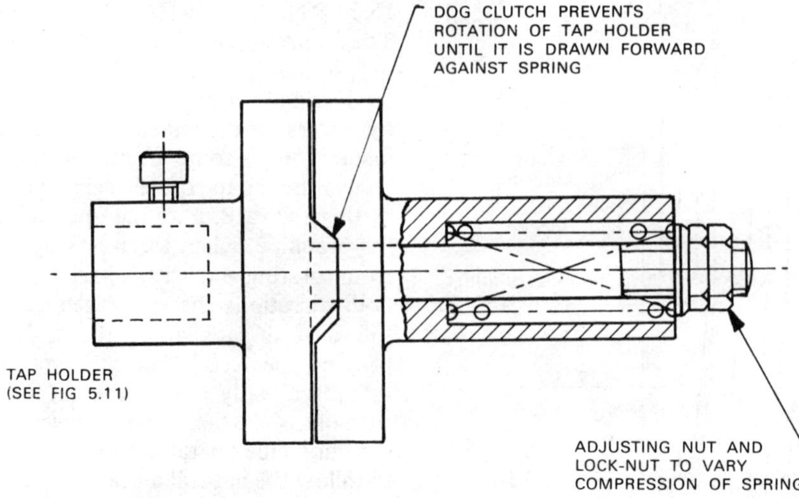

Fig. 7.53. Slipping clutch tap holder

carried out. This falls into two basic categories, work machined and parted-off from the bar, and work machined from a cut-off piece of material, or a casting or forging. In both cases first and second or subsequent operations must be considered separately as different conditions may apply and affect the work-holding method.

HOLDING BAR WORK

Much of the work machined on a capstan lathe consists of components machined from a bar and then parted off. The bar stock is usually held in a collet chuck with an automatic feed arrangement so that when the collet is opened the bar feeds forwards to the stop in the turret head. The type of collet used differs from that described on page 129, where the collet is drawn back into a taper to close on the work. Such a draw-in collet would tend to pull the bar away from the bar-stop if used on a capstan lathe and, instead, a *dead-length collet* is used. In this case the collet is fixed in its location in the machine spindle and the closing taper moves axially along the tapered nose of the collet to close the collet on the work. Thus the collet does not move and the bar remains securely against the stop. A diagram illustrating a typical dead-length collet and its method of operation is shown in fig. 7.54.

In conjunction with the dead-length collet, a means of controlling the bar protruding from the back of the machine is required as well as a method of feeding the bar when the collet is opened. The bar is usually gripped at its outboard end in a holder mounted in a bearing so that it is free to rotate. The holder slides along support bars, and when the collet is opened it is pulled forwards by a system of weights and pulleys. The weights are suspended in a triple-sheave pulley block giving a velocity ratio of six to one, the free end of the cable being attached to the bar holder. Thus if the weights fall 0.5 m the bar holder moves 3 m. As the available fall is usually greater than 0.5 m, it provides ample movement for the normal 3 m length of bar stock.

Hand operation of the bar feed is sometimes used, a handwheel and cable at the rear of the headstock drawing the bar holder forward. Another form of bar feed utilises an air cylinder.

SECOND-OPERATION COLLET WORK

Consider the component shown in fig. 7.55. Assuming that the first operation consisted of

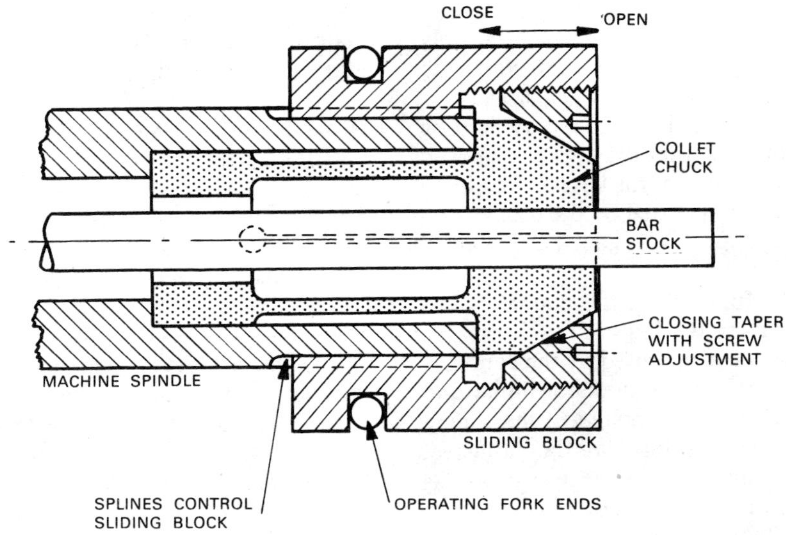

Fig. 7.54. Diagram of dead-length type collet chuck

feed to stop, centre drill, drill, tap, and part-off to length, the part must now be reversed in the collet to turn, undercut, chamfer and thread. It must be positioned so that the correct length is protruding from the collet, but in such circumstances the normal bar stop cannot be used. A soft steel or brass stop is therefore fitted into the back of the collet as shown in fig. 7.55.

Had the work been turned to a shoulder in the first operation the difficulty would not have arisen since the operator would then merely insert the work up to the shoulder before closing the collet.

HOLDING CUT-OFF BLANKS

Another common type of work is the machining of a part from a solid disc which has previously been cut off a bar of large diameter which, due to its size, could not be fed through the headstock spindle. Such work is gripped in a self-centring three- or four-jaw chuck, but such a chuck has different requirements from that used on a centre lathe:

(a) Its operation must be rapid.

(b) Only a small jaw movement is required, sufficient to release and grip work which, for long runs, is all of the same diameter.

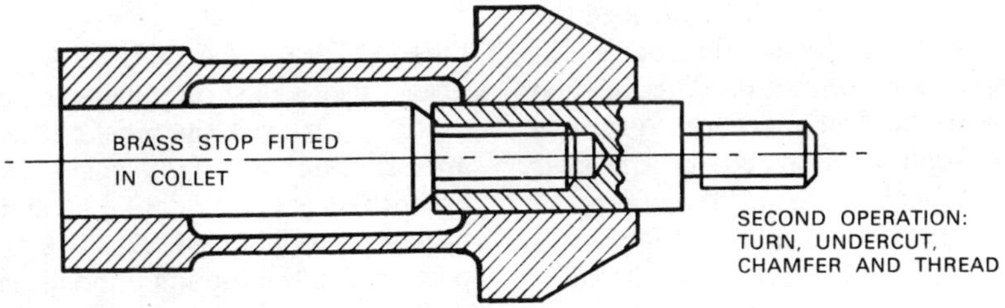

Fig. 7.55. Collet fitted with back stop for second operation work

(c) Although the jaw movement need only be small the jaws must be adjustable to accommodate a wide range of work sizes.

Rapid operation can be obtained by limiting the jaw movement so that the jaws are opened or closed by half a turn of the key. The opening and closing are not operated by the usual scroll but by a cam ring operating against a key on the back of each jaw.

Adjustment for different diameters is arranged by making the jaws in two parts, the sliding block which is operated by the cam ring and always moves between the same limits, and the gripping jaw which is bolted to the sliding block but whose position can be adjusted. Positive positioning of the gripping jaw is arranged by having serrations on the rear of the jaw engaging with similar serrations on the sliding block. These details are shown in fig. 7.56

By positioning the gripping jaws asymetrically on the sliding blocks, work of irregular shapes can be held, or rectangular blocks can be held for the machining of holes and other features off-centre.

SECOND-OPERATION CHUCK WORK

It is inevitable that much of the second-operation work performed in chucks is gripped on the surfaces machined in the first operation. For this reason soft jaws are used to avoid damaging the finished surface. It is also essential that the work is held so that turned diameters mate correctly without a step due to eccentricity. To ensure concentricity, a disc which is slightly less in diameter than that of the finished work is first gripped at the back of the jaws. The front of the soft jaws is then bored out to a suitable size to accommodate the finished outside diameter of the first operation. A typical component for such work would be the gear blank shown in fig. 7.57. The first operation would consist of face, face to step, centre-drill and turn, drill and chamfer, bore and chamfer. Before the second operation, soft jaws would be fitted and machined to suit the turned outside diameter, after which the second operation would be turn, face, chamfer hole and outside diameter. Figs. 7.57 (a) and (b) show the blank at the two stages of manufacture.

PLANNING FOR CAPSTAN OPERATIONS

Specific Objective: *The student should be able to list in logical order the sequence of operations for the manufacture of a typical component.*

The capstan lathe is essentially a tool of the production engineer. It can do nothing that cannot be done on a centre lathe but it can perform repetitive operations much more quickly and is used where the savings in labour costs more than outweigh the increased tooling and setting costs. However, the capstan lathe will only be economic if used efficiently and it will only be used efficiently if the work is planned correctly. Planning for capstan lathes falls into three stages:

(a) Selection of machine;
(b) Operation layout;
(c) Estimation of operation time.

MACHINE SELECTION

It is of little use proceeding with an operation layout if the work cannot be accommodated in the machine. For this reason manufacturers supply *capacity charts* for their machines which show the slide movements and adjustments available. A comparison of the work with the capacity chart enables the suitability of the machine for the job to be readily estimated.

OPERATION LAYOUTS

The cost of a part often depends on whether the machine set-up is the most efficient for the part and although in many cases the sequence of operations is self-evident it is often necessary to prepare a tool layout. This is a plan view of the part and all the tooling in position.

It enables the planning engineer to decide precisely what tools are required and to ensure that they are available before the job is begun. A comparison with the capacity chart will show

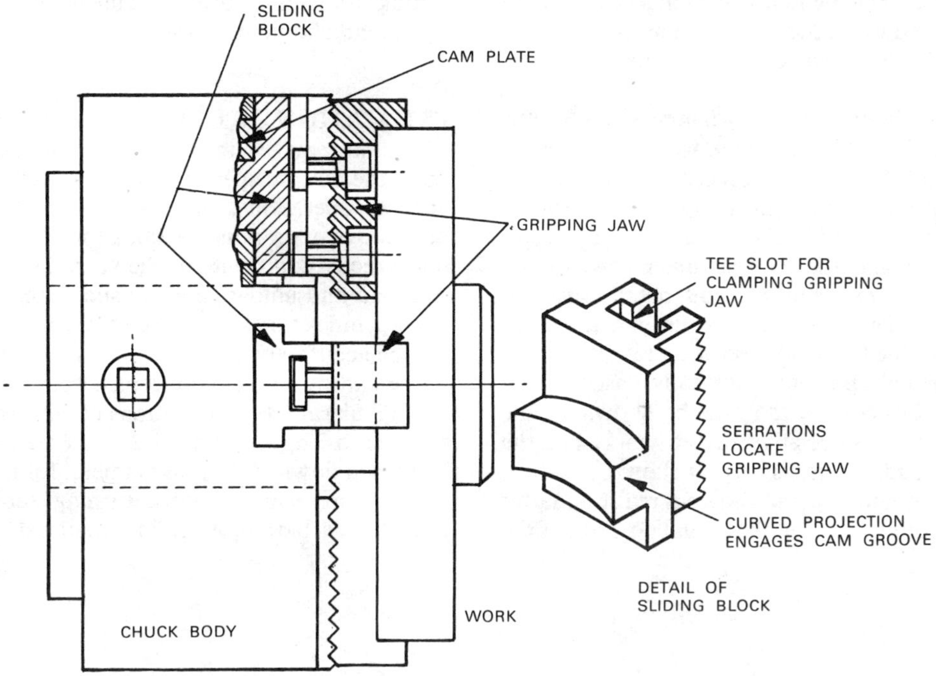

Fig. 7.56. Self-centring four-jaw chuck for a capstan lathe. One jaw assembly is shown in section

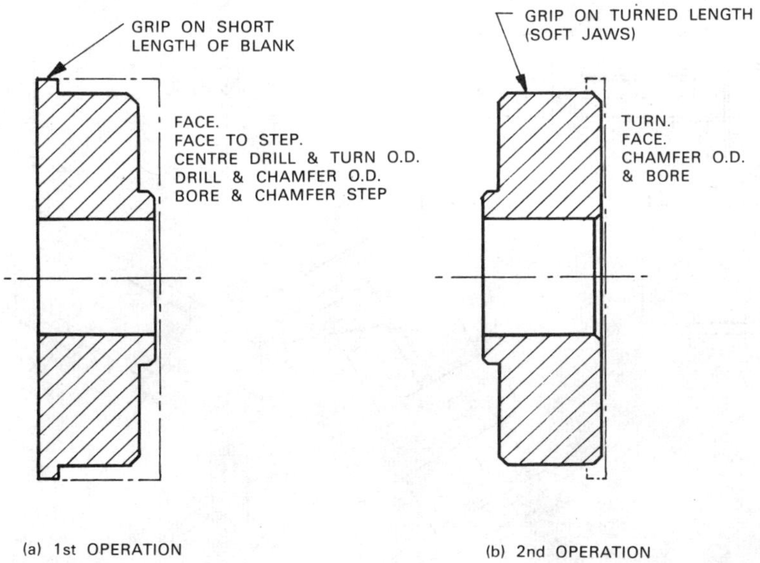

Fig. 7.57. Typical component for chuck work. The chain-dotted outline shows the metal removed in each operation

whether or not the machine will accommodate the job and when the machine and layout have been decided the operating time can be estimated.

Typical layouts are shown in fig. 7.58 for bar work, and in 7.59 for chuck work. The tooling shown on the layout is stylised, rather than a true representation. Thus in fig. 7.58 the rollers and tool in the roller box are correctly positioned and the overall dimensions of the box are correct but the full details are not shown. In fig. 7.59 the knee turning tools are shown in the horizontal position rather than in the vertical position normally used. This enables the cutting tools to be shown in the correct position relative to the work, but the knee tool guide bushes are not shown.

The amount of detail shown is really a matter of individual taste, but the actual position of the cutting tools and overall dimensions of the equipment should be correct.

ESTIMATION OF OPERATION TIME

The time required to machine a part depends upon the cutting speeds and feeds used for the various operations, and also upon the non-productive operations which, although vital, do not directly contribute to the value of the part, e.g. load and unload, change speed, feed stock to stop, index turret, etc. The actual cutting time can be accurately calculated from a knowledge of the cutting speeds and feeds used.

To illustrate the calculations of cutting times consider a length of bar of 25 mm diameter to be turned down to 20 mm diameter for a length of 75 mm, using a knee turning tool with tungsten-carbide tooling. The material is free-

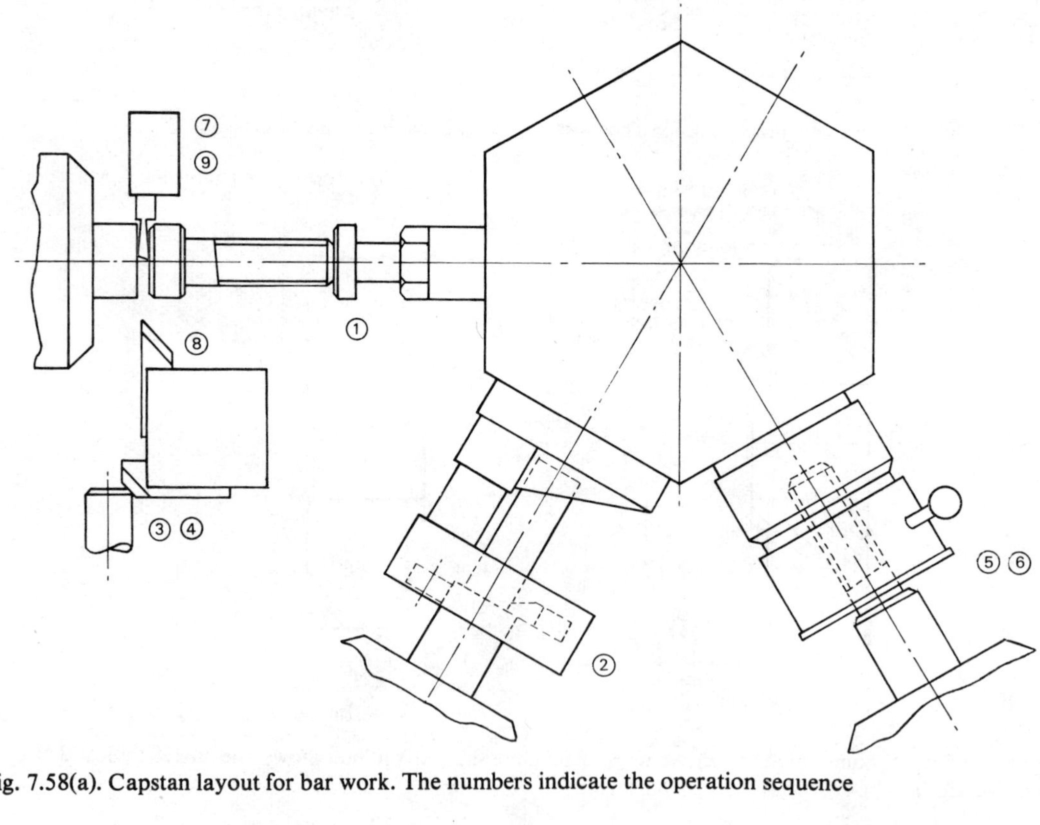

Fig. 7.58(a). Capstan layout for bar work. The numbers indicate the operation sequence

PART NO. ______

PART NAME: BOLT

1 OPERATN. NO.	2 OPERATION	3 TOOLING	4 TURRET OR X SLIDE	5 SPEED		6 FEED RATE CUTS/CM	7 TOTAL REVS	8 CUTTING TIME, MIN.	9 NON-PRODUCTIVE TIME, MIN.	10 REMARKS
				M/MIN	REV/MIN					
1	STOCK TO STOP	BAR STOP	T	-	-	-	-	-	0.105	
	INDEX & LOCK TURRET	-	-	-	-	-	-	-	0.100	
2	TURN 25mm x 75mm LONG	ROLLER BOX	T	50	600	30	225	0.375		ROLLERS LEADING H.S.S. TOOL
	INDEX & LOCK TURRET	-	-	-	-	-	-	-	0.100	
	POSITION X SLIDE & SET STOP	-	-	-	-	-	-	-	0.130	
3	CHAMFER TURNED END	45° R.H. TOOL	X	50	600	HAND	60	0.100		FRONT TOOL POST
	POSITION X SLIDE & SET STOP	-	-	-	-	-	-	-	0.130	READY FOR OP. 7
4	CHAMFER HEAD	45° R.H. TOOL	-	-	-	-	-	0.100		AS FOR OP. 3
	CHANGE SPEED	-	-	-	-	-	-	-	0.100	
5	ROUGH THREAD	S/O DIE HEAD	T	10	60	10	50	0.833		25mm x 1mm PITCH CHASERS
	WITHDRAW & RESET DIE HEAD	-	-	-	-	-	-	-	0.080	
6	FINISH THREAD	S/O DIE HEAD	T	10	60	10	50	0.833		AS FOR OP. 5
	INDEX TURRET 3 STATIONS	-	-	-	-	-	-	-	0.200	
	POSITION X SLIDE & SET STOP	-	-	-	-	-	-	-	0.130	
	CHANGE SPEED	-	-	-	-	-	-	-	0.080	READY FOR OP. 7
7	PARTIAL PART-OFF	PARTING-OFF TOOL	X	25	150	HAND	30	0.200		
	INDEX FRONT TOOL POST	-	-	-	-	-	-	-	0.110	
8	CHAMFER HEAD	45° L.H. TOOL	X	25	150	HAND	15	0.100		SET RELATIVE TO P/O TOOL - NO SLIDE ADJUSTMENT
	INDEX FRONT TOOL POST	-	-	-	-	-	-	-	0.110	READY FOR OP. 3
9	FINISH PART-OFF	PARTING-OFF TOOL	X	25	150	HAND	30	0.200		AS FOR OP. 8
	REMOVE PART	-	-	-	-	-	-	-	0.04	
	CHANGE SPEED	-	-	-	-	-	-	-	0.100	READY FOR OP. 2
							TOTALS	2.741	1.515	TOTAL TIME/PART 4.256 min

Fig. 7.58(b). Operation sheet for bar work

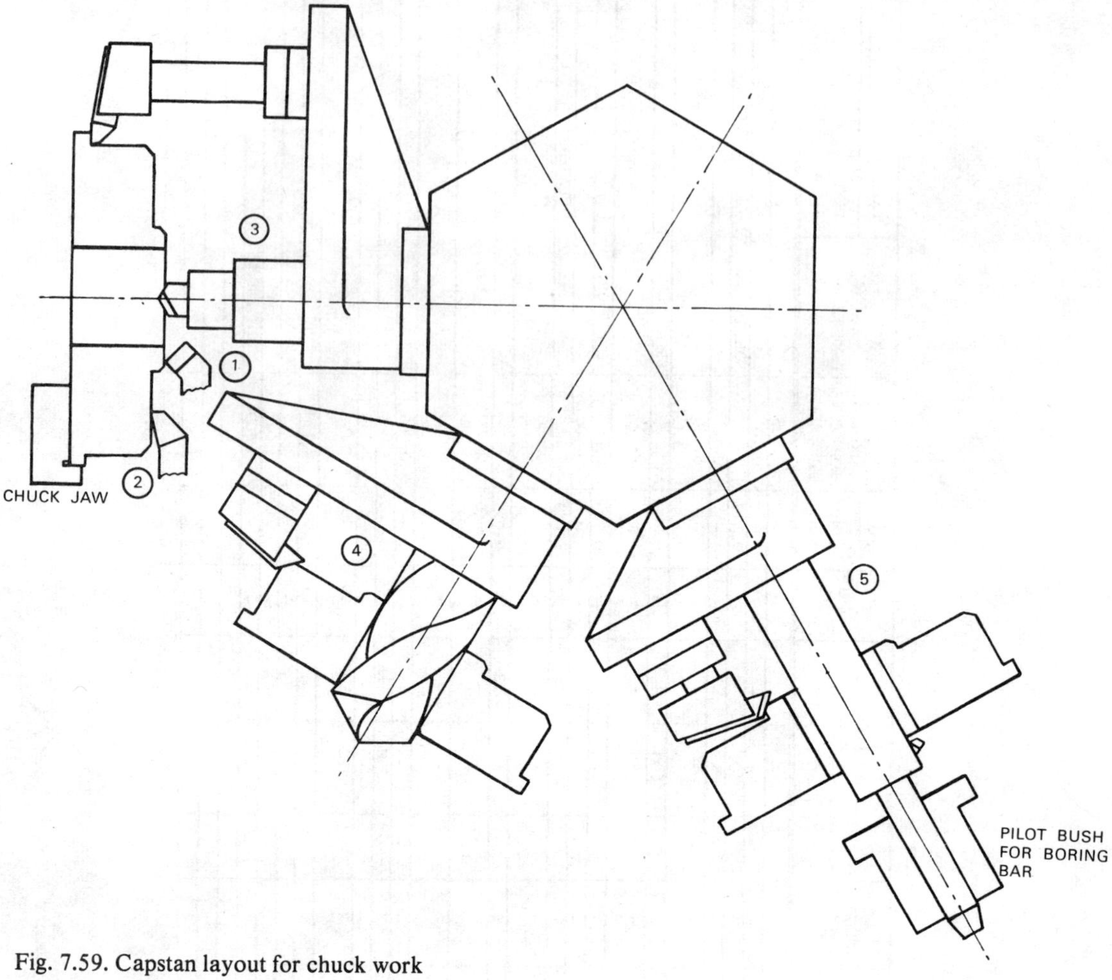

Fig. 7.59. Capstan layout for chuck work

cutting mild steel, allowing a cutting speed of 150 m/min and a feed rate of 30 cuts/cm.

Note that the cutting speed is the peripheral or surface speed in metres per minute and the feed rate indicates the number of revolutions the work must make as the tool advances 1 cm.

Circumference of bar

$$= \pi \times \text{diameter}$$

$$= \pi \times 25 \text{ mm}$$

$$= 78.5 \text{ mm}$$

Work speed in revolutions per minute

$$= \frac{\text{Cutting speed in metres per minute}}{\text{Circumference in metres}}$$

$$= \frac{150 \text{ m/min}}{78.5/1000 \text{ m}}$$

$$= \frac{150 \times 1000 \text{ rev/min}}{78.5}$$

$$= 1910 \text{ rev/min}$$

It is unlikely that the machine would have this precise speed available so for considerations of tool life the next nearest speed *downwards* is chosen. Assume a speed of 1800 rev/min.

Length of work

$= 75 \text{ mm} = 7.5 \text{ cm}$

Feed rate $= 30$ cuts/cm

Number of revolutions

$= \text{length in centimetres} \times \text{feed rate in cuts per centimetre}$

$= 7.5 \text{ cm} \times 30 \text{ cuts/cm}$

$= 225 \text{ rev}$

Cutting time

$$= \frac{\text{No. of revolutions}}{\text{Speed in revolutions per minute}}$$

$$= \frac{225 \text{ rev}}{1\,800 \text{ rev/min}}$$

$$= 0.125 \text{ min}$$

If the calculations for a series of operations were carried out as above, they would become hopelessly spread out and untidy. It is far better to set the calculations out in a tabular form. The table can include the times for non-productive operations and a total floor-to-floor time can be arrived at. Consider the bolt shown in fig. 7.58(a), for which the table, or operation sheet, would be as shown in fig. 7.58(b).

Columns 1, 2, 3 and 4 are self-explanatory.

Column 5 shows the cutting speed to be used and from this the work speed in revolutions per minute is calculated.

Column 6 is the feed rate in cuts per centimetre, which enables the number of revolutions required for the operations to be calculated.

Column 7 is the number of revolutions required for the operation and is obtained by multiplying the feed rate in column 6 by the length of work to be cut.

Column 8 is the time required for the cutting operations. The total revolutions are known and are divided by the revolutions per minute to give the cutting time in minutes.

Column 9 shows the time required for non-productive operations. These non-productive times are known for different machines by work study engineers from past experience, those used here being for a medium-sized capstan lathe. A full list of them is not given but from the values used in fig. 7.58(b) the student could compile a useful basic table of such values. These he could augment by estimating times for other non-productive operations with which he is familiar.

Column 10 is a useful place to include any explanatory comments, as shown.

From the operation sheet, fig. 7.58(b), it would appear that a capstan operator, working under the conditions shown, could produce this part in 4.256 min, or 14.1 parts per hour or 112.8 such parts in an eight-hour shift. Allowances must be made, however, for unavoidable stoppages. Since the part is 100 mm long plus a parting-off allowance, about 105 mm is required per part. Thus, if the operator is using bars of 3 m length he must stop after every 28 parts and insert a new length of bar. Time must be allowed for this. Allowance must also be made for tool maintenance and for the operator to attend to the needs of nature. Unfortunately, these two contingencies rarely coincide.

Various organisations have their own arrangements and standards for these allowances but detailed descriptions of them form an important part of work study and would be out of place in this book. However, the student should be aware of the need for such allowances, and that the simple time-estimation sheet is not the whole story.

SUMMARY

The capstan lathe is designed to eliminate the skill required to produce a part on a centre lathe. It has a greater number of tooling stations and enables repetition work to be carried out without frequent tool changes and resetting. The skill largely rests on the tool setter who sets the machine up with suitable tools for the work in hand.

Turret lathes are larger and will accommodate a wider range of work. They are often used for smaller batches and the operator is a skilled man usually doing his own setting.

Cost estimating for capstan and turret lathes requires a considerable skill in planning and can be carried out to a high degree of accuracy of estimation of floor-to-floor time or cycle time.

The next logical development is to make the machine fully automatic and dispense with the operator. It is interesting to note that many single-spindle automatic lathes are indeed similar to capstan lathes but all the operations are controlled by cams and timing trips. The estimating for these machines is important to enable the cams to be designed and the timing settings calculated. The operation and cam design sheets are very similar to fig. 7.58(b) but are calculated in revolutions per operation for both cutting and non-productive operations so that the total revolutions can be calculated. The proportion of each cycle devoted to each operation can be calculated and the cam-layouts made accordingly.

QUESTIONS

1. Sketch a flat form tool with zero rake and show that the depth of form ground on the tool will not be the same as the depth of form produced on the work. Show how the height of the cutting edge can be adjusted for a radial type tool and a tangential type tool.

2. Sketch (a) a knee turning tool holder, and
   (b) a combination tool holder.

Show how the rigidity of the set up is increased when these tool holders are used.

3. Explain the difference between 'rollers leading' and 'rollers following' type roller box tool holders, and state the circumstance under which each is used.

4. Describe suitable tool holders for the following operations on capstan lathes.
   (a) Centring
   (b) Drilling
   (c) Recessing
   (d) Profiling.

5. Explain why self-opening die-heads are preferred to die holders when producing external threads on capstan or turrent lathes. With the aid of diagrams show the difference between radial and tangential die-heads and list the advantages and limitations of each. What is the advantage of thread-rolling over each of these methods of producing external threads?

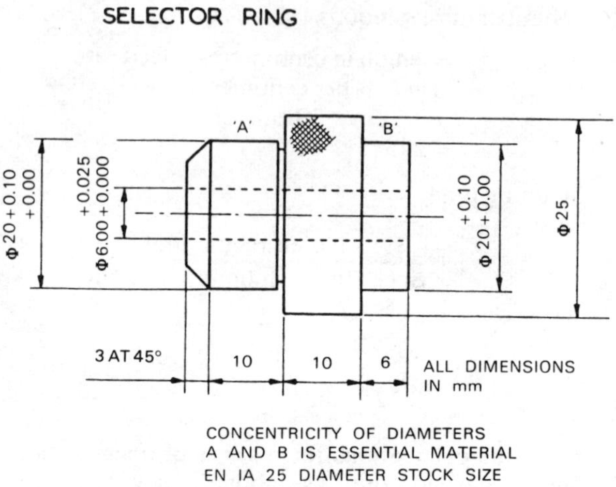

6. The component shown above is to be produced on a capstan lathe and about 800 are required.
   (a) Give a suitable sequence of operations.
   (b) Show diagrammatically a suitable tooling layout.
   (c) Explain what is meant by the synthetic production time and why it is obtained.

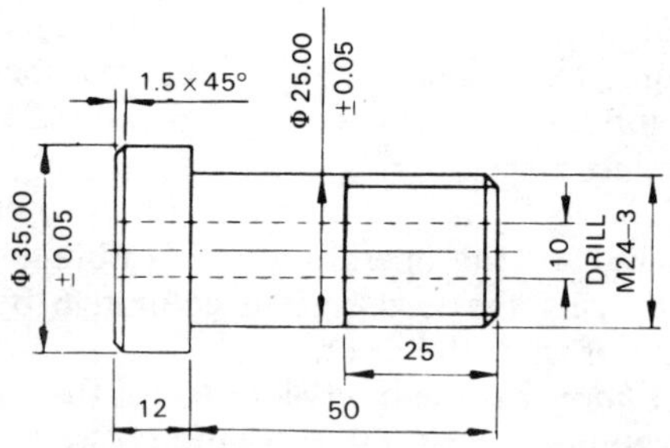

7. The component shown above is to be produced in batches of 100 from bright mild steel bar of 40 mm diameter. Draw up an

operation layout, make a simple tooling diagram and estimate the cycle time per component. Assume that high-speed steel cutting tools are to be used and allow 20% for all contingencies. The available speeds and feeds are:

Speeds:
rev/min 1531 834 453 241 131 71 46
Feeds:
mm/rev 0.15 0.22 0.30

8. Using sketches or diagrams, indicate the general arrangement and set-up of **three** of the following items of capstan tooling equipment. Show the kind of work feature each would produce and what settings are required.
   (i) knee tools
   (ii) roller steady box
   (iii) recessing tool
   (iv) a floating holder
   (v) a knurling head.

9. (a) Make an operation layout for producing the bolt shown opposite from 50 mm diameter bright mild steel bar on a capstan lathe. Draw a line diagram of the tool set up.
   (b) Calculate the time for machining the 20 mm diameter portion and the 12.5 mm diameter H8 hole. Use the cutting speeds and feeds shown below.

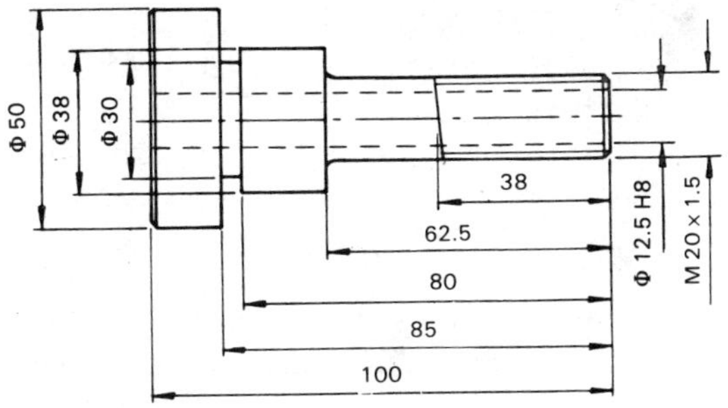

ALL DIMENSIONS IN mm

Operation	Surface speed m/min	Feed mm/rev
Turning	50	0.13
Drilling	50	0.10
Reaming	10	0.50

ANSWERS

9. (b) 0.6 min; 0.785 min.

8 The milling machine

General Objective: *The student should be able to compile block diagrams of the commonly used milling machines and describe typical milling operations.*

CUTTING AND FEED MOTIONS

A milling machine may be defined as one whose main function is to produce flat surfaces with a rotary cutter, by either a copying or a generating process. This statement is true of both horizontal and vertical-spindle machines, and for each the basic table motions of the machine are the same, as shown in fig. 8.1(a).

The motions are:

1. *Rotation of cutter*

As stated in Chapter 5, the largest force applied to a milling cutter is the tangential force F_t, which is therefore the motion through which the power is applied.

2. *Longitudinal table motion*

This is the feed motion which, in conjunction with motion no. 1, produces the cutting action.

3. *Vertical table motion*

This motion is used to set the depth of cut. On vertical-spindle machines it is also usually possible to feed the cutter axially downwards, an additional motion, no. 5.

4. *Transverse table motion*

This crossward motion of the table is used to position the work under the cutter. It may occasionally be used as a cross-feed.

During cutting, only motions 1 and 2 are normally used, the other motions, i.e. adjustments, being locked.

It might appear at first sight that horizontal machines always utilise a copying process and vertical-spindle machines a generating process, but this is not so. With a face milling cutter mounted in the spindle, or a side and face cutter on the arbor, the horizontal machine can be employed for a generating process. Correspondingly, the periphery of an end mill copies a vertical surface when used in a vertical-spindle machine as shown in fig. 8.1(b).

MACHINE ALIGNMENTS

Specific Objective: *The student should be aware of the errors induced by incorrect machine alignments.*

Certain relationships must exist and be maintained between these motions if the geometry of workpieces produced is to meet requirements. These relationships are the machine alignments and the important ones are given below. The diagrams in figs. 8.2 to 8.6 illustrate the errors induced if the alignments are incorrect.

1. TABLE SURFACE MUST BE PARALLEL TO ITS OWN MOTION

This alignment applies to all machines in which the feed motion is produced by moving the table, e.g. surface grinders and planing machines. Any error produces non-parallelism of the work as shown in fig. 8.2.

2. CENTRE-TEE SLOT MUST BE PARALLEL TO TABLE MOTION

The centre tee-slot is used to locate various milling fixtures such as the vice, rotary table and dividing head. If this alignment is not correct an incorrect datum is being used for the various operations.

With this in mind it is obviously necessary to ensure that the tee-slot and the fixture tenons which locate in it do not suffer damage, since a

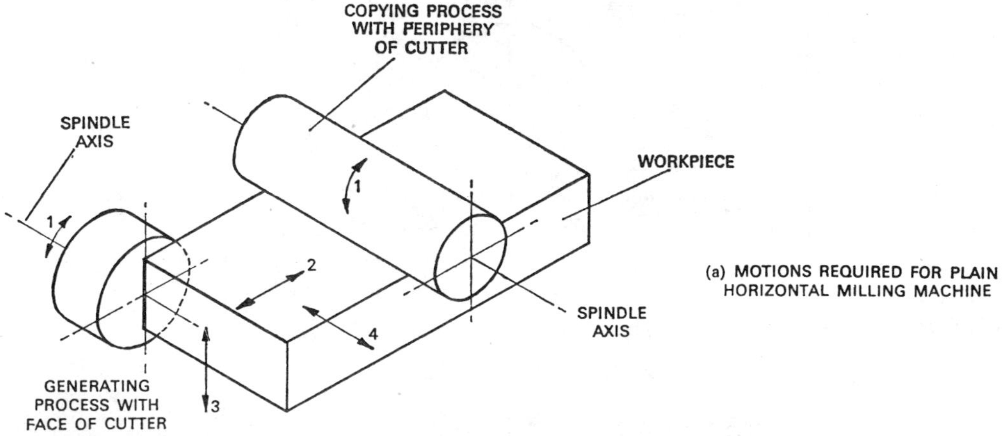

Fig. 8.1(a). Basic motions of tool and work on milling machine

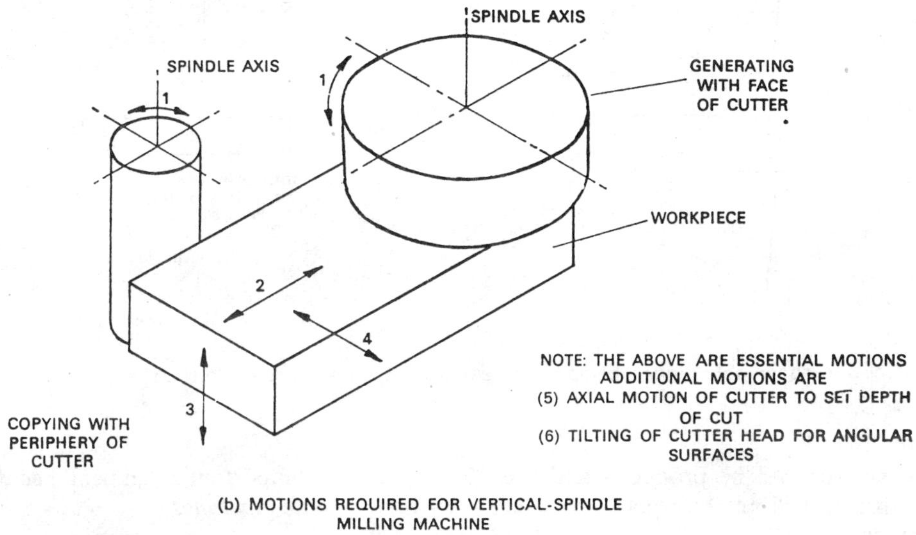

Fig. 8.1(b)

burr can destroy alignment locally even though the alignment is elsewhere correct.

3. VERTICAL TABLE MOTION MUST BE AT RIGHT ANGLES TO TABLE SURFACE

If all other alignments are correct, an error in this alignment will produce a side component of motion as the table is raised, and errors in depths of cut and widths of slot, as shown in fig. 8.3.

The above alignments are essential in all milling machines. Those peculiar to *horizontal machines* are:

4. SPINDLE AXIS MUST BE PARALLEL TO TABLE SURFACE

If this alignment is incorrect, non-parallel work

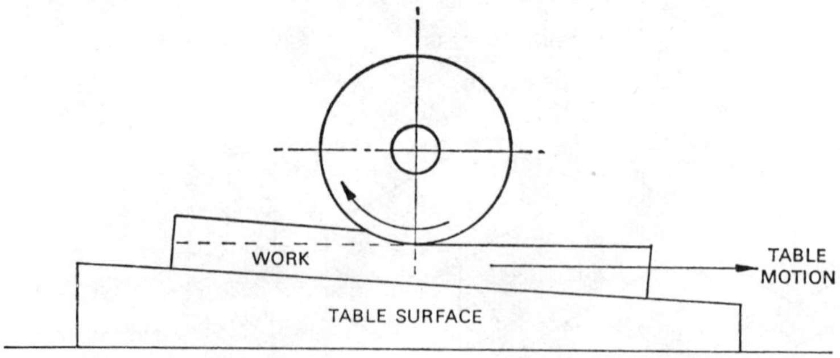

Fig. 8.2. Taper produced by table motion being non-parallel to table surface

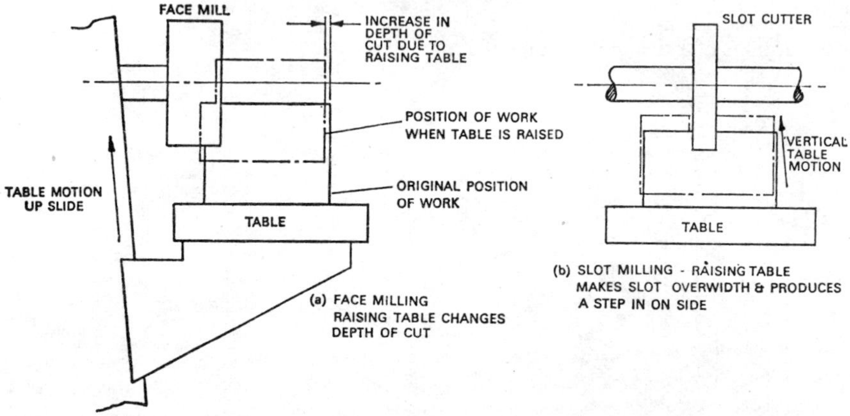

Fig. 8.3. Effect of vertical table motion being out of square to table surface spindle axis

across the surface will be produced and the machined surface will not be square to other faces, as shown in fig. 8.4.

5. SPINDLE AXIS MUST BE AT RIGHT ANGLES TO TABLE MOTION

If this alignment is incorrect, slots produced by side-and-face cutters or by slotting cutters and saws will be too wide. The width of slot will be equal to the projected width of the cutter across the line of table movement, as shown in fig. 8.5. Note that the slot produced will not necessarily be out of alignment.

The most important alignment peculiar to *vertical-spindle machines* is:

6. SPINDLE AXIS MUST BE AT 90° TO TABLE SURFACE IN BOTH DIRECTIONS

This alignment is particularly important across the table; if it is in error, work produced will be non-parallel in that direction, or non-square if the cutter periphery is used to produce a vertical face, as in fig. 8.6.

In most vertical-spindle machines the cutter head can be swung in the other vertical plane and is often deliberately misaligned slightly in

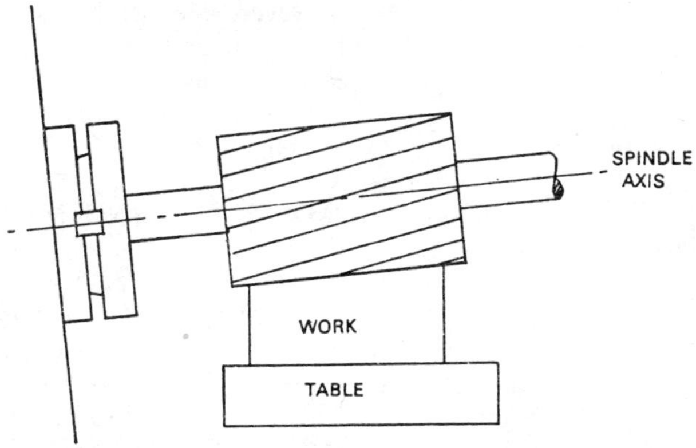

Fig. 8.4. Non-parallel work produced when spindle axis is not parallel to table surface

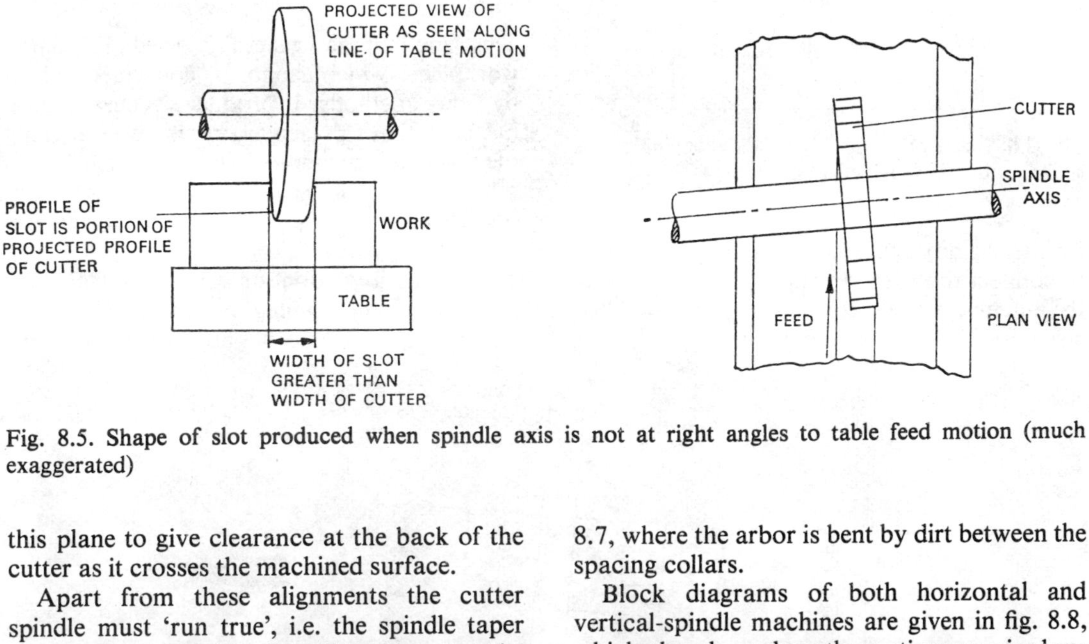

Fig. 8.5. Shape of slot produced when spindle axis is not at right angles to table feed motion (much exaggerated)

this plane to give clearance at the back of the cutter as it crosses the machined surface.

Apart from these alignments the cutter spindle must 'run true', i.e. the spindle taper must not run out of truth and the arbor must be straight. If the cutter teeth are not concentric with the axis of rotation, all the cutting load has to be borne by the teeth running at the greater radius, with possible cutter damage or breakage or with a reduced cutter load to avoid such damage. In a horizontal machine a bent arbor will cause the cutter to oscillate from side to side and produce an overwidth slot, as shown in fig. 8.7, where the arbor is bent by dirt between the spacing collars.

Block diagrams of both horizontal and vertical-spindle machines are given in fig. 8.8, which also shows how the motions required are obtained and aligned.

The arbor of a horizontal machine is prone to misalignment through bending and its unsupported length should be kept as short as possible. This is generally achieved by keeping the cutters as near to the column as practicable, and then moving an overarm support as close to the cutters as possible, as shown in fig. 8.9.

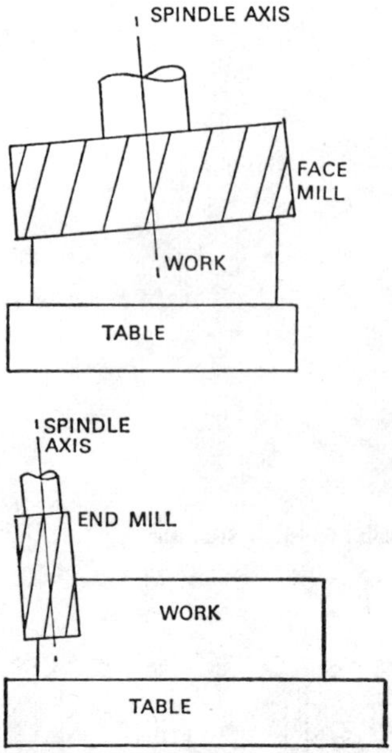

Fig. 8.6. Effect of vertical spindle axis being out of square with table surface across table

Further rigidity can be obtained by using braces to connect the overarm support to the knee of the machine, thereby forming a rigid closed-box structure. The braces have to be released, however, when the cut is changed and removed to change the cutter, which is often inconvenient, but for production work with long runs and heavy cuts the increase in rigidity is well worthwhile.

WORK-HOLDING METHODS

Specific Objective: *The student should be able to select suitable workholding methods for given operations on a given workpiece.*

The method used to hold the work depends on the type of workpiece, its size and shape, and on the operations to be carried out.

1. CLAMPING DIRECTLY TO THE TABLE

This method is generally used for large workpieces which cannot be conveniently held by other methods. It produces accurate work but sometimes presents a problem in aligning the work: two examples will be given.

(a) *Cleaning up the faces of 100 mm by 100 mm angle iron*

This can easily be done on a horizontal machine by a shell end milling cutter. The work is clamped to the table, slightly overhanging the

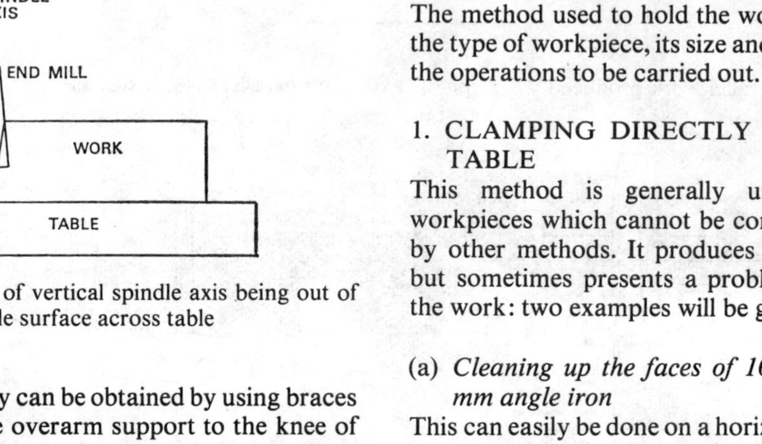

Fig. 8.7. Effect of cutter running out of true because of bent arbor (much exaggerated)

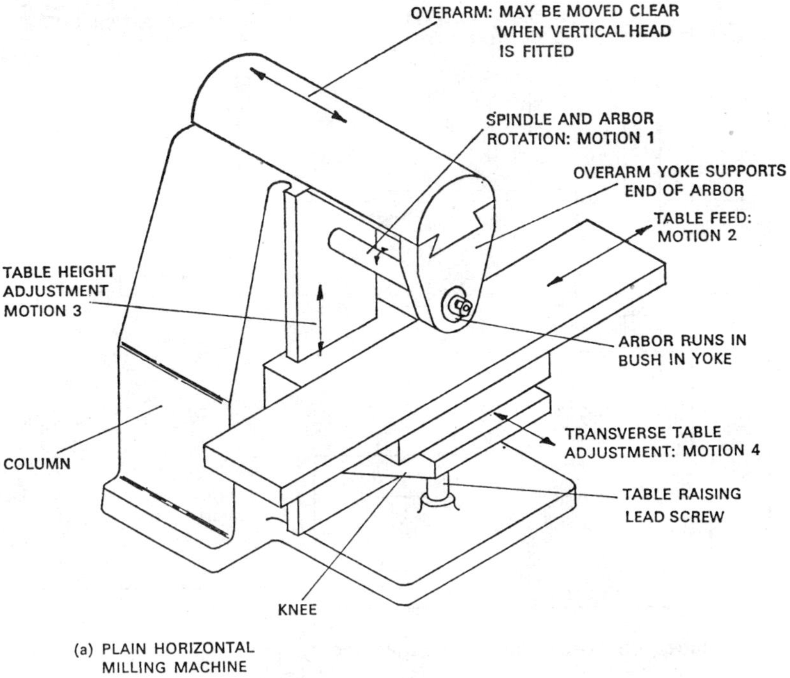

Fig. 8.8(a). Block diagrams of column and knee-type plain milling machines. The motions and adjustment numbers refer to the motions in fig. 8.1

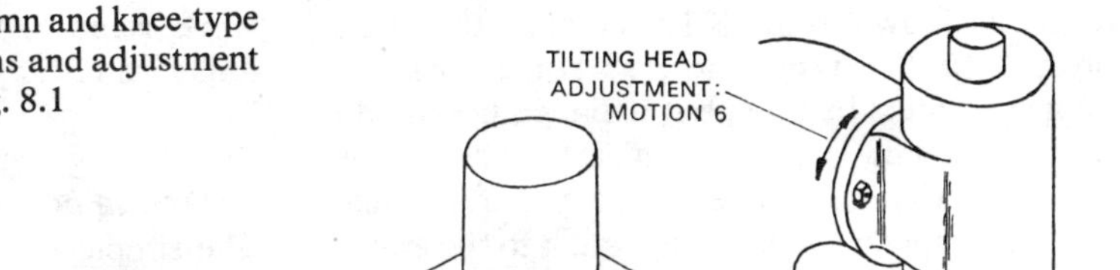

Fig. 8.8(b). Fixed-head vertical-spindle machine. The small diagram shows an alternative spindle head which swivels and allows the spindle axis to be tilted. The axial spindle feed motion 5 is still provided by a similar mechanism to that used on a drilling machine, within the head itself

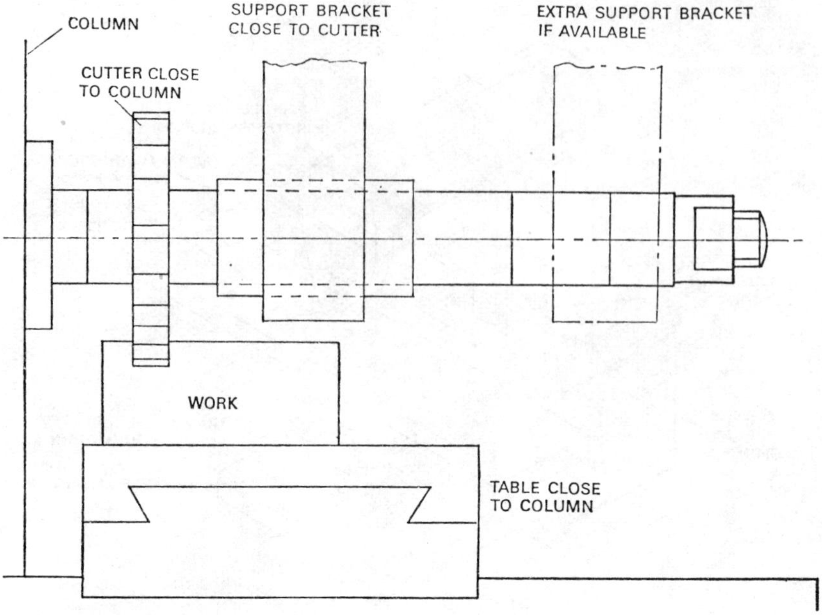

Fig. 8.9. Support of cutter on horizontal milling machine

edge, as shown in fig. 8.10. To align the angle iron with the table motion, some form of indicator must be used, but a dial gauge should not be used on such a rough surface as this. A simple device known as a 'sticky pin' is used. This is simply a small scriber stuck to the cutter with a blob of plasticine. The table is adjusted to bring the top edge of the angle iron close to the pin, and then traversed by hand along the length of the iron to see if the gap between pin and work remains the same. The position of the angle iron is adjusted as necessary.

(b) *Cutting keyways in line at each end of a round bar*

The simplest way of doing this is to locate and clamp the bar along a tee-slot of the table, mill the keyway at one end and, at the same setting, traverse the table and mill the keyway at the other end of the bar. The set-up is shown in fig.

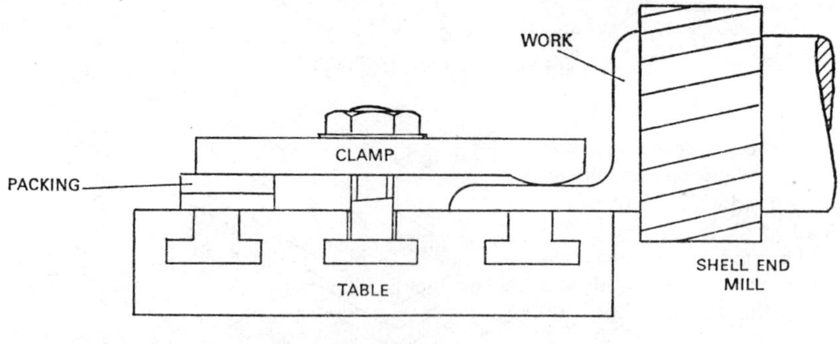

Fig. 8.10. Work clamped direct to table. Facing angle iron

8.11(a). The bar could alternatively be clamped in vee blocks as shown in fig. 8.11(b), and a dial gauge used to align it with the table motion, since the bar would either be machined or bright-drawn.

2. THE MACHINE VICE

The plain (non-swivelling) machine vice consists of a base with a fixed jaw at one end and a moving jaw which slides along the base of the vice and is operated by a leadscrew. It is a piece of precision equipment and is sometimes located by tenons in the centre tee-slot so that the fixed jaw is at right angles to the table motion. In that case, a face or slot milled in a workpiece should be square to the face clamped against the fixed jaw. The actual gripping faces of the jaws are usually ground and are perpendicular not only to the table motion but also to the table surface. Thus, if a vice is well maintained and properly used it should produce work of correct geometry.

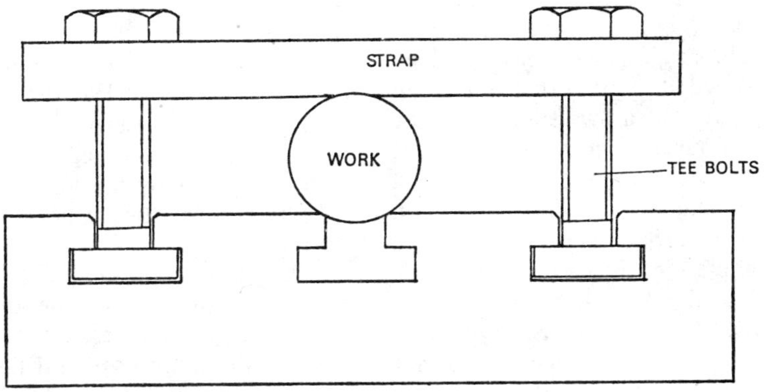

Fig. 8.11(a). Work clamped direct to table. Round bar located in centre tee-slot

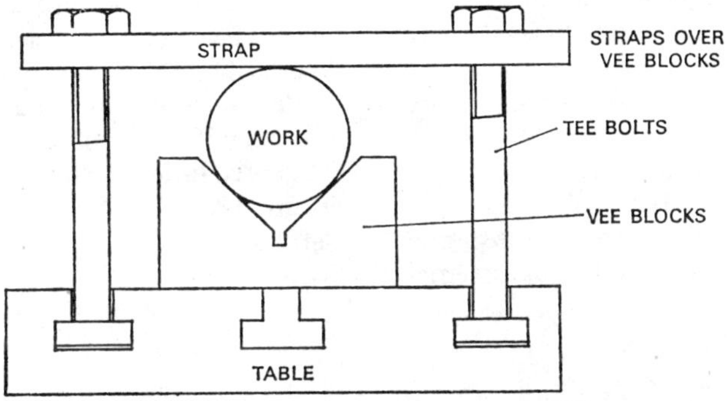

Fig. 8.11(b). Alternative method of locating and holding round bar

Note that the clamps and vee blocks would have to be clear of the cutting zones and, to reduce vibration and avoid deflection from the cutting load, it might be necessary to put supports under the ends of the bar.

If the keyways had to be at 90° to each other, one could be cut at one end with a slot cutter on the arbor, and the other with a slot drill or end mill, mounted in an adaptor in the spindle nose, without disturbing the setting of the workpiece.

The vice should normally be set so that the cutting forces act against the fixed jaw and not against the moving jaw, where the cutting forces would be resisted only by the leadscrew thread which might become strained or loosened by the vibration.

The plain vice is normally limited to workpieces whose faces and slots are to be machined at right angles. Workpieces which have to be machined at angles other than right angles are often held in a swivelling machine vice, which is mounted on a rotary base, graduated in degrees, and is, in fact, almost identical with the shaping machine vice shown in fig. 8.11. However, this facility of swivelling through an angle is in some ways a disadvantage since the fixed alignments of the plain vice are lost. Although the rotary base is provided with a scale of degrees, it is not accurate enough to be relied upon for squareness or parallelism. For squareness, the fixed jaw should be checked with a dial gauge against the transverse table motion and, for parallelism, against the longitudinal table motion. This type of vice is shown in use for cutting a slot angled at 60° to the datum face of the work in fig. 8.12.

3. THE ROTARY TABLE

A vertical-spindle milling machine equipped with a rotary table can be used for an extremely wide variety of jobs. The rotary table is bolted on to the machine table and is usually located by tenons in the centre tee-slot. It has tee-slots cut in its own upper surface so that work can be cramped or bolted to it. A taper hole is accurately positioned at its centre so that, with the aid of a short mandrel, work can be accurately positioned over the centre of the table. If this is not possible the 'sticky pin' mentioned on page 190, the scriber of a surface gauge or a dial guage can be used to set the work central, according to circumstances. The work is adjusted until its marked-out circle or circumference runs true when the table is rotated.

The table rotation is produced by a hand-operated worm drive. The angle of rotation is measured in degrees by a graduated scale around the table base and then to a further 3 minutes or 5 minutes of arc, depending on the table, by graduations on the handwheel scale. Note that the base scale reads directly and does not take backlash into account. But the worm and wheel have a clearance which causes backlash and settings on the handwheel minute scale must always be made in the same direction to eliminate the effect of backlash. This precaution is unnecessary if the table scale is fitted with a vernier for fine readings and settings.

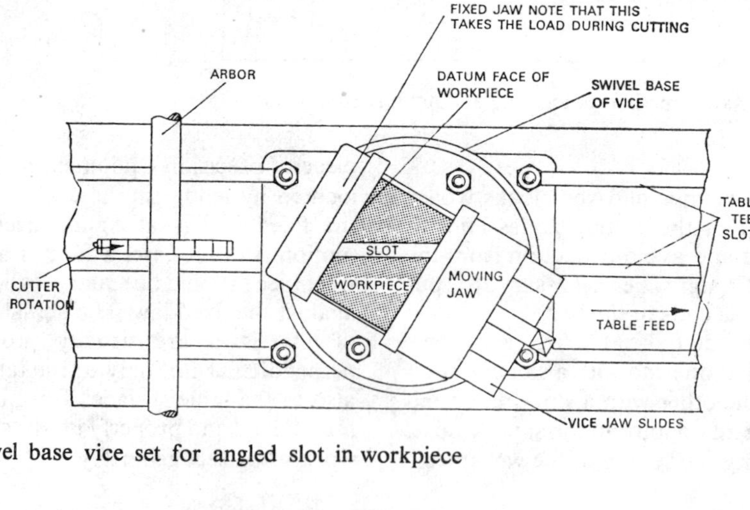

Fig. 8.12. Swivel base vice set for angled slot in workpiece

Apart from allowing accurate angular positioning on the table to be achieved, the worm drive also provides a hand feed which can be used for milling arcs of circles, etc. A typical workpiece for production on a vertical-spindle machine equipped with a rotary table is shown in fig. 8.13 and the task may be resolved into two parts:

1. setting-up the work on the rotary table:
2. setting the cutter to the work.

Let us assume that the plate has been machined to a square of size 180 mm × 180 mm and the 25 mm diameter hole has been drilled and reamed. The procedure could then be as follows.

1. Fit a suitable locating plug, of diameter to suit the hole in the workpiece, in the centre of the table and fit the workpiece on it, but resting on parallel strips to give clearance for the end of the cutter. Lightly clamp the workpiece.
2. Set the rotary table to zero (0°) and mount a dial gauge on the vertical head so that it bears against side AA of the workpiece.
3. Use the longitudinal table motion and adjust the work position on the rotary table until the dial gauge reading is constant. Tighten the clamp on the workpiece and recheck the setting.

The work is now positioned so that the centre

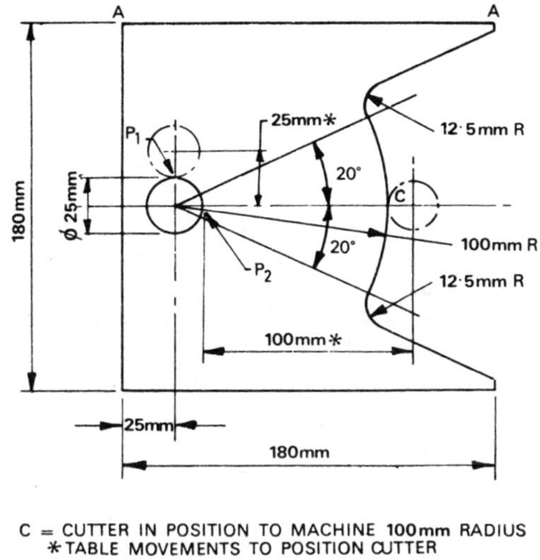

Fig. 8.13. Workpiece to be produced on a vertical spindle milling machine equipped with a rotary table

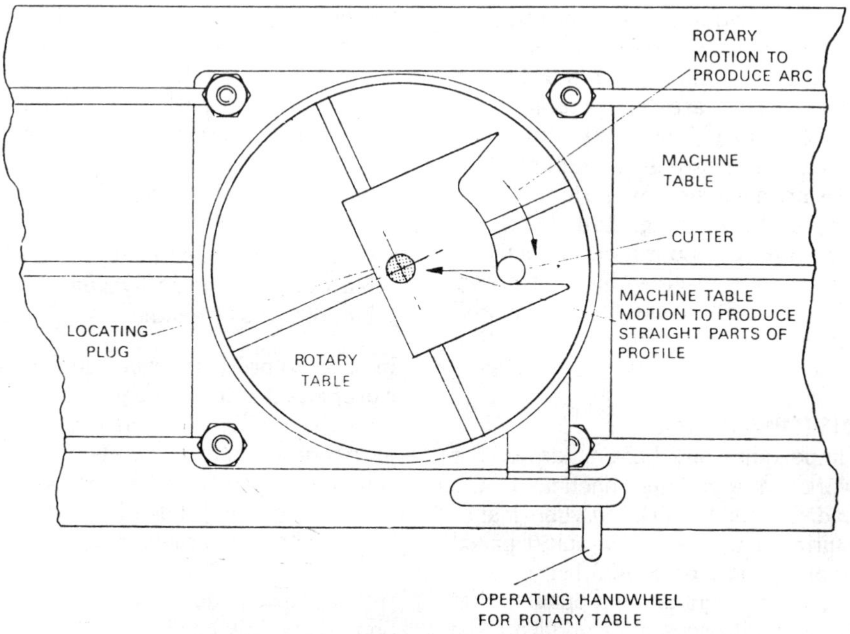

Fig. 8.14. Workpiece of fig. 8.13 set up on rotary table. Clamps and parallel strips not shown

of the 100 mm radius is over the centre of the rotary table, as shown in fig. 8.14, and the edges are parallel or square to the table motion.

4. Fit a 25 mm diameter end mill in the spindle and adjust the table movement until the cutter registers at point P1 on the locating plug. Set the cross-slide leadscrew dial to zero, traverse the cutter clear of the plug and move the table a distance of half the cutter diameter plus half the plug diameter to bring the cutter centre over the centre line of the plug.

$$\text{Table movement} = \frac{25\ \text{mm}}{2} + \frac{25\ \text{mm}}{2} = 25\ \text{mm}$$

Lock the cross slide. The cutter is now positioned over the centre line of the work.

5. Traverse the table until the cutter registers at point P2 on the datum plug. Set the leadscrew dial to zero and move the table *more than* 100 mm to bring the cutter *beyond* point C in fig. 8.13. When it is moved back to point C for the finishing cut the backlash in the table leadscrew will have been eliminated, since the movements to point P2 and position C will have been in the same direction.

6. Set the cutter position about 2 mm beyond the line of the finished radius, pierce the workpiece and rough out the arc and the two tapering sides. The arc is generated by traversing the rotary table 20° in either direction from the centre line, and the two tapering sides are generated by the longitudinal table motion, as shown in fig. 8.14. When the profile has been roughed out it can be finished by setting the table leadscrew dial to 100 mm from the datum position and taking suitable finishing cuts.

4. THE DIVIDING HEAD

The most important work-holding device for indexing work on a milling machine is the dividing head. So much work is cylindrical in nature, requiring a number of equally-spaced slots, teeth or splines, or a number of flats positioned at stated angular relationships that the dividing head has become a standard piece of equipment on the general-purpose machine.

Direct indexing

The simplest form of dividing head consists of a work-holding attachment, usually a three-jaw self-centring chuck, attached to a spindle on which is mounted an index plate with a number of equally-spaced holes into which a taper plunger fits. The number of holes chosen is usually one with many factors, e.g. 48. This enables the chuck to be turned through 48, 24, 16, 12, 8, 6, 4, 3, and 2 equal divisions. A 24-hole plate is almost as useful, giving 24, 12, 8, 6, 4, 3, and 2 divisions. When the pin has been engaged in the correct hole the spindle is locked and the cut taken. Such a dividing head is shown in fig. 8.15(a) in its vertical form and in fig. 8.15(b) in its horizontal form, with a catch plate instead of a chuck.

Simple indexing

Although quick and easy to operate, the direct-indexing head is limited in scope and most dividing heads consist of a work-holding device on a spindle which carries a 40-tooth worm wheel driven by a single-start worm and handle, i.e. 40 turns of the handle produce one turn of the work. Thus if 5 equal divisions of the work are required, $\frac{40}{5} = 8$ turns of the handle are required for each division; if 4 equal divisions are required, $\frac{40}{4} = 10$ turns of the handle per division are necessary, or generally:

$$\text{No. of turns of handle} = \frac{40}{n}$$

where n = no. of equal divisions of workpiece

Now let $n = 7$ equal divisions

$\therefore$ No. of turns of handle $= \frac{40}{7} = 5\frac{5}{7}$ turns.

In order to produce seven equal divisions of the workpiece it is therefore necessary to subdivide a complete turn of the handle. To achieve this the handle contains a spring-loaded plunger and can be adjusted for radius of rotation so that it can engage in any one of a number of rings of holes in an index plate, as shown in figs. 8.16 and 8.17.

The index plates supplied with Browne and Sharpe dividing heads have three plates, viz.:

Plate No. 1: 15, 16, 17, 18, 19 and 20 holes

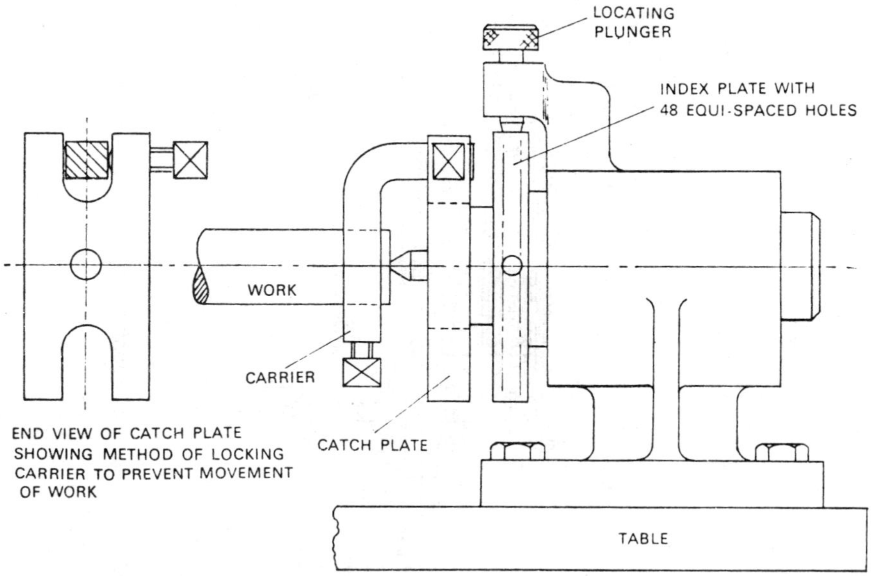

Fig. 8.15(a). Direct-indexing dividing head (vertical type). This type is often used in production work for milling flats on cylindrical heads, as shown

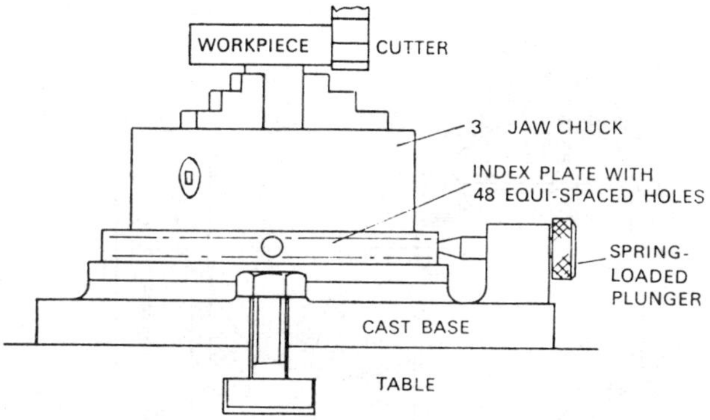

Fig. 8.15(b). Direct-indexing dividing head (horizontal type)

Plate No. 2: 21, 23, 27, 29, 31 and 33 holes
Plate No. 3: 37, 39, 41, 43, 47 and 49 holes

Parkinson and Cincinnatti dividing heads have larger, double-sided plates with (a) on one side:

24, 25, 28, 30, 34, 37, 38, 39, 41, 42 and 43 holes

and (b) on the reverse side:

46, 47, 49, 51, 53, 54, 57, 58, 59, 62 and 66 holes.

For seven equal divisions of the workpiece, we adjust the radius of the plunger arm so that the plunger engages a circle of holes which are a multiple of seven. Using a Browne and Sharpe head we could select 21 holes.

$$\tfrac{5}{7}\text{ turns} = \tfrac{5}{7} \times 21 = 15\text{ holes.}$$

Thus to index 7 equal divisions would require

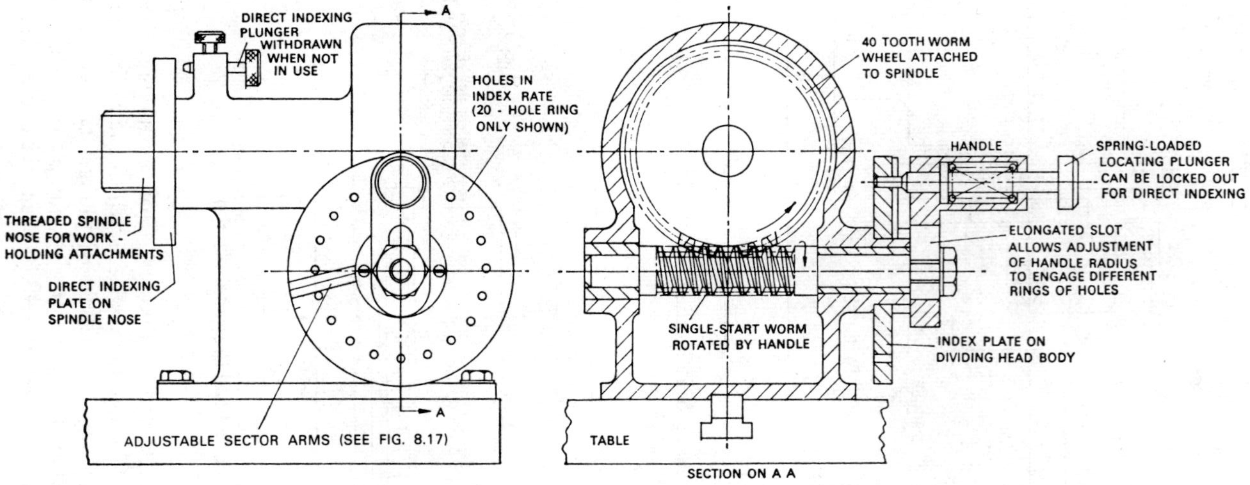

Fig. 8.16. Vertical dividing head equipped for both direct and simple indexing. This is the simplest type. More complex types can be swung to any angle between vertical and horizontal

Note: When used for direct indexing the worm is normally disengaged from the worm-wheel

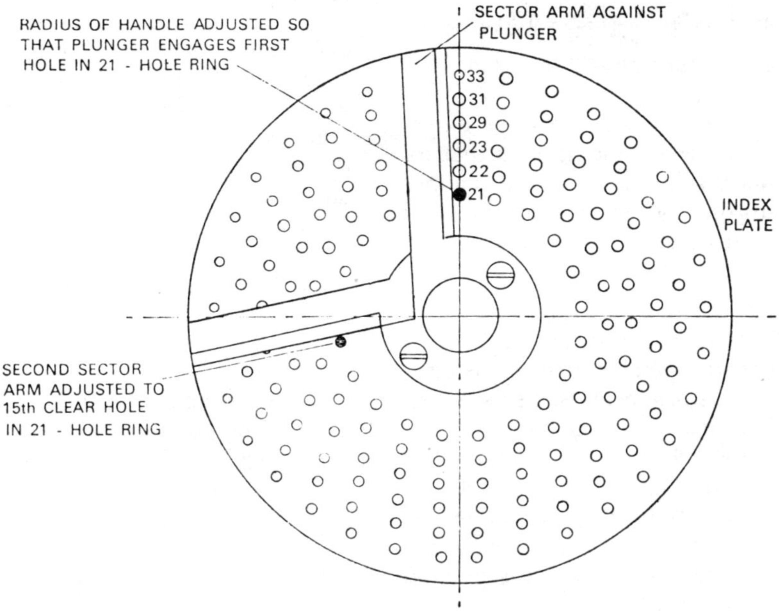

Fig. 8.17. No. 2 index plate set for indexing 7 equal divisions. The handle is turned 5 revolutions and a further 15 holes in the 21-hole ring

5 complete turns and 15 holes in a 21-hole circle.

To enable the 15 holes to be indexed without continual counting, the index plate has adjustable sector arms as shown in fig. 8.17. With the plunger in a correct hole, one sector arm is brought up to it and the second locked to indicate 15 *clear* holes. To index $\frac{1}{7}$ rev, the handle is turned 5 complete turns and on to the hole indicated by the second sector arm. The arms can be moved relative to the plate and are now swung round so that the first arm contacts the plunger ready for the next index.

SAFETY NOTE

Backlash must be avoided by turning the handle always in the same direction. If you 'overshoot' the required hole, turn the handle back one complete revolution and forward again to the correct position. This avoids the awkward position of asking the foreman if he wants one thick tooth or two thin ones at the end of a gear-cutting operation!

Angular indexing

It may be that a shaft is required with two keyways at an angle of 36° to each other. Since 40 turns of handle $= 360°$

$\therefore$ 1 turn of handle $= \frac{360}{40} = 9°$ work

or 36° of work $= \frac{360}{9} = 4$ turns of handle

Or generally:

$$\text{Turns of handle} = \frac{\text{angle required}}{9°}$$

Consider an angle of 43°

$$\text{Turns of handle} = \tfrac{43}{9} = 4\tfrac{7}{9} \text{ turns}$$

That is, 4 turns plus 14 holes in an 18-hole circle on a Browne and Sharpe No. 1 plate, or 42 holes in a 54-hole circle on a Parkinson plate.

If the angle required is quoted in minutes of a degree, of one revolution of the handle $= 9° \times 60 = 540$ minutes of the work or:

$$\text{Turns of handle} = \frac{\text{angle required in minutes}}{540'}$$

Thus for an index of 15° 40′ we have:

$15° 40' = 940'$

Turns of handle $= \frac{940'}{540} = \frac{94}{54} = \frac{47}{27} = 1\frac{20}{27}$ turns.

Thus the index is 1 turn plus 20 holes in a 27-hole circle.

Most dividing heads also carry a direct-indexing plate on the spindle nose so that for common numbers of divisions no calculations are required and the quicker direct-indexing can be used. This plate is shown in fig. 8.16.

Work-holding in the dividing head

Work may be held in the dividing head in a three-jaw self-centring chuck, between centres or on a mandrel between centres. As usual, the accuracy of concentricity of self-centring chucks is not too reliable, particularly when wear has occurred. Thus work held between centres is to be preferred, but the chuck has the advantage that it can be used at any angle up to the vertical, whereas centres can only be used parallel to the table surface or nearly so (a height adjustment is provided on the tailstock).

When work which has to be machined parallel is held between centres or supported by the tailstock it should be checked with a dial gauge to ensure that it is parallel to the table motion, particularly if the tailstock alignment has been altered, e.g. to machine the flutes of a taper-pin reamer. The same applies to chuck-held work when the chuck is returned to the horizontal after use in some other position.

CUTTER SETTING

Specific Objective: *The student should be able to explain how to set a milling machine to produce a workpiece to a given specification.*

Just as drilled holes are positioned relative to datum faces, so milled features are also positioned relative to some datum, either a datum face, reference line or reference point on a workpiece. To produce the required feature it is necessary to position the cutter relative to the datum. Furthermore, milled features often have prescribed widths, e.g. slots, and the width of slot produced may not be equal to that of the cutter, particularly if the cutter is running slightly out of truth. It is therefore often better deliberately to select a cutter thinner than the width desired and measure the slot produced after the first cut. The correct width can then be obtained with a finishing cut. Problems such as this will be dealt with in the following examples, all of which presume that the work is already set correctly, relative to the machine alignments.

1. SETTING A CUTTER TO CUT A STEP (fig. 8.18)

Cutter: side-and-face, 20 mm wide = 100 mm diameter.

(a) With machine running, adjust cross-slide until side of cutter just contacts datum face AA. Set cross-slide dial to zero.

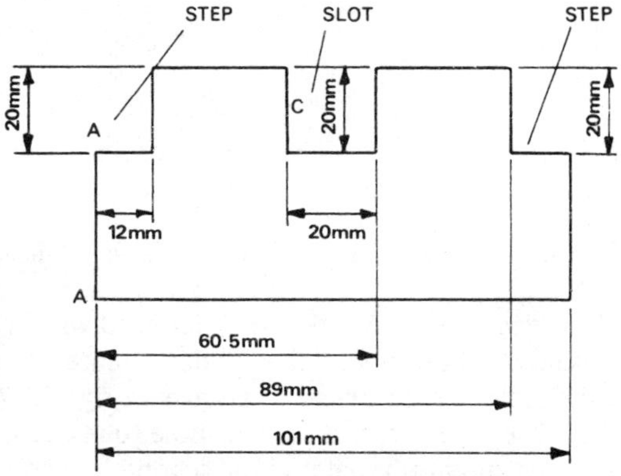

Fig. 8.18. Profile of workpiece used to illustrate cutter-setting techniques

(b) Lower table until cutter clears and traverse across 12 mm on cross-slide. Lock cross-slide.

(c) Raise table until cutter just touches work and set vertical dial to zero.

(d) Back-off table and raise to take first cut. Continue to cut until vertical dial reads 20 mm.

(e) Lower table and move cross-slide until cutter is beyond other side of work. Bring table back and take a cut to mark a shoulder of greater width than 89 mm from datum face AA.

(f) Measure shoulder width and move cross-slide the remaining amount to reduce it to 89 mm. Check width and lock cross-slide.
(g) Take a cut at this setting and check width from datum face.
(h) Continue cutting at this settting until vertical dial reads 20 mm.

Note how backlash is eliminated by this procedure to obtain the 89 mm dimension.

On a vertical-spindle machine the procedure would have been similar but end mills would have been used.

2. SETTING A CUTTER TO MACHINE A SLOT (fig. 8.18)

Cutter: side-and-face, 15 mm wide × 100 mm diameter.

In this case we assume the cutter to be unreliable for width, so an undersize cutter is used.

(a) Take a touch on face AA. Set cross-slide dial to zero.
(b) Lower table until cutter clears and traverse across 60.5 mm on cross-slide, using cross-slide dial.
(c) Cut slot to depth.
(d) Move table back to take a light cut on face C. This eliminates backlash.
(e) Measure width of slot using workshop-grade gauge blocks and subtract this measurement from 20 mm.
(f) Move table required remainder and finish slot. Again note elimination of backlash.

3. SETTING A CUTTER TO CUT A KEYWAY IN A ROUND BAR (fig. 8.19)

Cutter: slot cutter, 120 mm diameter × 12 mm wide.

In this case we assume the cutter to be reliable enough to produce the width of keyway to the required accuracy.

(a) Raise table and use cross-slide to touch side of cutter on diameter at A.
(b) Lower table and traverse cross-slide for half work diameter plus half cutter width.

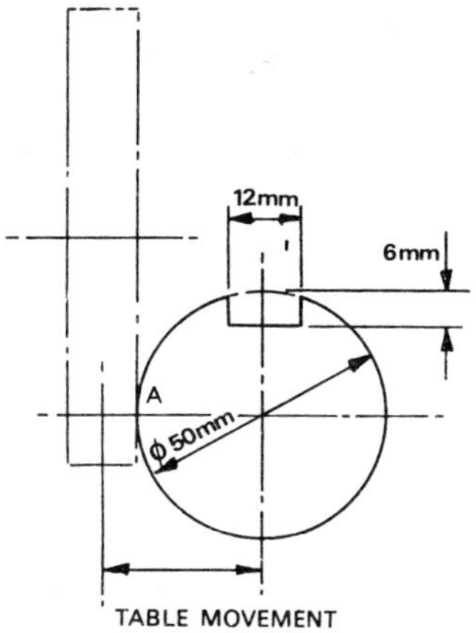

Fig. 8.19. Setting slot cutter to a small diameter bar

$$\text{Table movement} = \frac{50\text{ mm}}{2} + \frac{12\text{ mm}}{2} = 31\text{ mm}$$

Lock cross-slide.

(c) Raise table until cutter touches work. Traverse table until cutter clears work and raise table a further 6 mm. Lock vertical slide and proceed to cut slot.

4. SETTING A CUTTER TO CUT A SLOT IN A LARGE DIAMETER DISC (fig. 8.20)

In this case the diameter of the work is too large to permit a touch by the cutter at the side (this would require a cutter about 200 mm diameter).

Cutter: slot cutter, 100 mm diameter × 6 mm wide.

(a) Set a square on the machine table or on parallel strips if necessary to obtain the height. *Lightly* clamp the square in contact with the work as shown.
(b) Wring-up workshop-grade blocks equal to half work diameter *minus* half width of cutter.

$$\text{Gauge block width} = \frac{150\text{ mm}}{2} - \frac{12\text{ mm}}{2}$$
$$= 69\text{ mm}$$

(c) Adjust table until gauge blocks fit

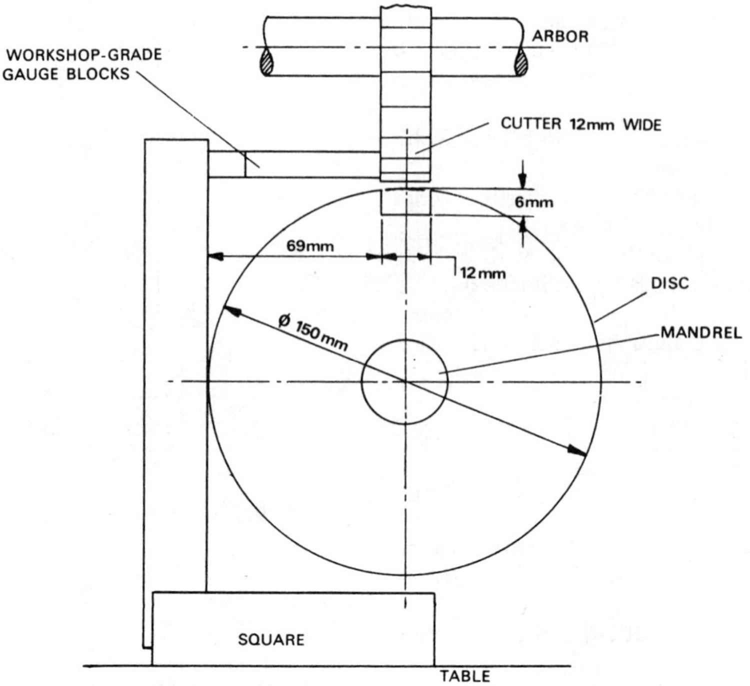

Fig. 8.20. Setting a cutter to a large-diameter disc. Note: an alternative method is to set the cutter to the mandrel before mounting the disc on it

without shake when held between cutter face and try square.

(d) Check cutter setting from other side of disc and adjust as necessary.

(e) Machine slot to size.

All the above operations have been based on the use of slot cutters or side-and-face cutters. They can also be performed by end mills and as a series of exercises the student can give corresponding descriptions of such operations.

MORE SPECIALISED CUTTERS AND OPERATIONS

Specific Objective: *The student should know the applications of specialised types of milling cutters.*

Form-relieved cutters

All the examples described have dealt with flat surfaces, except for the use of the rotary table to generate an arc. However, some shapes or forms need to be *copied* by specially-made cutters, such as that shown in fig. 8.21. These cutters are usually form-relieved, i.e. the form of each tooth is continued and preserved for the full depth of the tooth, clearance being obtained by making each tooth a small part of a spiral whose cross-section is the form of the tooth. Thus however far the teeth are ground back on the rake face their form will remain correct. This is true only if the rake angle is correct and care must be exercised in sharpening such cutters to preserve the correct rake angle.

Dovetail cutters

A dovetail cannot be produced by a normal right-angled cutter and a special type of cutter is used, as shown in fig. 8.22. It machines the base surface with its end teeth and the angled surfaces with its inclined cutting face.

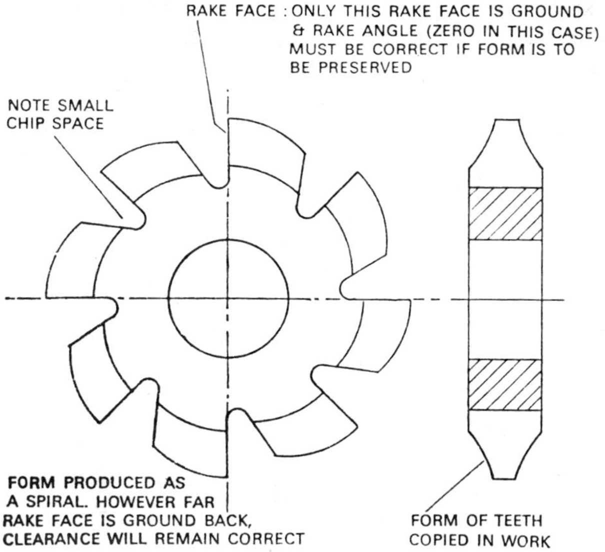

Fig. 8.21. Form-relieved cutter

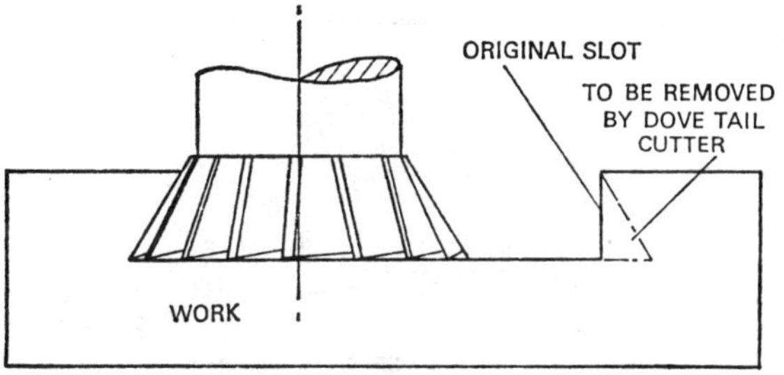

Fig. 8.22. Action of dovetail cutter

Tee-slot cutters
Tee-slots are first machined by cutting the vertical slot to full depth with an end mill or slot drill. The tee of the slot is then cut by a special cutter as shown in fig. 8.23. It consists essentially of a small side-and-face cutter mounted on a shank, with a neck small enough to clear the vertical slot machined in the first operation. With the aid of a rotary table, a tee-slot can be machined to an arc of a circle.

Woodruff cutters
These cutters are used for producing Woodruff key-seats and must conform to the shape of the key, which is a segment of a circle. The cutter is too small to be mounted on the machine arbor and it is made in the form of a slot cutter on a solid shank, side clearance being obtained by dishing. Unlike a tee-slotter it has no side-cutting teeth and can further sometimes be identified by the fact that its end is not flat but carries a centred spigot which allows a support

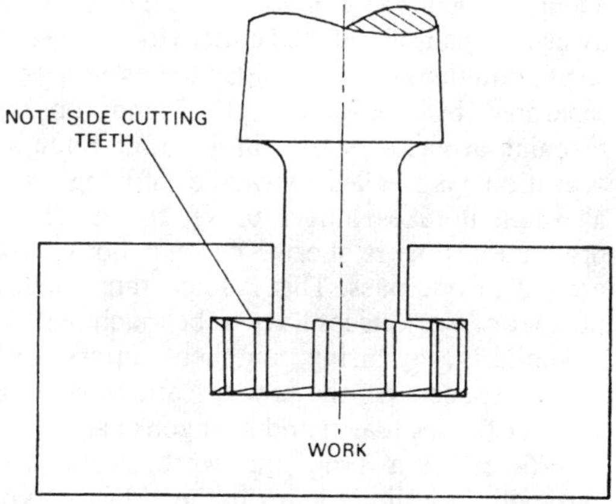

Fig. 8.23. Action of tee-slot cutter

centre to be used as shown in fig. 8.24. To produce the key-seat, the cutter is positioned over the work and sunk directly to the required depth. Different sizes of Woodruff keys, and therefore of Woodruff cutters, are used.

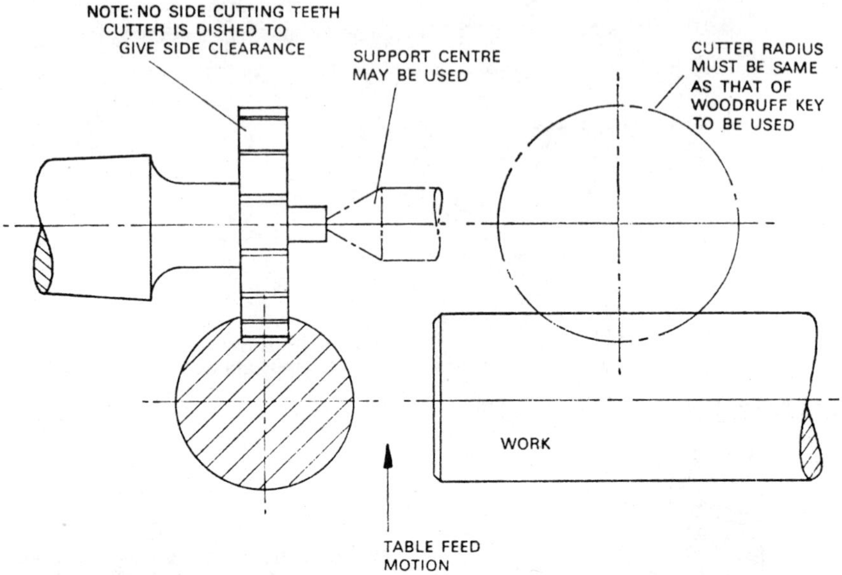

Fig. 8.24. Action of Woodruff key-seat cutter

STRADDLE MILLING

Referring to milling example no. 1 on page 198 and its associated fig. 8.18, it would be possible by using a pair of matched cutters to cut the two steps simultaneously, provided the cutters were separated by a distance of 77 mm on the machine arbor, as shown in fig. 8.25. Such an operation is called straddle milling and although it takes longer to set the machine, production time is shorter because both steps are cut at one pass. This is important when a number of workpieces have to be machined.

Similarly, by using matched cutters with correct spacers on the arbor, features such as spanner flats, squares and hexagons can be cut two faces at a time, the work being held vertically in a direct indexing head of the type shown in fig. 8.15(a).

GANG MILLING

Referring again to milling example no. 1 on page 198 and its associated fig. 8.18, by selecting suitable cutters it should be possible to machine the whole of the top surface in one

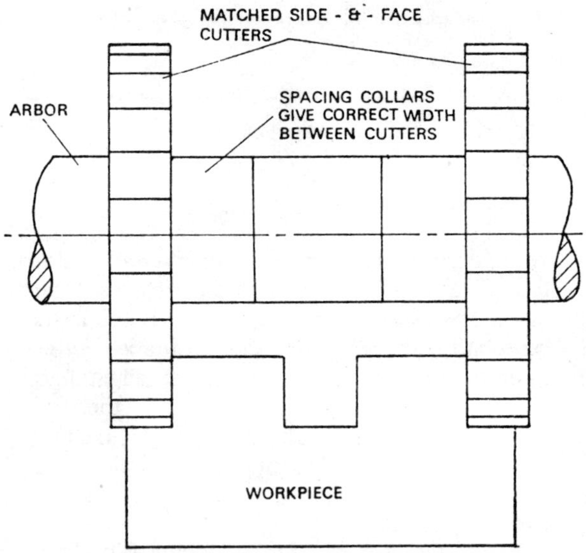

Fig. 8.25. Straddle milling

pass, the cutters required being two side-and-face cutters, two slab mills, and one slot cutter. These cutters must be of correct appropriate diameters and thicknesses since the shape of the surface is a copy of the form produced by the cutters in combination, as shown in fig. 8.26. Such a process often calls for special cutters and is only economical where a considerable number of parts are to be machined.

of a slab of mild steel, 340 mm long by 100 mm wide. We will assume that a helical-tooth slab mill, 100 mm diameter by 125 mm long, is used to remove the metal at one pass, with the same cutting speed of 20 m/min and a feed per tooth of 0.25 mm, the cutter having 4 teeth.[1]

$$\text{Circumference of cutter} = \pi \times 100\ \text{mm} = 314.2\ \text{mm}$$

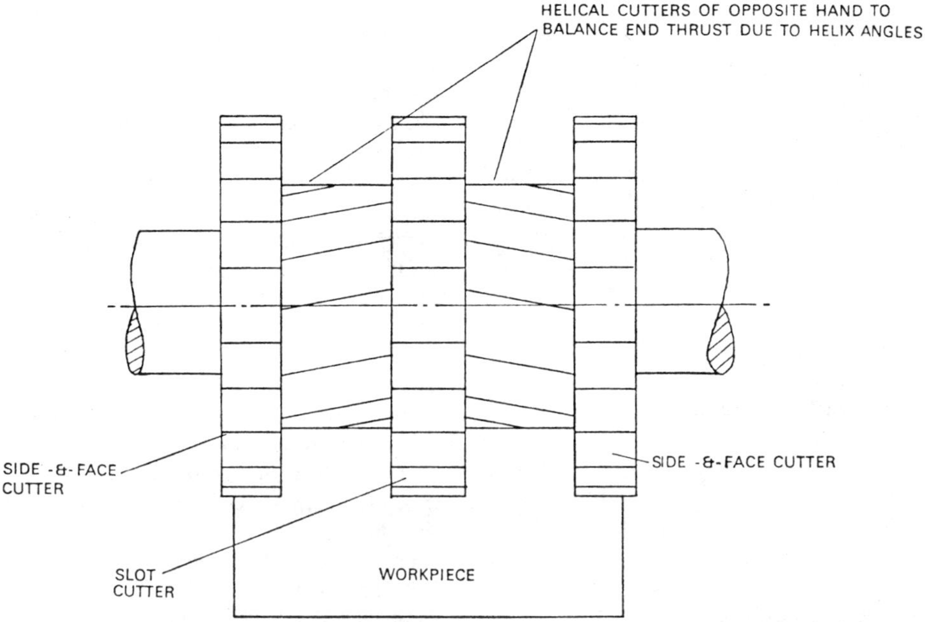

Fig. 8.26. Gang milling. Matched set of cutters machines complete and correct profile on work

Note that the slab milling cutters shown in fig. 8.26 are of opposite hand so that the end thrusts produced by the helix angles cancel each other out.

ESTIMATION OF CUTTING TIMES

Specific Objective: *The student should be able to compare cutting times for a milling operation with those for producing similar results by other methods.*

Consider the removal of 5 mm from the surface

$$\text{Rev/min} = \frac{\text{Cutting speed (m/min)}}{\text{Circumference (m/rev)}} = \frac{20}{0.314} = 64\ \text{rev/min}$$

$$\text{Feed/tooth} = 0.25\ \text{mm}$$

$$\text{Feed/rev} = \text{Feed/tooth} \times \text{no. of teeth} = 0.25 \times 4 = \underline{1\ \text{mm}}$$

[1] In Chapter 10, page 233 the same problem is considered for a shaping machine.

$$\text{Feed/min} = \text{Feed/rev} \times \text{rev/min}$$
$$= 1 \times 64$$
$$= 64 \text{ mm/min}$$

$$\text{Time taken} = \frac{\text{Length (mm)}}{\text{Feed (mm/min)}}$$

$$\text{Time taken} = \frac{340}{64} \text{ min}$$

$$\text{Time taken} = \underline{5.32 \text{ min}}$$

This is over twice as fast as with the shaping machine. In fact, the operation would take rather longer than 5.32 minutes since we have not allowed for the cutter to feed in to full depth of cut and to run out. However, with the shaping machine we used an average cutting speed of 20 m/min, knowing that the maximum speed at the midpoint of the stroke would be greater. With the milling machine we should in practice use a higher cutting speed, because the speed is constant during the whole cutting time.

Thus, under similar conditions, we have a higher metal removal rate with the milling machine than with the shaping machine, for the following reasons:

(a) For each revolution of the mechanism of the shaping machine there is only one cutting edge in operation. For each revolution of the milling machine spindle, four teeth engage the work.

(b) On the shaping machine nearly half the operating time is wasted on the return stroke. This is not so on the milling machine, which has an almost continuous cutting action.

The question then arises, 'Why use shaping machines?' The answer is given if a similar comparison is made for a surface 375 × 200 mm. The milling machine would require two passes and the total time would be more than doubled because of the time needed to feed back, cross-traverse the table and restart the cut. The shaping machine would still be cutting and thus as surface areas increase the shaping machine becomes more economic. Furthermore, the shaping machine uses relatively simple hand-ground tools, while the cutters of a milling machine are expensive to buy and maintain. The milling machine is, however, capable of much more sophisticated work, while the shaping machine is mainly suited to the machining of large flat surfaces. Each has its own special functions for which it is purchased and used.

SAFETY NOTES

The following safety notes, which have again been left until the end of the chapter, not only deal with the safety of the operator but also with safety of the equipment.

1. Always use the guard supplied for the machine. The time taken in fitting and adjusting the guard may save a lifetime as a one-handed operator.

2. Milling attachments are heavy. Never try to lift a vice, dividing head, rotary table or other attachment unless you know it is well within your capabilities, and be modest rather than conceited. Get assistance or use a portable hoist, or you may do yourself an injury.

3. Some operations are difficult to guard. Keep your hands away from cutters and never brush swarf away when the cutter is running. Use a coolant feed, not a brush.

The following safety notes are concerned more with the machine than with the operator:

4. The machine attachments should be regarded as instruments rather than just work-holding attachments. Treat them with care to maintain the built-in alignments. Ensure that the bases and tenons do not get damaged, by always resting the attachments on wood blocks, not on a concrete floor.

5. The tee-slots of the machine are important, since they provide locations for the attachments. Maintain the tee-slots in good condition and, if a burr is raised scrape or file it off carefully.

6. Ensure that the table and the underside of any attachment are clean before setting-up and bolting down the attachment.

7. Ensure that the spindle-nose taper and the arbor tapers are clean before fitting an arbor and pulling it home with the drawbolt.

8. Ensure that collars and cutter faces are clean when mounting or changing a cutter.

QUESTIONS

1. Show with the aid of diagrams the errors induced in the geometry of workpieces by the following alignment errors in milling machines.
   (a) Table surface not parallel with table motion.
   (b) Spindle axis not at right angles to the table motion when milling a slot using a side and face cutter.
   (c) Spindle axis not parallel to table surface when slab milling a flat surface.
   (d) Effect of arbor being bent as a result of dirt between collars.

2. The outside faces of a piece of angle iron 100 mm × 100 mm are to be cleaned up square to each other on a horizontal milling machine. This could be done using a side and face cutter or a shell-end mill. State which you would use, giving reasons, and explain how you would hold the workpiece for this operation.

3. A round bar 50 mm diameter × 1.5 m long is to have two slots milled in its end. The slots are to be 6 mm wide, cut on the centre line of the bar, that at one end being at right angles to that at the other.

Explain how the workpiece would be located and clamped, the cutter set and the two slots located at right angles to each other.

The machine to be used is a horizontal milling machine.

4. Explain the advantages and limitations of using an end mill compared with a side and face cutter for machining:
   (a) straight slots
   (b) profiles.

In the case of profiles illustrate your answer with a diagram of a typical component to be machined assuming that a rotary table is available.

5. Calculate the indexing movements necessary for the following jobs to be done using a dividing head:
   (a) Milling 30 equi-spaced splines on a shaft end.
   (b) Milling two slots in a shaft, the slots to be at an angle of 85° relative to each other.
   (c) Milling three slots in a shaft with angles of
   118° 40′ between slots 1 and 2
   115° 20′ between slots 2 and 3.

In each case assume that the calculation is to be done for both types of index plate.

6. A block of metal 100 mm wide is to have a slot 25 mm wide cut across it so that the near edge of the slot is 50 mm from the datum face of the work. The cutter available is a side and face cutter which was 20 mm wide but which has been re-ground and is no longer reliable for width. Explain how the slot would be cut to ensure that its position and width are to specification.

7. Slots are to be cut on the centre lines of:
   (a) a shaft 45 mm diameter
   (b) a disc 185 mm diameter.

If the slots are to be 15 mm wide and cut with a side and face cutter 150 mm diameter explain for each case how the cutter would be set to bring it over the centreline of the work.

8. (a) Use simple sketches to show the differences between an end-milling cutter and a slot drill.

b = 14 mm
h = 9 mm
t_1 = 5·5 mm
t_2 = 3·8 mm

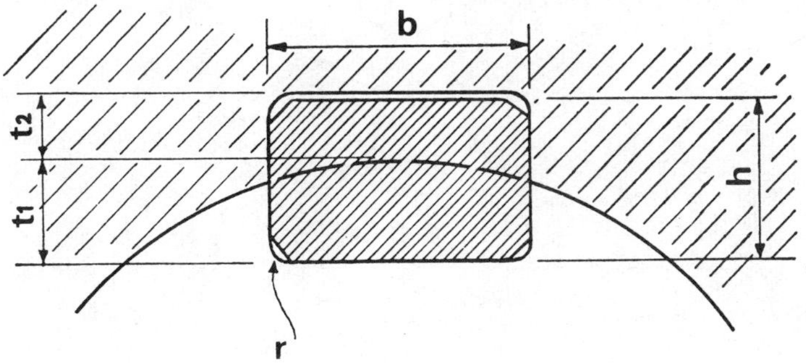

Fig. 8.27

(b) Explain why the slot drill can be used with an axial feed and the end mill cannot.

(c) B.S. 49 (Keys and Keyways) gives the information shown in fig. 8.27 for a 50 mm diameter shaft. Explain carefully what settings would have to be made on the milling machine, using the index dials, in order to cut the keyway correctly.

ANSWERS

5. (a) Browne and Sharpe: 1 turn and 6 holes in 18 hole circle; Cincinnati: 1 turn and 8 holes in 24 hole circle

(b) Browne and Sharpe: 9 turns and 8 holes in 18 hole circle; Cincinnati: 9 turns and 24 holes in 54 hole circle

(c) For 118° 40′:
Browne and Sharpe: 13 turns and 5 holes in 27 hole circle; Cincinnati: 13 turns and 10 holes in 54 hole circle
For 115° 20′:
Browne and Sharpe: 12 turns and 25 holes in 27 hole circle; Cincinnati: 12 turns and 50 holes in 54 hole circle.

9 The drilling machine

General Objective: *The student should be able to compile block diagrams of the commonly used drilling machines and describe typical drilling operations.*

INTRODUCTION

We have seen that in order to cut metal successfully any cutting tool must overcome the resistance of the metal. In the case of a drill, two forces are required, a torque to provide the tangential cutting force and a direct thrust normal to the work surface to feed in the drill. Thus any drilling machine must provide two motions to the drill:

1. A rotary motion whose speed is adjustable to give the required cutting speed and whose power is sufficient to provide the necessary tangential force for cutting.
2. A linear motion for the feed.

If the machine is to produce accurate holes without exerting bending forces on the drill:

(a) the drill must rotate on its own axis, i.e. 'run true' without eccentricity, which demands an accurate drill-holding device;
(b) the drill must cut equally on both cutting edges;
(c) the feed motion must be along the line of the axis of the drill.

Furthermore, a rigid base or table must be provided on which the work can be fixed for the drilling operation. This table must normally be at right angles to the axis of rotation of the drill and therefore at right angles to the direction of feed.

These requirements are called the basic alignments of the machine and are illustrated in fig. 9.1. They are common to all drilling machines, and machines differ only in size, power and the way the drill is positioned over the work.

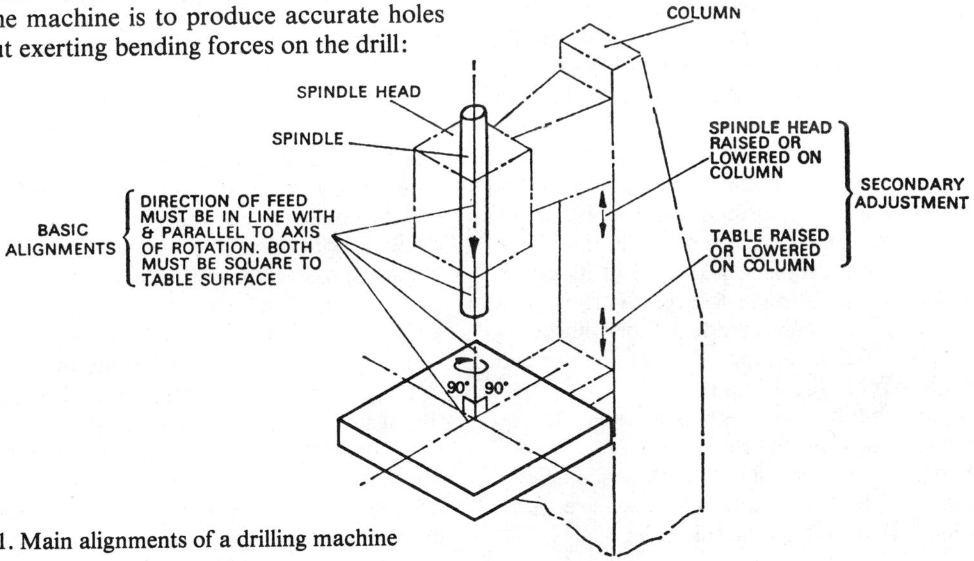

Fig. 9.1. Main alignments of a drilling machine

DRILL-HOLDING DEVICES

1. *Taper shanks and sockets*

Drills, and indeed other tools requiring similar mounting, can be obtained with straight shanks, to be gripped in a chuck, or with taper shanks which fit in a taper socket in the end of the machine spindle. The straight-shank drills are driven by friction between the shank and the chuck jaws and may slip if the torque required is greater than that provided by the frictional grip.

Taper-shank drills, on the other hand, have a positive drive. The shank of the drill is made as shown in fig. 9.2, with a shallow taper of approximately 50 in 1000 on diameter. This is known as a *self-locking taper* because when the drill is pushed into a mating taper hole it stays in position and a considerable force is required to loosen it. However, this alone would provide only a friction drive, so the end of the taper bears a *tang*, as shown, which fits in a corresponding slot at the end of the tapered hole in the spindle. The tang thus provides a positive drive. Since the drill is self-locking in its taper, provision must be made for removing it. The slot is carried through the sides of the spindle, as shown, to enable a taper drift to be tapped through to force the drill out.

Taper shank drills are generally to be preferred, not only for their positive drive but also for the greater accuracy of concentricity obtained with them. However, they are more expensive than straight-shank drills.

2. *Drill chucks*

The type of drill chuck most commonly used is a three-jaw self-centring one in which the outer sleeve is threaded and, when rotated, moves axially along the chuck body. The inner surface is conical to correspond with the jaws, whose axial position is fixed inside the chuck body. Thus, when the sleeve is screwed up, the conical surface causes the jaws to close on the drill, and vice versa. Closing and releasing the chuck may be either by hand, on smaller sizes of chuck or, more commonly, by means of a key which carries a small bevel gear which engages with similar teeth on the chuck.

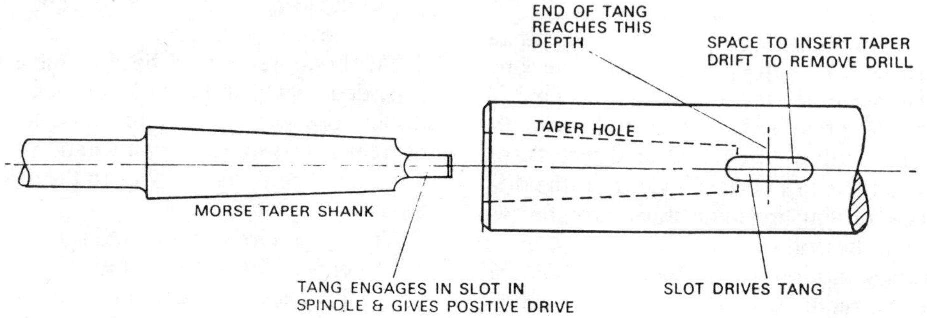

Fig. 9.2. End of drilling machine spindle and taper shank of drill

If frequent drill changes are required, for example when a radial drilling machine is used to drill a large number of different-sized holes in one component, a quick-change chuck may be used. In this case the outer sleeve of the chuck is freely rotating and when it is moved up the chuck body the drill is released. The drill is fitted with a collet to suit the chuck and a positive drive is obtained by a tang on the collet engaging a driving slot in the chuck.

THE MACHINE SPINDLE

The spindle must not only rotate but must also provide a linear feed motion with the necessary thrust for the feed. The drive to the spindle does not normally move down with the feed so provision must be made for transmitting the drive regardless of the movement of the spindle downwards and upwards.

A simple machine spindle is shown in fig. 9.3. This has a ball thrust bearing at top and bottom, any end play being taken up by adjusting a pair of locknuts at the top. The radial bearings are a pair of plain phosphor-bronze bushes and the complete spindle is carried in a sleeve. The sleeve slides axially in the spindle housing, which is attached to the machine frame or column. A gear or pinion mounted in the spindle housing engages with a rack, or length of teeth, cut in the back of the sleeve. The operating, or feed, lever is attached outside the spindle housing to the shaft of the pinion, and movement of the lever turns the pinion, thus moving the sleeve and the spindle downwards or upwards.

The spindle drive pulleys or gears are mounted in bearings in a fixed part of the frame above the spindle housing, and drive the spindle by a sliding key or splines, so that the spindle can move axially and still be driven in any position.

This technique is common to almost all drilling machines. Various refinements in the bearings are used, and power feed is commonly provided in addition to manual feed, but the basic features are the same.

TYPES OF DRILLING MACHINE

Specific Objective: *The student should know the different types of drilling machine and their uses.*

A simple method of classifying drilling machines is to consider two categories, all machines having the basic alignments already mentioned:

1. those in which the work is moved into position under a fixed drill axis which are the most common;
2. those in which the work is fixed and the drill axis is moved into position over the hole.

FIXED-AXIS MACHINES

1. *The sensitive drilling machine*

These machines are usually bench mounted,

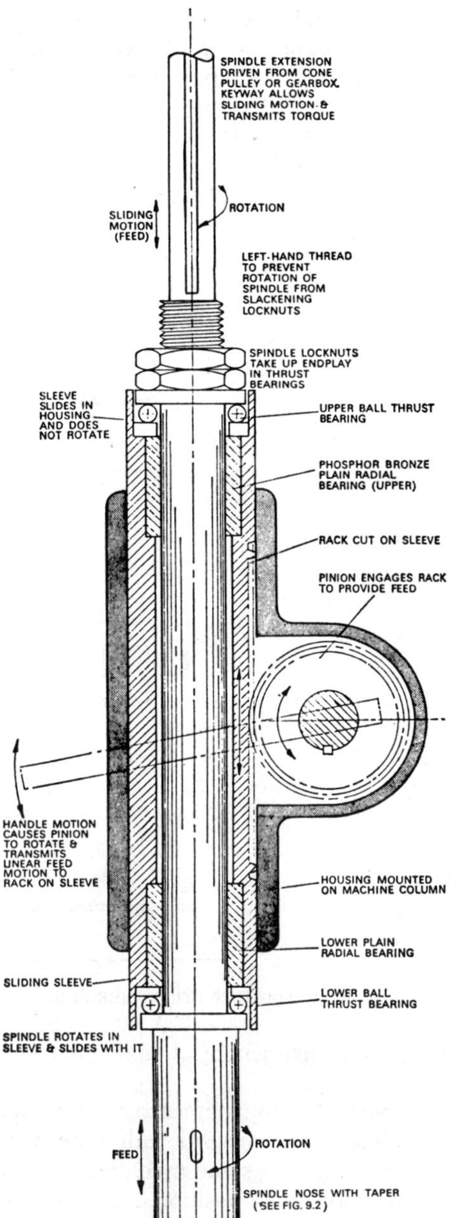

Fig. 9.3. Cross section of drilling machine spindle assembly showing how both rotary cutting motion and sliding feed motion are obtained

although pedestal machines are available. They are of small capacity, i.e. maximum size of drill, and do not have power feed. Thus the operator, by hand feeding, can 'feel' the drill cutting, particularly as it breaks through – hence the name 'sensitive'.

An outline of such a machine is given in fig. 9.4, and the adjustments available are as follows:

(d) spindle feed stop to allow holes to be drilled to a set depth (see drilling operation, pages 215–16).

2. *The pillar drilling machine*
This is simply a much more robust version of the sensitive drill, floor-mounted and provided with power feed, usually at three rates, and with a wide range of spindle speeds provided by a gearbox rather than by cone pulleys. As an example let us consider a machine with a drill-size range from 6 mm diameter to 25 mm diameter, to be used for all materials from brass, requiring a cutting speed of 70 m/min, to high-carbon steel, requiring a cutting speed of 10 m/min.

Obviously the fastest speed will be required when drilling brass with the smallest drill and therefore:

$$\text{Maximum rev/min} = \frac{\text{Cutting speed}}{\text{Drill circumference}}$$

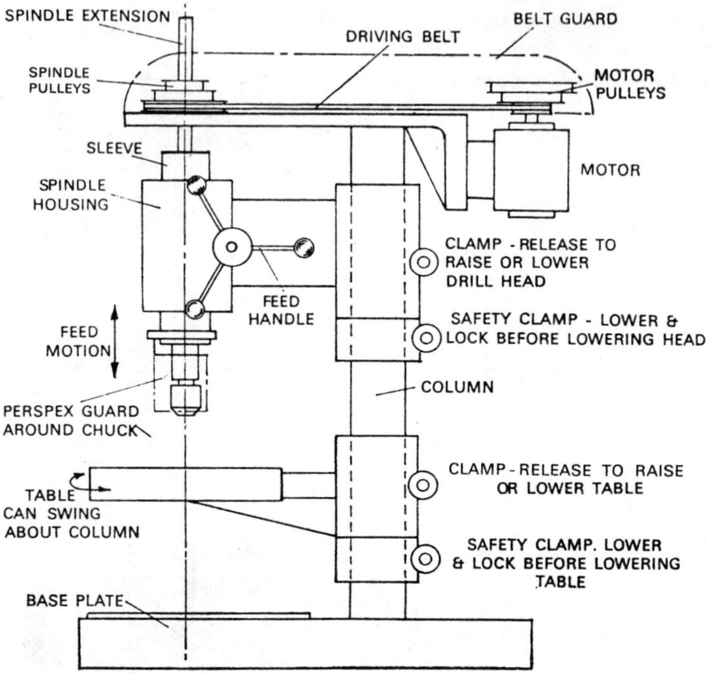

Fig. 9.4. Outline of sensitive drilling machine

(a) speed adjustment by changing the belt on the pulleys;
(b) table height adjustment to accommodate different work heights within the spindle feed range. In some machines, as in the one shown, the position of the drill head and spindle housing on the column can also be adjusted;
(c) table position adjustment about the column. This enables the table to be swung clear to accommodate larger workpieces;

$$= \frac{70}{\pi \times \frac{6}{1000}}$$

$$= 3700 \text{ rev/min}$$

The lowest speed will be used when drilling high-carbon steel with the largest drill and similarly:

$$\text{Minimum rev/min} = \frac{10}{\pi \times \frac{25}{1000}}$$

$$= 127 \text{ rev/min}$$

Thus the speed range needs to be from 127 to 3700 rev/min and would usually be provided in nine steps.

Note that the table can be swung clear of the base to allow large castings to be accommodated, as with a sensitive drilling machine. This has the disadvantage that although the machine is more massive it loses rigidity since the table cannot be supported from underneath. A similar machine using a box column is much more rigid, since the table moves on slides on the front of the column as shown in fig. 9.5, and is supported under the

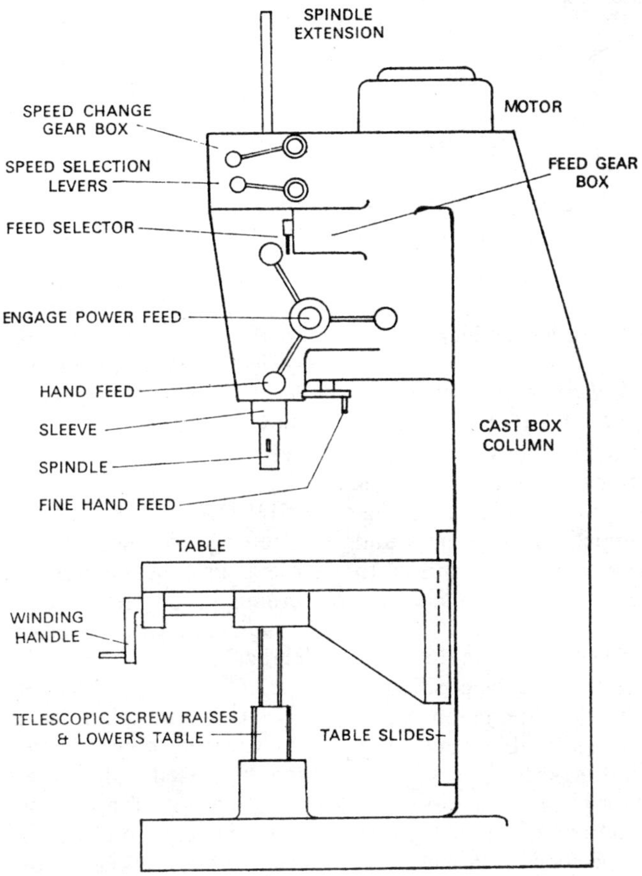

Fig. 9.5. Vertical drilling machine of the box column type. Compare this machine with a sensitive drill or pillar drill for rigidity

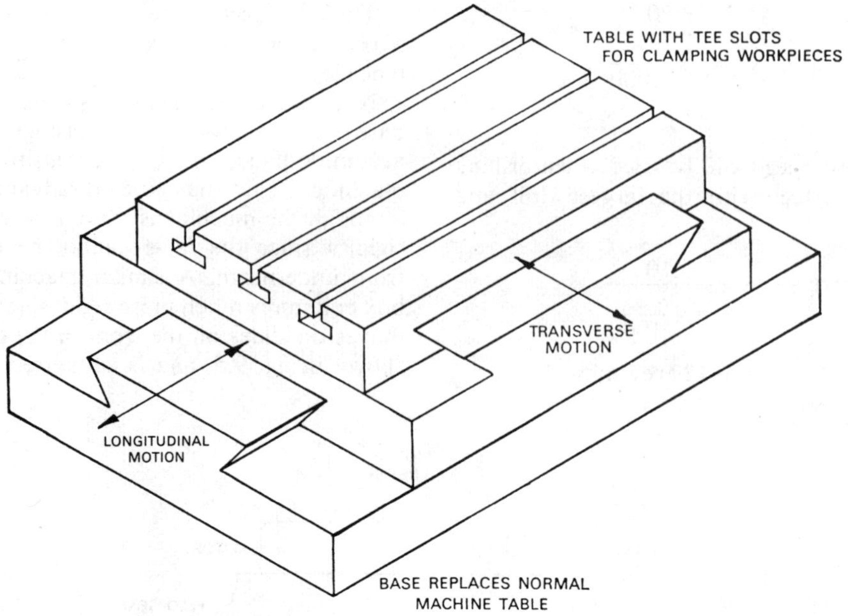

Fig. 9.6. Compound table for drilling machine. This is particularly useful where a large number of holes is to be drilled in a single component. Operating lead screws not shown

point of drill thrust by the table-raising screw. However, the table cannot be moved to position the work under the drill, nor swung clear.

3. *The compound-table drilling machine*
This is a development of the box column machine with a table mounted on two slideways to give horizontal motions at 90° to each other and controlled by lead screws, as shown in fig. 9.6. The work is clamped to the compound table which can then be accurately adjusted to bring each hole centre under the spindle axis in turn for drilling.

If the work is located accurately on the table, by locators in the tee slots, once the position of the first hole has been fixed, the remaining holes can be positioned by means of table movements without recourse to marking-out.

Ideal jobs for this type of machine are plates requiring a large number of holes whose coordinate centres are specified, as shown in fig. 9.7. The method is used where the number of components does not justify making a special jig.

Looking ahead to more advanced work, it is worthy of note that these machines lend themselves to fully-automatic operation by tape control. The operator simply loads the workpiece on the table at the correct location and presses the button. The machine does the rest.

MOVING-AXIS MACHINES

The only common machine of this type is the radial drilling machine, so called because the drilling head is mounted to move on slides along a radial arm which itself can be swung about the end column.

Such a machine is shown in outline, in front elevation and plan, in fig. 9.8, so that the adjustments can be shown. The base of the machine itself is the work table and is provided with tee-slots for clamping large workpieces. An auxiliary table is usually provided to which smaller workpieces can be clamped and in some cases, where utilisation is high, two such tables are used. One is placed on the machine while drilling is in progress and the other is at one

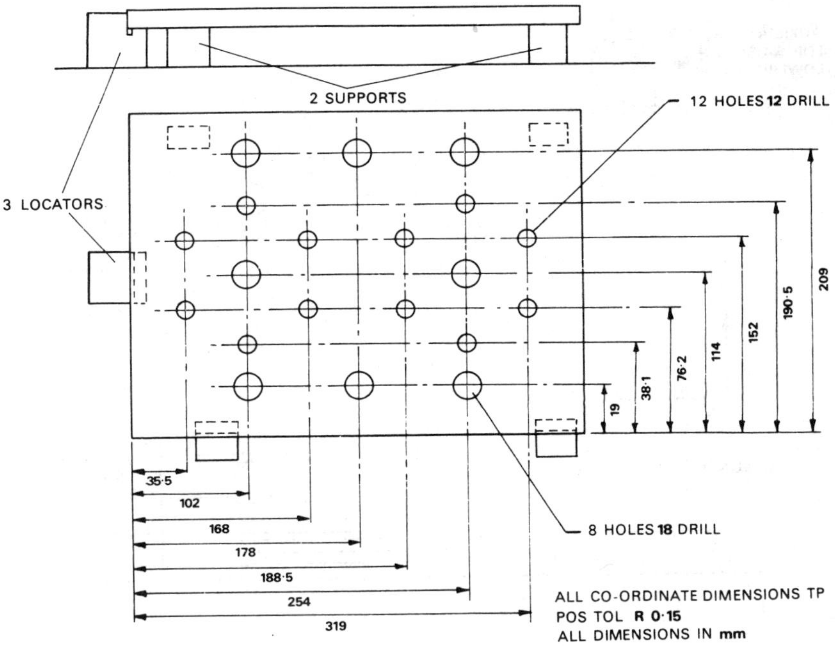

Fig. 9.7. Typical component for drilling on a compound-table machine. The 12 mm holes could have been specified twice as '6 holes ϕ 12 drill equi-spaced on 152.5 PCD' but, for this method of drilling, co-ordinate dimensions are necessary

side, having the previous work removed and new parts positioned. When the work is completed the tables are interchanged by hoist, the radial arm being swung clear for the purpose.

DRILLING OPERATIONS

The obvious purpose of a drilling machine is to drill holes but, in fact, the machine is much more versatile than this and is used for other operations. However, drilling itself is worthy of comment since there are several points to be made clear for correct and safe procedures to be followed.

DRILLING

The machine table is machined flat and accurately mounted square to the axis of rotation and feed. Dirt or burrs will destroy this alignment and the table must therefore be kept clean and in good condition. Great care must be taken never to drill into the table surface.

WORK-HOLDING

The holding of work by hand is dangerous. If the drill 'picks up' the workpiece, a bad cut may result as the workpiece rotates with it. Work should always be held in a vice or clamped to the table. In some cases work is best clamped to an angle plate as shown in fig. 9.9. Note that it should then normally be clamped on the inner face of the plate so that the feed force does not tend to tilt the plate.

DRILLING A HOLE TO A SET DEPTH

Specific Objective: *The student should know how to set a machine to drill holes to a specified depth.*

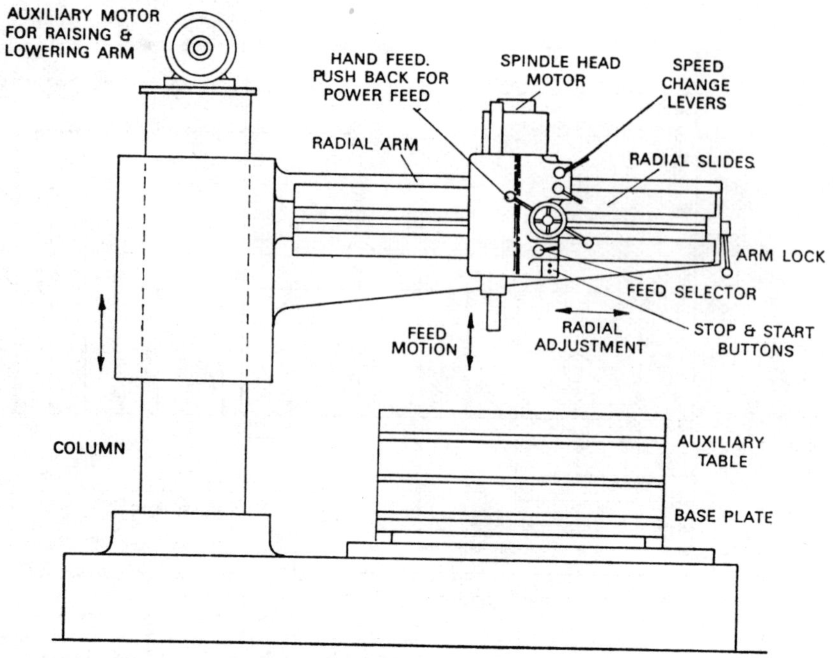

Fig. 9.8. Radial drilling machine: note that all controls are on the spindle head within easy reach of the operator

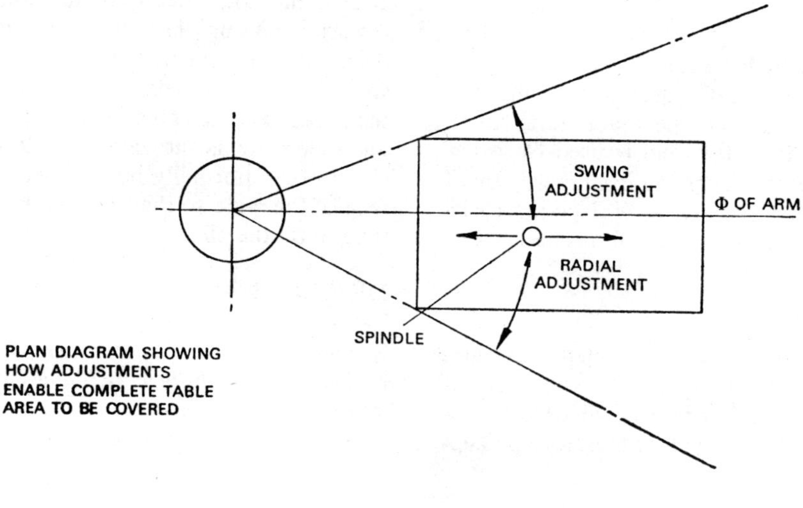

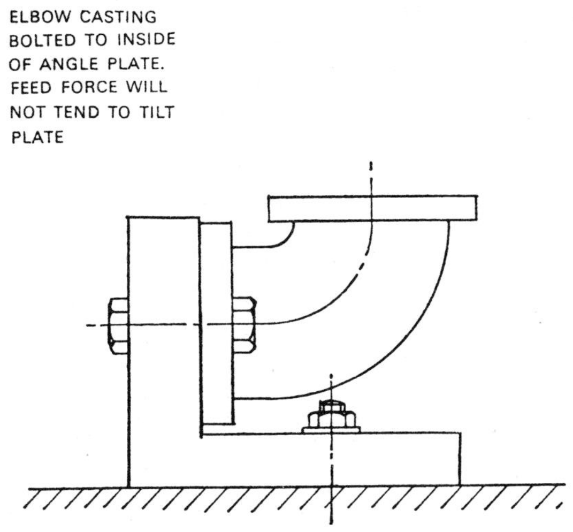

Fig. 9.9. Use of angle plate in holding workpiece for drilling

If a blind hole is to be drilled, some means other than guessing is required to gauge its depth. The drill sleeve motion is usually provided with adjustable stops which can be set in two ways.

1. The stationary drill is brought down lightly on to a pile of grade 2 gauge blocks of the correct height, as shown in fig. 9.10(a), with protective blocks top and bottom, to give the total depth of hole. With the drill in this position, the feed stop is appropriately adjusted and locked.
2. If the depth of full diameter is specified, drilling is started until full diameter is just obtained and the machine is then stopped. The stationary drill is brought down until it contacts the conical depression, and the feed stop is set so that the drill penetrates only to the depth required. This method is shown in fig. 9.10(b).

In both cases the drill should be withdrawn as soon as the stop is reached since, if it is allowed to rub, rapid overheating and wear will result.

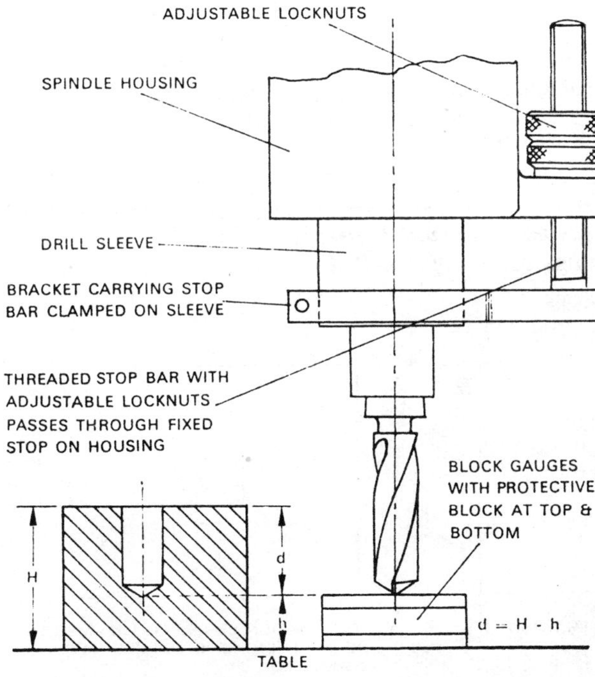

Fig. 9.10(a). Setting total depth of hole by gauge blocks

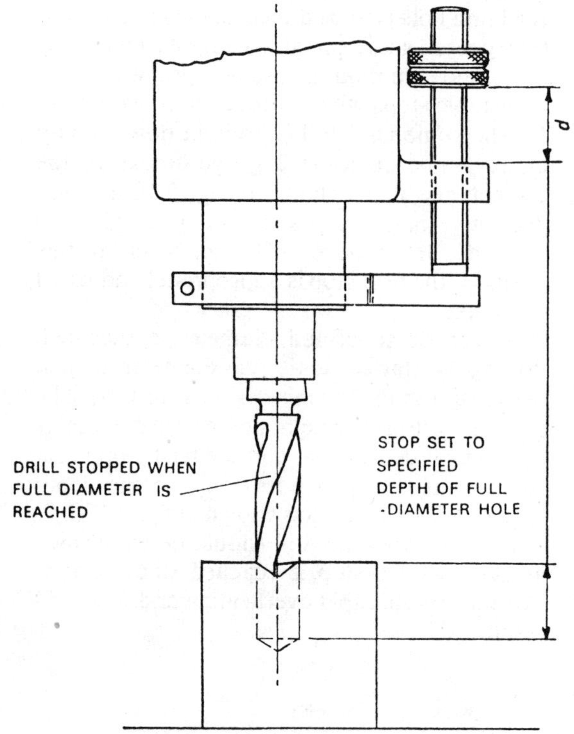

Fig. 9.10(b). Setting depth of full-diameter hole

CORRECTING HOLE POSITIONS DURING DRILLING

Specific Objective: *The student should know how to correct the starting position of a drilled hole.*

In fig. 9.11 the technique of 'boxing' holes when marking them out for drilling is illustrated. However accurately the centre punch mark is made the drill may not position itself quite correctly, which can be detected by comparing the initial cone cut by the drill with the box *before the drill has attained its full diameter.* If the cone is off-centre a groove is chipped down the side of the cone to the apex, as shown in fig. 9.11. This draws the drill in the direction of the groove so that, with care, it is brought back to cut centrally.

Note: This 'drawing' of the drill cannot be done once the drilled cone has reached full diameter.

SPOT-FACING

Specific Objective: *The student should know of operations other than drilling which are carried out on a drilling machine.*

When a bolt hole is drilled in a boss on a casting the boss surface, on which a bolt has to pull down, is rough and does not provide a good seating for the bolt. The boss is therefore cleaned up, or machined flat, by a spot-facing cutter as shown in fig. 9.12(a).

The cutter is an interchangeable blade-fitted in a cutter bar whose nose diameter is slightly less than that of the drilled hole, in which the nose acts as a pilot for the cutter. The blade can also be fitted in the reverse position and used to spot-face an inside face after the bar has been

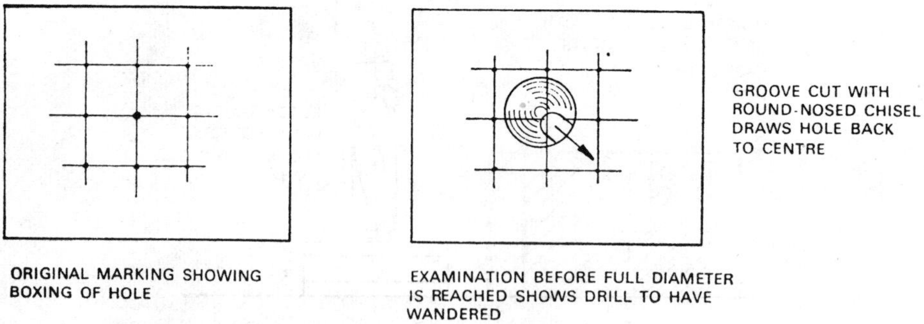

Fig. 9.11. Correcting hole centre position after initial error in starting drill

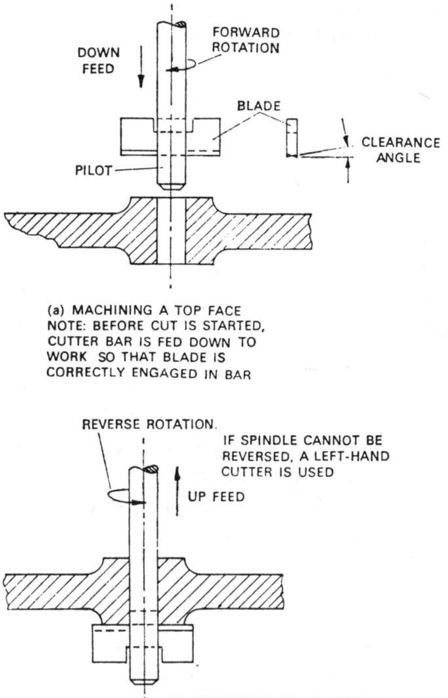

Fig. 9.12. Spot-facing with blade-type spot-facing tool

fed through a previously drilled hole, as shown in fig. 9.12(b).

COUNTERBORING

If a hole has to take a socket-head screw it must be opened out to a depth slightly greater than the depth of the screw head. This operation is performed with an end-cutting tool similar to an end mill but with a pilot to guide the tool and keep it concentric with the hole, as shown in fig. 9.13. Larger counterboring tools are sometimes used for spot-facing.

TAPPING

Unless the machine has special equipment for power tapping, the tapping of screw threads by it is still a hand operation, but the machine alignments are used to ensure that the tap is correctly aligned with the hole. When the hole to be tapped has been drilled, a centre is fitted in the drill spindle. The tap is entered into the hole and the spindle is fed down until the centre engages with the centre hole in the shank of the tap. The tap is then turned by hand in the normal way to produce the thread and is kept square to the work by maintaining the contact of the centre.

Many sensitive drilling machines have the spindle weight counterbalanced by spring loading. If the spring loading is disconnected, the spindle will follow and support the tap by its own weight, leaving both hands of the operator free for the tapping operation. The counter-balance must be reconnected when the tapping has been completed.

REAMING

Reaming may be performed on a drilling machine with a machine reamer. A very slow speed must be used, with a plentiful supply of cutting fluid, mainly to wash away the swarf which might mar the surface finish if allowed to build up in the hole.

DRILLING PLASTIC MATERIALS

Plastic materials may be drilled on standard drilling machines using high speed steel drills which have been ground in the flutes to remove the positive rake angle at the cutting edge. The standard 118° point angle is suitable for most materials but in nylon a better breakout is

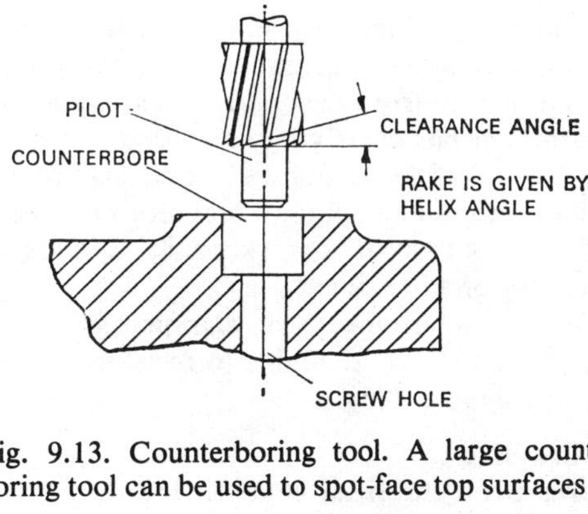

Fig. 9.13. Counterboring tool. A large counter-boring tool can be used to spot-face top surfaces

obtained if the angle is reduced to as little as 80° if the material is thick enough. In general when drilling thin material the point angle should be large enough to ensure that the full drill diameter is engaged before the point breaks through.

Two problems are peculiar to plastics, they are poor conductors of heat and are sensitive to heat. Thus coolants are required to carry away the heat which will not be conducted away and the method chosen should not allow too much heat to be generated. This calls for a high speed with a fine feed and frequent clearing of the drill. This is called a 'woodpecker' method. This also prevents clogging which also increases the heat generated.

Thermoplastic materials are particularly heat sensitive and the powder produced may fuse into a solid if allowed to overheat.

SAFETY NOTES

Many incidents occur on drilling machines but the machines themselves are rarely to blame. The fault usually lies with the operator, but the causes of the accidents are easily enumerated and knowledge of the dangers enables them to be avoided.

1. Work picked up and revolving with the drill.
The drilling machine is one of only two machines in general use (the other is the offhand grinder) on which the operator can work without clamping or fixing the workpiece. No one thinks of holding work by hand in a shaping or milling machine, yet similar cutting forces are at work in a drilling machine. If work picks up and is torn from the operator's grasp at 360 rev/min, he has one-sixth of a second to remove his hand before it comes round again and strikes him. Since human reaction times are of the order of two-fifths of a second, the odds are in favour of an accident.

Work should always be clamped, or a stop should be fixed on the table to resist the torque set up by the drilling.

2. Scalping
Never bend the head close to the drill to see what is happening, or blow away swarf. The rotating spindle can pick up hair and tear away a part of the scalp with very painful results. When the drill is held in a chuck, the chuck should be covered by a guard attached to the spindle sleeve so that the guard screens all but the drill itself. This reduces the danger but does not eliminate it. Keep your head out of the way.

3. Swarf cuts
A well-ground drill will produce two equal-length spirals of swarf, one from each flute, which rotate with it and tend to fly outwards under centrifugal force. If these become too long and lash the back of the hand, bad cuts can result. Before the swarf streamers become too long, ease the feed so that they break off. Never try to break them off with a brush or piece of wire.

4. Coolant brushes
These are an abomination, for they can easily be picked up by the drill and cause an accident. If no coolant feed is available use a pressure oil-can to squirt the coolant into the work area.

Like any other machine tool, the drilling machine is quite safe as long as the operator is its master. It can, however, exact swift retribution from the careless and the foolish.

QUESTIONS

1. Explain why it is preferable for small drills to be held in a chuck and larger drills to be held directly in the spindle nose taper.
2. Make a block diagram of a sensitive drilling machine and identify clearly:
   (a) table and explain how it is adjusted on the:
   (b) column
   (c) spindle
   (d) speed adjustment.
3. Explain with the aid of diagrams two methods of setting a machine for drilling holes to a predetermined depth. State which you consider to be the more accurate.
4. List three hazards associated with the use of

drilling machines and state what steps should be taken to avoid accidents happening due to each cause.

5. State the precautions to be taken when drilling plastic materials. Explain what is meant by 'woodpecker' drilling. Why is it particularly used on plastics.

6. Show with the aid of a diagram how a sensitive drilling machine may be used to align a tap correctly although the tapping operation is carried out by hand.

7. Name the tools used to provide the following conditions on a component:

(a) Clean seating on a cast surface for a bolt head to pull down on.

(b) Recess to take an 'Allen' screw head.

In each case sketch the tool to be used to perform the operation.

10 The shaping machine

General Objective: *The student should be able to compile block diagrams of the commonly used shaping machines and describe typical shaping operations.*

THE GENERATING PROCESS IN SHAPING

The shaping machine may be defined as a machine tool designed to produce flat surfaces by a generating process which uses a reciprocating single-point tool. These surfaces may be horizontal, vertical or at some angle between the two.

The essential motions for the machine to fulfil this function are as shown in fig. 10.1 and are as follows.

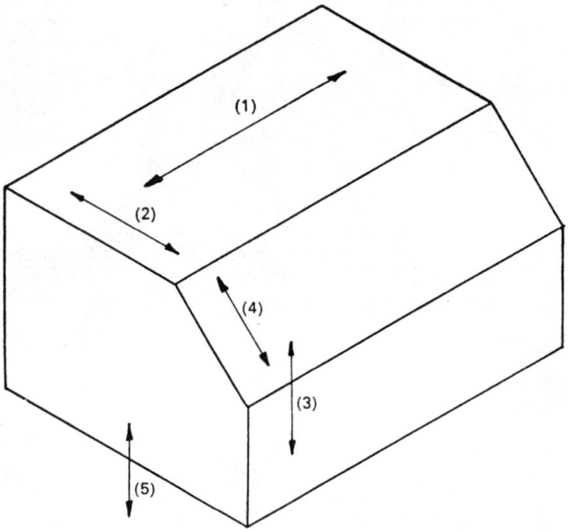

Fig. 10.1. Basic motions of tool and work on a shaping machine

1. RECIPROCATING MOTION OF TOOL

This motion supplies the power for cutting and must be adjustable for speed, length of stroke and position of stroke over the workpiece.

2. FEED MOTION FOR HORIZONTAL SURFACES

This is the motion of the table across the front of the machine, which may be hand or power fed.

3. FEED MOTION FOR VERTICAL SURFACES

This is normally a hand-fed motion of the toolpost in a head slide mounted on the front of the ram head, but it may be power fed.

4. FEED MOTION FOR INCLINED SURFACES

This is a variant of 3 above; the head slide is set over at the angle of inclination required and the tool feeds down at this angle.

5. VERTICAL TABLE ADJUSTMENT

This is an additional motion not normally used for feed purposes but to adjust the height of the table to accommodate wide variations in the height of the work.

The flat horizontal surfaces generated by the machine must normally be parallel with opposite surfaces, and vertical surfaces must normally be at right angles to adjacent horizontal surfaces. If these conditions are to be realised, other requirements in the alignments of the basic motions must be satisfied. For flat, parallel surfaces, motions (1) and (2) must be straight and parallel to the table surface. If the vertical surfaces are to be square then motion (3) must be at right angles to the table surface.

This statement presumes that the work-holding methods adopted ensure that the underside of the work rests on or is parallel to the surface of the table, and we shall discuss how this is ensured in a later section. Basically, the motions and alignments listed are built into the machine.

As was emphasised in Chapter 6 a generating process requires two motions, the cutting motion and the feed motion, with all other adjustments locked while the cutting proceeds. In the case of horizontal surfaces, motions (1) and (2) produce the surface, and motion (3) provides adjustment for depth of cut. Vertical surfaces are produced by motions (1) and (3), and motion (2) provides adjustment for depth of cut. Inclined surfaces use motion (1) and motion (3) inclined to give motion (4), and the depth of cut is again adjusted by motion (2).

Fig. 10.2 shows how these alignments are produced on the actual machine. The ram, which carries the head slide at the front, moves forwards and backwards in long slideways which are at right angles to the vertical slide in the front of the machine. The head slide gives the vertical tool feed for motion (3) and is mounted on a rotary head so that it can be inclined to give motion (4). The table is mounted on a horizontal feed slide, which in turn is mounted across the main vertical slide, so that the table surface is parallel to the ram slide in both directions. The horizontal feed motion (2) is provided by moving the table across the horizontal slide, which is parallel to its own upper surface (this requirement for parallelism is also found in the milling machine; see Chapter 8).

Finally, the vertical adjustment of the table, motion (5), is provided by the vertical slide but this is an adjustment, not normally used for feeding.

TESTING ALIGNMENTS

The shaping machine is a notable example of the advantage of checking the machine alignments before starting to use the machine, rather than checking machined work and finding too late that it is in error. Two of the important alignments for parallelism and

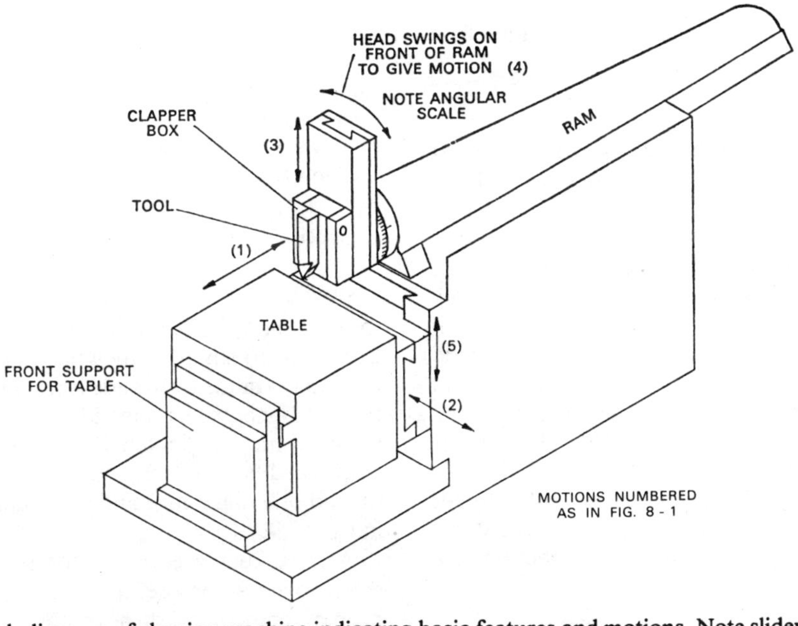

Fig. 10.2. Block diagram of shaping machine indicating basic features and motions. Note slideways provided for motions

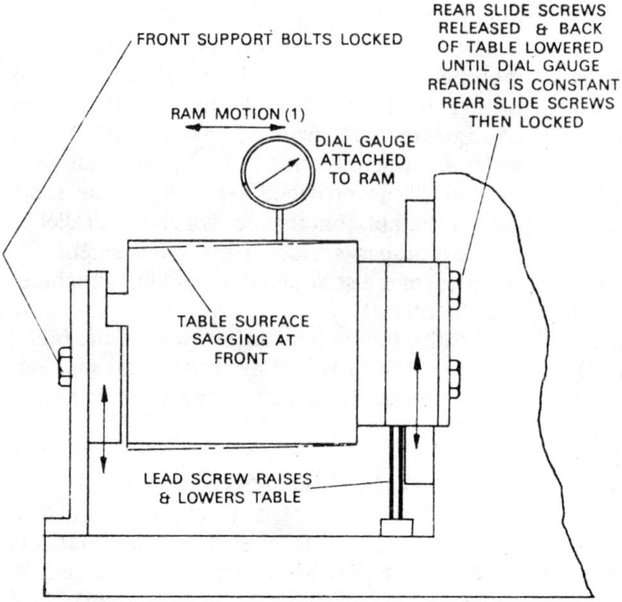

Fig. 10.3. Checking and correcting table sag after height adjustment

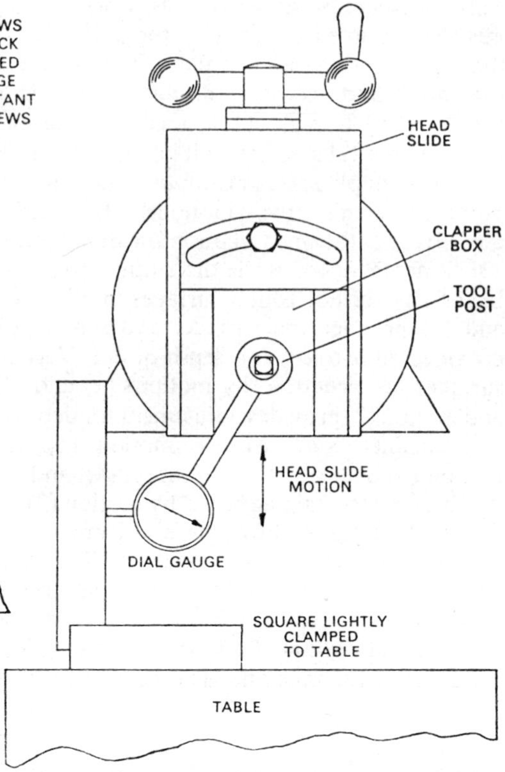

Fig. 10.4. Motion of head slide for squareness to table. The dial gauge reading will remain constant only when the tool-slide motion is perpendicular to the table surface

Note: A cylindrical square may be used instead of an engineer's square as shown

squareness can be adjusted and may therefore be incorrect.

(a) PARALLELISM

The cross-slide motion of the table cannot normally be adjusted, but the work will only be parallel in a fore-and-aft direction if the table surface is parallel to the ram motion. Through wear in the front vertical slide and the weight of the table, there is a tendency for the table to sag at the front every time the vertical adjustment (5) is used. The amount of sag can be checked by mounting a dial gauge in the tool post and noting the reading as it is traversed over the table. If the front is low the bolts on the front table support should be locked and the rear of the table lowered by the vertical adjustment until the reading on the dial gauge is constant. This alignment error and its correction are shown in fig. 10.3.

(b) SQUARENESS

When adjacent horizontal and vertical faces are to be shaped at one setting, the vertical face will be machined square only when the vertical feed motion (3) of the head slide is truly square to the table surface. When this slide has been 'set over' to machine an inclined surface, it can be readjusted to the perpendicular position with the aid of a dial gauge and try square, as shown in fig. 10.4.

A mistake sometimes made is to run a dial gauge over a surface, in position, after it has been machined and then wonder why, although the dial guage reading is constant, the surface is later found to be not parallel or square to other surfaces previously machined. The offending surface was, of course, parallel with the tool motion, and since the dial gauge had the same motion as the tool this 'test' shows nothing except that the operator does not realise or know what he is doing. Too many people believe that a dial gauge has magic powers and give the advice 'Clock it up' with complete abandon and disregard for what may be involved.

THE DRIVE MECHANISM OF THE SHAPING MACHINE

Specific Objective: *The student should know how to adjust the length and position of the stroke of a shaping machine and appreciate the need for a quick return motion.*

For its reciprocating motion the ram of the shaping machine must have:

(a) a speed adjustment to cater for different materials and lengths of stroke;
(b) an adjustment for length of stroke;
(c) an adjustment for position of stroke;
(d) a quick-return action.

(b) and (c) are necessary to reduce the time spent in 'cutting air'. The tool does no useful work on the return stroke and the drive is arranged so that the return stroke is quicker than the forward stroke. The quick-return mechanism is incorporated in the drive, which is shown in fig. 10.5.

(a) SPEED OF STROKE

Common with many reciprocating motions, the linear ram motion is obtained from the rotary motion of a driving wheel or *bull wheel* which rotates once per stroke. The number of strokes per minute therefore depends on the revolutions per minute of the bull wheel, which are controlled by a gear box driven from the motor. It follows that for a given number of strokes per minute the cutting speed will change with the length of stroke which must be allowed for in calculating cutting speeds (see p. 231).

(b) LENGTH OF STROKE

The drive from the bull wheel to the ram is by means of a *slotted link*, as shown in fig. 10.5. A driving pin in the bull wheel engages a block which slides in the slotted link so that, as the wheel rotates, the link is caused to rock about its pivot in the base of the machine. The top end of the slotted link moves forwards and backwards in an arc and thus cannot be connected directly to the ram, so a *compensating link* is introduced. To alter the length of stroke, the radius at which the pin on the bull wheel rotates must be adjustable from outside the machine. This is achieved by mounting the pin on an inner block which fits in a radial slide in the bull wheel and is moved along the slide by a leadscrew. The leadscrew is rotated by a pair of bevel gears, as shown, and the spindle which operates them passes through the centre of the bull wheel spindle to the outside of the machine where it can be adjusted by a detachable handle.

SAFETY NOTE

Always ensure that this stroke-adjusting handle is removed before starting the machine. It will rotate at the same rate as the strokes of the ram and can do considerable damage if left on its spindle – and possibly more if it flies off.

(c) ADJUSTING THE POSITION OF THE STROKE

The length of the stroke is adjusted to the length of cut required but it is also necessary to adjust the position of the cut to suit the position of the workpiece, i.e. to adjust the position of the stroke with respect to the table. The ram carries a longitudinal slot through which a stud passes, screwed into a block beneath and with a locking handle on top, as in fig. 10.5. This block is connected through the compensating link to the end of the slotted link of the main drive. When the locking handle is tightened, the ram is secured firmly to the block and thus moves with the end of the slotted link.

When the locking handle is freed and the

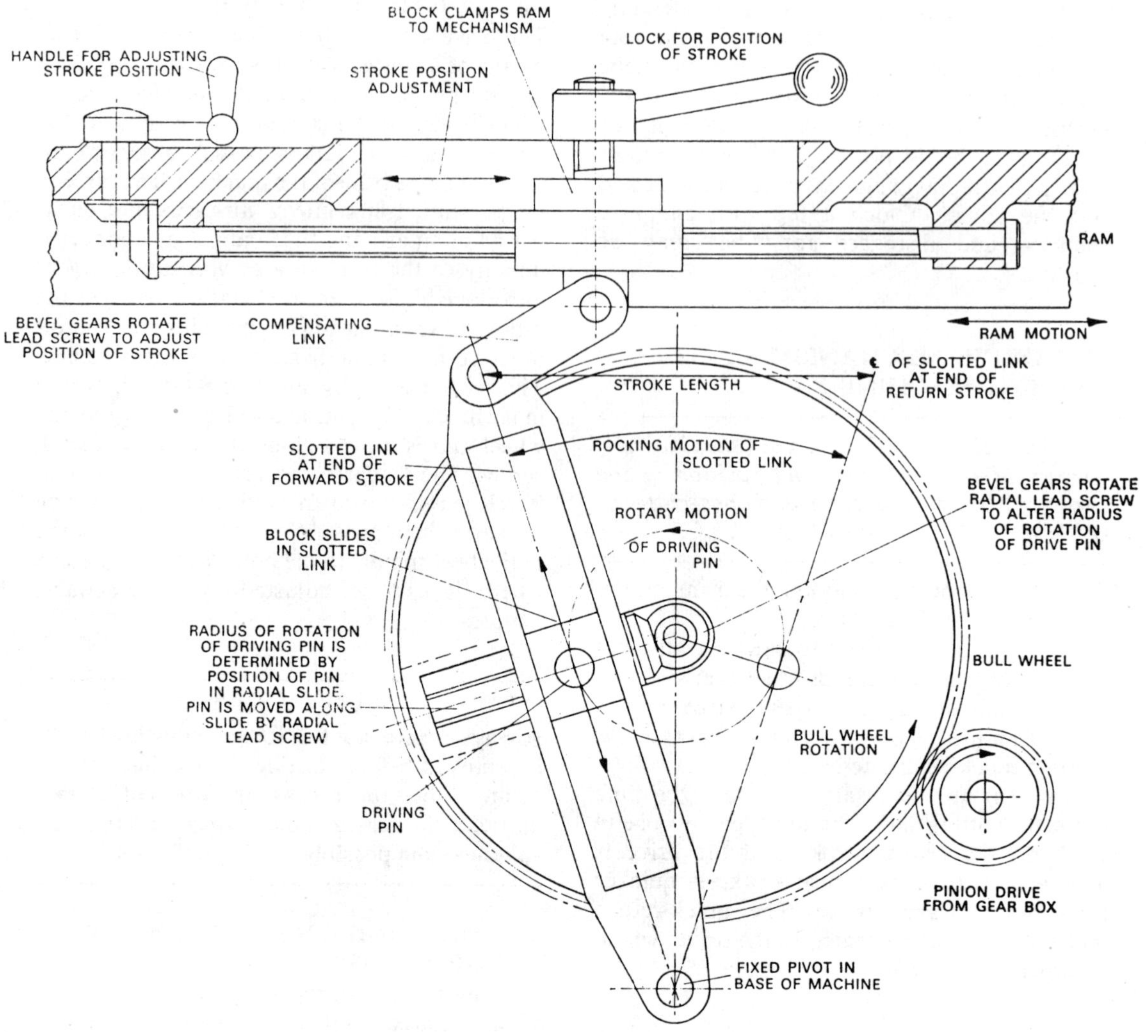

Fig. 10.5. Driving mechanism for shaping machine, showing slotted link quick-return motion

block is released, the block can be moved to a new position in the ram slot by a leadscrew within the ram, which is operated through a pair of bevel gears driven by a rotating handle on top of the ram. This allows the ram position to be adjusted as required. After the adjustment, the block is re-locked.

(d) THE QUICK-RETURN ACTION

In order to save time on the return stroke, when no work is being done, the drive mechanism provides a quick-return action, the geometry of which is shown in fig. 10.6(a). The driving pin on the bull wheel moves at a constant speed but, because the slotted link is pivoted at the bottom end, the pin travels through a larger angle *A* on the forward stroke than angle *B* on the return stroke. Since the pin is moving at constant speed it thus takes longer to traverse the forward stroke than the return stroke and, in

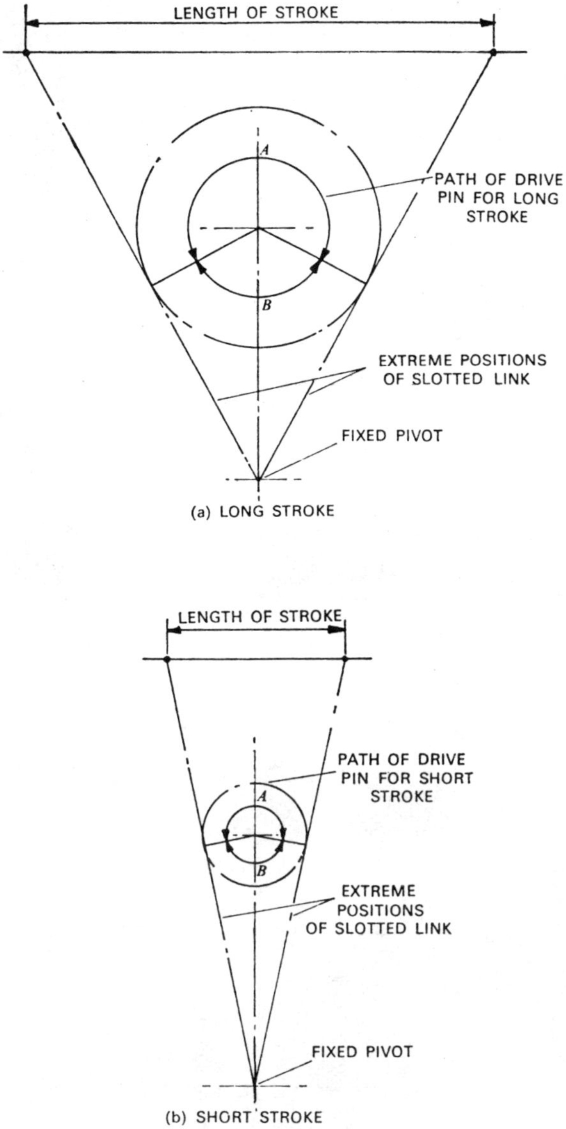

Fig. 10.6. Geometry of quick-return motion

fact the ratio

$$\frac{\text{time for cutting stroke}}{\text{time for return stroke}}=\frac{\text{angle for cutting}}{\text{angle for return}}=\frac{A}{B}$$

If we examine the conditions for a shorter stroke, shown in fig. 10.6(b), we see that these angles, *A* and *B*, are more nearly equal and therefore as the length of the stroke is reduced the quick-return action is much less marked.

It is of interest to note that the Butler Machine Tool Co., Ltd, will provide a perspex cover plate for the drive of a shaping machine, with interior illumination, which enables the details of the drive mechanism to be seen clearly.

THE FEED MECHANISM OF THE SHAPING MACHINE

Specific Objective: *The student should know how to vary the table feed on a shaping machine.*

Automatic feed is provided to the table cross-slide of a shaping machine, since this is usually the motion required to feed across the greatest width of work face. Unlike that of a lathe, where a continuous feed is used, the shaper feed must be intermittent, occurring during each return stroke of the tool, and it is therefore necessary to turn the cross-slide leadscrew a part of a turn on each return stroke. If we require a feed of 0.1 mm per stroke and the table cross-slide leadscrew has a pitch of 5.0 mm, the screw must rotate $\frac{0.1}{5.0}=\frac{1}{50}$ revolution per stroke. If a toothed wheel of 50 teeth is fitted to the end of the leadscrew, a rotation of the amount of one tooth will give the required feed of 0.1 mm per stroke, rotation to the extent of two teeth will give a feed of 0.2 mm per stroke and so on.

This feed of a few teeth at a time can be arranged by mounting a ratchet pawl on the toothed wheel in a bracket which rocks about the leadscrew axis, the rocking motion being provided by a connecting link attached to a crankpin on a disc which rotates at the same speed as the bull wheel. This arrangement is shown in fig. 10.7 and it will be noted that the crankpin radius can be adjusted to give a greater or lesser amount of rock to vary the feed as required.

The ratchet pawl is square on one side and bevelled on the other, so that it carries the

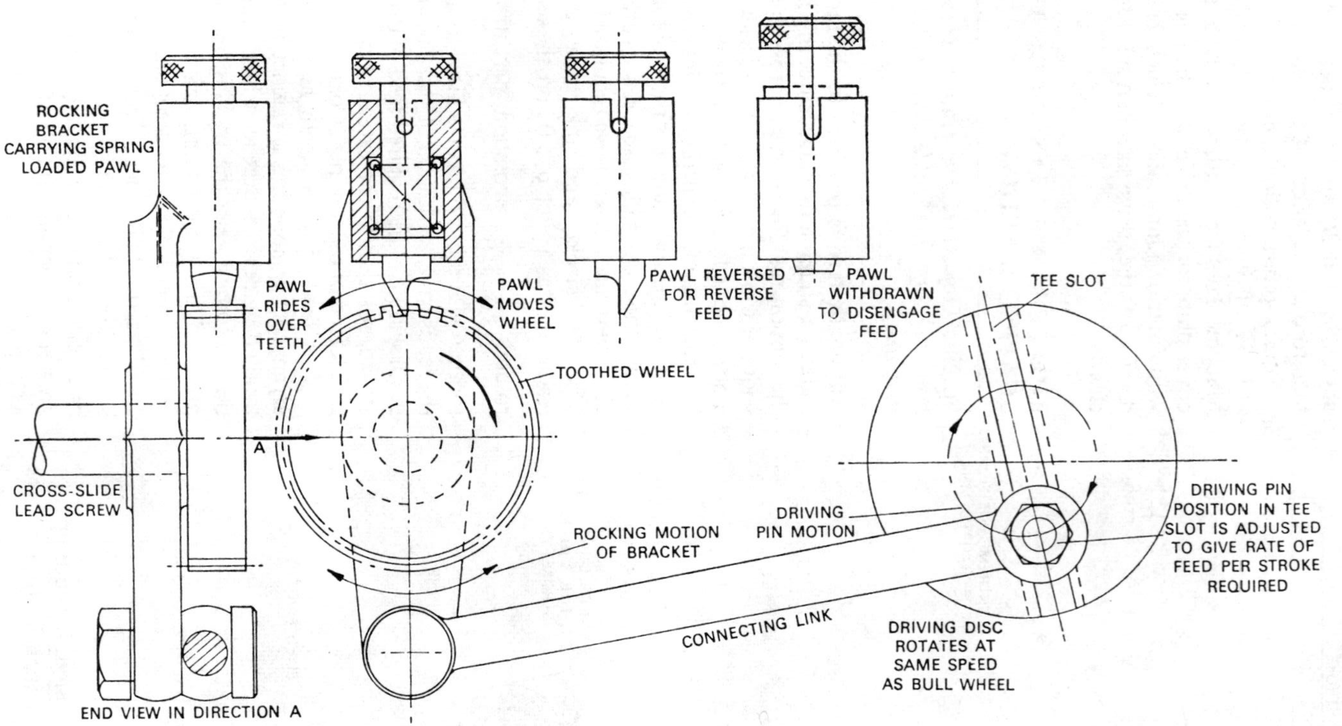

Fig. 10.7. Feed mechanism for shaping machine

toothed wheel as it rocks in one direction but rides over the tooth as it rocks in the other direction. Thus feed can be obtained in either direction by turning the pawl through 180°, or can be disengaged by withdrawing the pawl so that the cross-pin rests on top of the bracket and lifts the pawl clear from the toothed wheel.

TOOL MOUNTING IN THE SHAPING MACHINE

Specific Objective: *The student should appreciate the need for a clapper box and know how to set it for a given operation.*

The shaping machine is one of the few machine tools in common use which is based on a reciprocating action, others being the slotting machine, the punch shaping machine, the gear shaping machine and the planing machine. In all these machines the tool would rub on the work during the return stroke and lose its edge if clamped rigidly. On the shaping machine, rubbing of the tool is avoided by use of the clapper box. On the front of the head slide is a *back block* which is mounted on a pivot screw near the bottom and fastened at the top by a screw which passes through a quadrant slot, as shown in fig. 10.8. This enables the back block to be swung at an angle on either side of the vertical. The *clapper block* is fitted in front of the back block between two side cheeks to which it is attached by a cross pin, or pivot pin, near the top, thus forming a *clapper box* in which the clapper block can swivel. The *toolpost*, which carries the cutting tool, is attached to the front of the clapper block.

On the cutting stroke, the clapper block sits firmly on the back block in its box, but on the return stroke, the tool, touching the workpiece, is lifted slightly, the clapper block being swung about its pivot pin. When the tool drops clear of the work the clapper block falls back into its box with a clapping sound, which gives the box its name. Some modern machines have a tool lifting device operated by a Bowden cable, which clears the tool on the return stroke so that neither the finished surface nor the tool is damaged by its falling on the back stroke.

As stated above, the clapper box can be swung about its pivot screw to either side of the line of the tool slide and it is this facility which enables vertical and inclined surfaces to be machined. *The top of the clapper box is always swung away from the surface being machined*, as shown in the various diagrams of fig. 10.9. An examination of the geometry shows that this swings the tool clear of the surface on the return

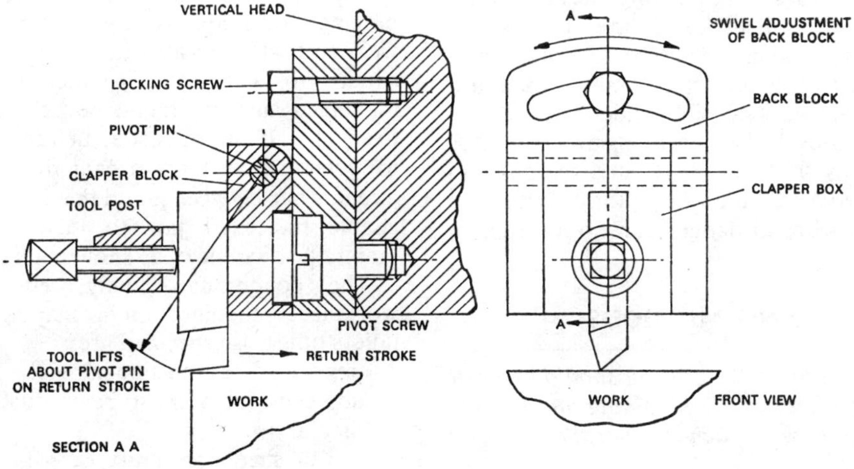

Fig. 10.8. General arrangement of clapper box

stroke instead of allowing it to dig in, as it would if the box was not correctly adjusted.

In fig. 10.9(a), which shows the clapper box set for cutting a vertical surface, line AB on the set-up is the radius from the centre line of the pivot about which the tool point swings. The action is shown at the left-hand side of the drawing as a geometric layout, and the end view of the geometry shows the line AB in the position AB_1 as the tool swings up on the back stroke. Projecting this position back to give the point B_1 in the front view reveals the clearance of the tool from the surface. Fig. 10.9(b) shows a corresponding set-up and geometry for the cutting of an inclined surface.

THE SWAN-NECKED SHAPER TOOL

Specific Objective: *The student should be able to explain the reason for the use of a swan-necked tool on shaping and planing machines.*

The clapper box is a device to prevent the tool rubbing into the work on the back stroke. The swan-necked tool, shown in fig. 10.10, is sometimes used to prevent digging-in and chatter on the forward stroke. If the cutting load becomes excessive the tool bends slightly about point A and a straight tool bending about this point cuts deeper and increases the cutting load. This causes more bending, deeper cutting and increased load, and the process is progressive, making conditions worse until something gives or the depth of cut becomes too great and the work possibly undersize.

The swan-necked tool bending about point A moves away from the work, and the cut and load are reduced until an equilibrium situation is reached with no danger and no increase in depth of cut.

WORK-HOLDING METHODS

Specific Objective: *The student should know of different methods of work-holding and be able to explain how to achieve parallelism and squareness.*

Large workpieces may be clamped directly to the machine table if their shape lends itself to this method, with clamp bolts fitted in the table tee-slots. The clamps should always be clear of the surface to be machined so that a complete cut can be taken over it: never clamp a workpiece so that the cut has to be stopped and the clamp removed to allow the cut to proceed. Incidentally it is surprising how many students suggest this during an examination, though they probably would never consider it if confronted with a shaping machine and the workpiece.

Suitably-shaped parts can be clamped or bolted to an angle plate which is itself bolted to the table.

THE MACHINE VICE

The commonest method of work-holding is probably by the machine vice, as shown in fig. 10.11. The machine vice is a piece of precision equipment which maintains the machine alignments and should be treated with care. It usually locates by a tenon in one of the table tee-slots and is mounted on a rotary base, graduated in degrees, so that it can be swung to any angular position, between jaws parallel to the stroke and jaws at right angles to the stroke. When the vice is in good condition, the fixed jaw is set accurately at right angles to the table and will remain so if the vice is used carefully. The moving jaw may lose its alignment through wear so that when a surface is to be machined square to a previously-machined one, the machined surface should be held against the fixed jaw. If the opposite surface is rough or irregular it may tend to cant the work and a circular bar between it and the moving jaw will enable the fixed jaw to take control and maintain squareness, as shown in fig. 10.12.

The conditions usually required of a workpiece machined on a shaping machine, unless otherwise specified are:

1. opposite faces must be parallel;
2. adjacent faces must be mutually perpendicular;
3. the finished part must be within specified limits of size.

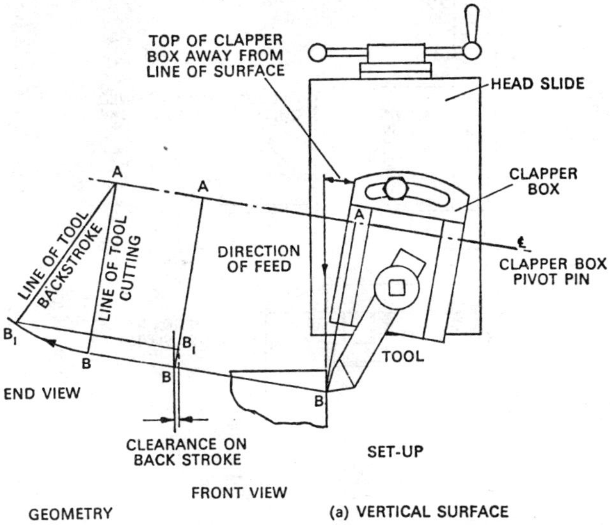

Fig. 10.9(a). Clapper box inclination and its geometry

Fig. 10.9(b)

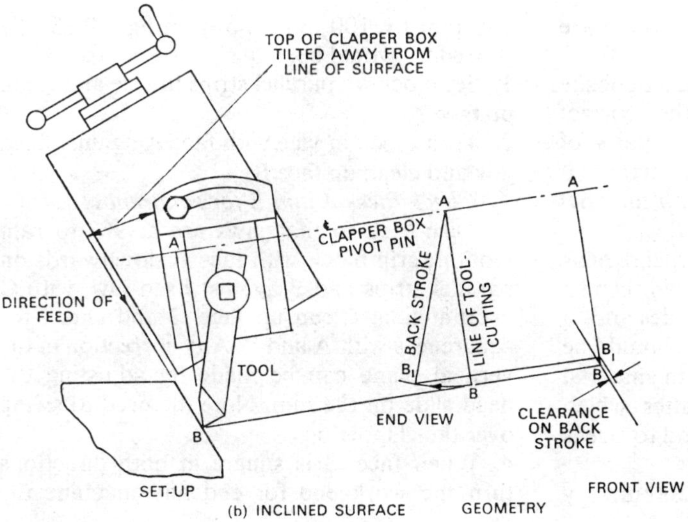

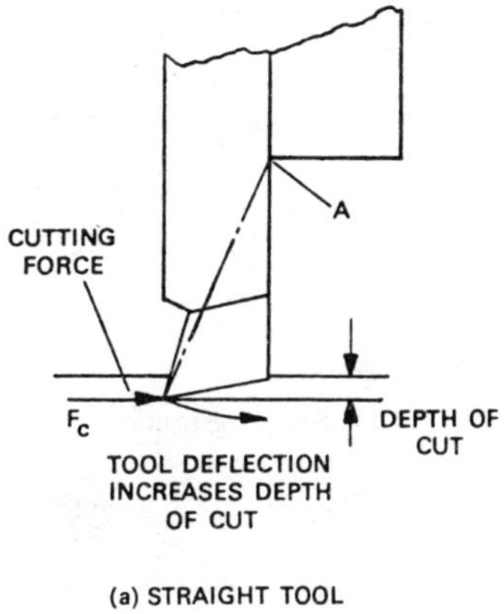

Fig. 10.10. Effect of tool deflection in straight and swan-necked tools

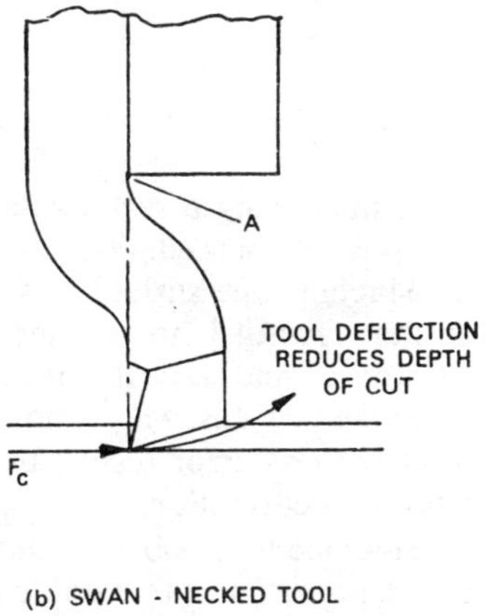

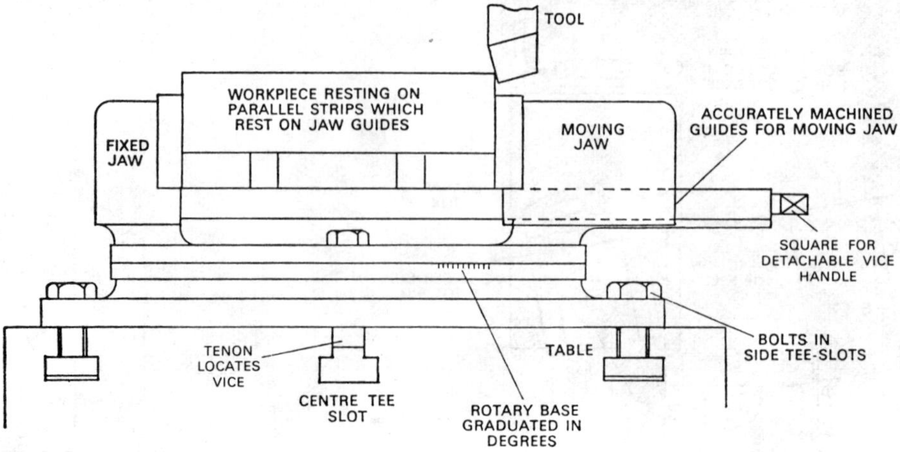

Fig. 10.11. Shaping machine vice

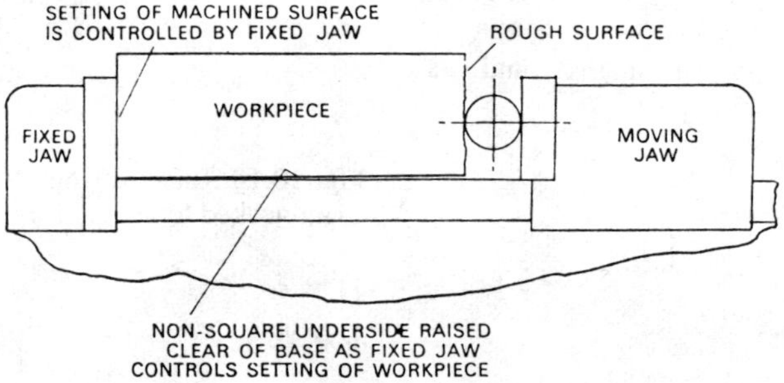

Fig. 10.12. Use of round bar to obtain a square setting in machine vice

In machining a rectangular block there are two possible procedures.

1. Machine one surface, machine the opposite surface parallel to it and to the correct thickness, and repeat for all three pairs of opposite surfaces. This is not a good method; if a squareness error results there is no material left for its correction.
2. Machine three sides mutually perpendicular to each other. The remaining three can then be machined parallel to their opposite sides and to the correct size. This method should be adopted, since it provides maximum material for obtaining initial squareness, after which machining the other sides parallel and to size is relatively easy.

Thus, for a block to be finished 200 mm by 150 mm by 100 mm, shown in fig. 10.13, the procedure would be:

1. Set block on parallel strips in vice and clean up face A.
2. Turn block in vice with face A against fixed jaw and clean up face B.

Check faces A and B for squareness

3. Turn vice so that jaws are at 90° to ram motion; grip block with face A downwards on parallel strips and B against fixed jaw, with C overhanging. Clean up face C and check for squareness with A and B. Any correction in the vertical plane can be made by adjusting the head slide on the ram. Note the need to swing over the clapper box.
4. When face C is square in both directions turn the work end for end and machine the

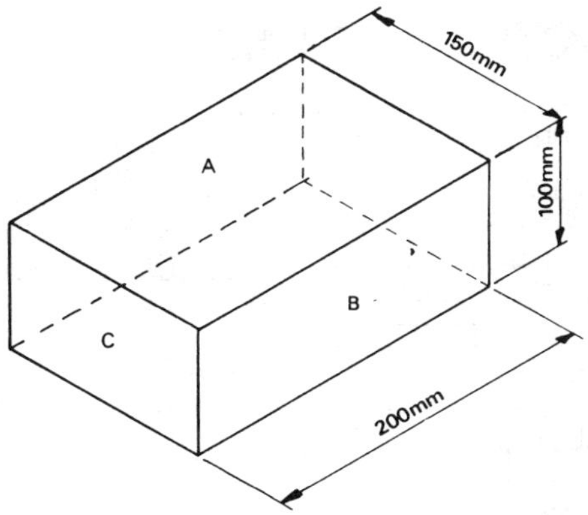

Fig. 10.13. Rectangular block to be machined all over to size and to be square and parallel

other end of the length of 200 mm.

5. Without removing work from vice, turn vice through 90° and machine top face parallel to face A and to 100 mm size.

6. Turn work so that face B is downwards on parallel strips and machine its opposite face to the 150 mm size, checking parallelism.

Note that the workpiece is held on parallel strips and if it is lightly tapped down until the strips are securely held between the vice base and the underside of the work, a parallel workpiece should result, unless the machine alignments are incorrect. A method of correcting error in the alignment for parallelism was given on page 222.

A method of ensuring that the work is held down tightly on the parallel strips is to use hold-down strips, as shown in fig. 10.14(a). These are wedge-shaped strips of metal, resting on parallel strips, between the vice jaws and the work. The angle A is greater than 90° so that the thrust of the jaw is directed obliquely downwards along the line between the points of contact at the work and the jaw. This force, being inclined, has a vertical component which pushes the work down on to the parallel strips, and a horizontal component which grips the work. The actual clamping force is no less than that with a full vice jaw since the frictional force, which resists the tool cutting force, does not depend on the area of contact.

If the workpiece is to be stepped as shown in fig. 10.15, parallelism of the vertical faces can be achieved by attaching a dial gauge to the toolpost and running it along the fixed jaw of the vice, adjusting the setting of the vice until a constant dial gauge reading is obtained.

A component which has to have a face machined at a compound angle is shown in fig. 10.16. The compound angle can be produced by settting the vice at 15° from its square alignment and the head slide to 30°, with the clapper box tilted correctly away from the surface to be machined.

When practicable, the vice should be used with its jaws along the line of stroke. This generally allows a longer stroke to be used which is more economic than a greater number of short strokes because, as already shown, the quick-return action is more marked with a long stroke and, additionally, less time is spent 'cutting air'. Apart from this, the layout of the machine demands that when the jaws are across the line of stroke the cutting force is generally taken by the moving jaw and leadscrew of the vice. This should normally be avoided on any machine tool vice, if the jaw alignments are to be maintained. It must again be emphasised that a machine vice is a piece of precision equipment and should be treated with care for its accuracy to be maintained.

SHAPING MACHINE SPEEDS AND FEEDS

The cutting speed on a shaping machine should normally be less than the figures quoted in Chapter 9, for two reasons. Firstly, the cut is an interrupted one and the tool undergoes a shock load each time the cut is taken. Secondly, the ram does not move at a constant speed but reaches maximum speed at the mid-point of each stroke, the speed being zero at the beginning and end of the stroke. Thus the maximum speed reached by the tool is much greater than the average speed, on which calculations are based. Another factor to consider is the quick-return action which, as

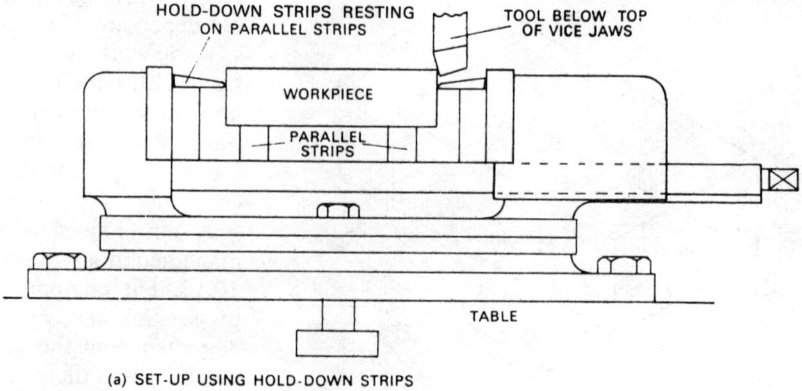

Fig. 10.14(a). Use of hold-down strips

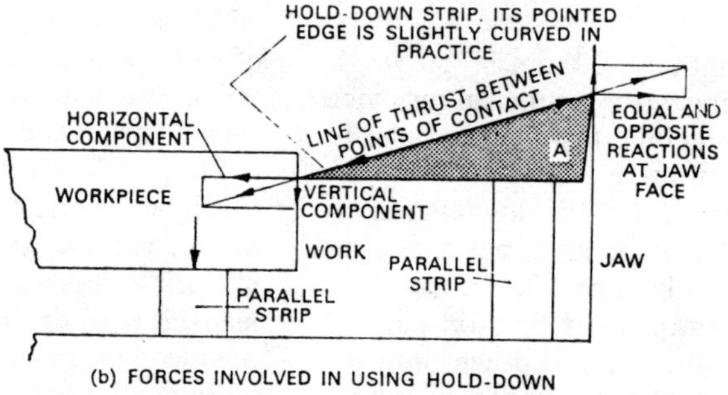

Fig. 10.14(b)

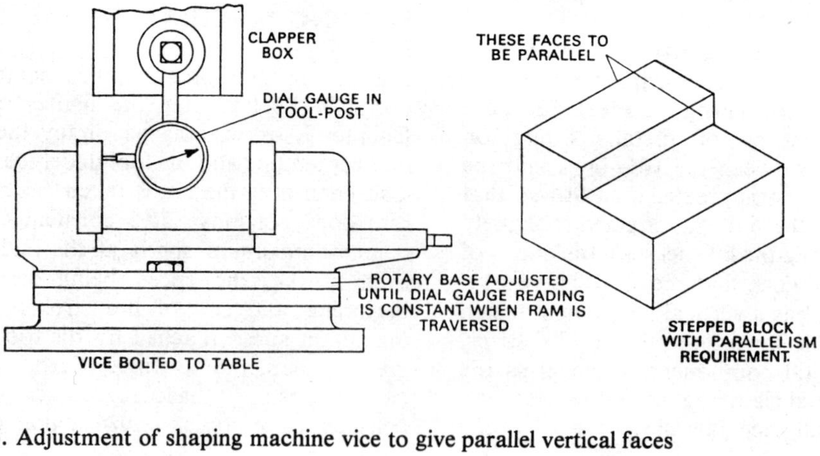

Fig. 10.15. Adjustment of shaping machine vice to give parallel vertical faces

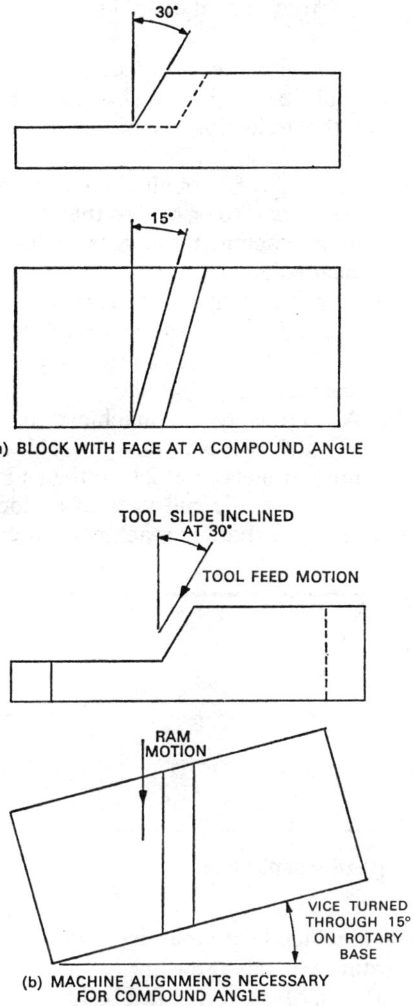

Fig. 10.16. Shaping a face at a compound angle

previously shown, is much more marked for a long stroke than a short one.

Bearing these factors in mind, consider a machine which is roughing a mild steel workpiece. Assume that the average cutting speed is 20 m/min and that the return stroke is twice as fast as the forward stroke[1] If the workpiece is 340 mm long and allowing 30 mm overrun at each end:

Length of stroke

$$= 400 \text{ mm}$$

∴ Time for cutting

$$= \frac{400 \text{ mm}}{20 \text{ m/min} \times 1000} = \frac{1}{50} \text{ min}$$

∴ Time for return

$$= \frac{1}{2} \text{ time for cutting} = \frac{1}{100} \text{ min}$$

∴ Total time per stroke

$$= \frac{1}{50} + \frac{1}{100} \text{ min} = \frac{3}{100} \text{ min}$$

∴ Strokes per minute

$$= \frac{100}{3} = 33 \text{ strokes/min approx.}$$

A suitable feed for a roughing cut would be of the order of 0.25 mm per stroke, reduced to 0.10 mm per stroke for finishing unless a special broad-nosed tool is used, with a shallow cut and feed equal to the tool width.

Assume the workpiece described above is 100 mm wide and that a depth of cut of 5 mm is taken with a feed of 0.25 mm per stroke.

Over a 100 mm width, a feed of 0.25 mm per stroke would require 400 strokes, the approach and runout being neglected.

At 33 strokes/min the time required would be

$$\frac{400 \text{ strokes}}{33 \text{ strokes/min}} = 12.1 \text{ min}$$

This should be compared with the same situation in milling on pages 203–4.

GENERAL SAFETY NOTE

It is difficult to guard the cutting tool on a shaping machine because of its motion, and hands must be kept clear. The swarf produced can fly all over the place and be a hazard if it hits anybody, apart from being uncomfortable to walk on. It can be collected in a simple tray, made as shown in fig. 10.17, the tongues of which slide into the table tee-slots.

Above all it should be remembered that the

[1] This quick return ratio is an assumption for this example only. It must not be assumed that the speed of the return stroke is generally twice that of the forward stroke.

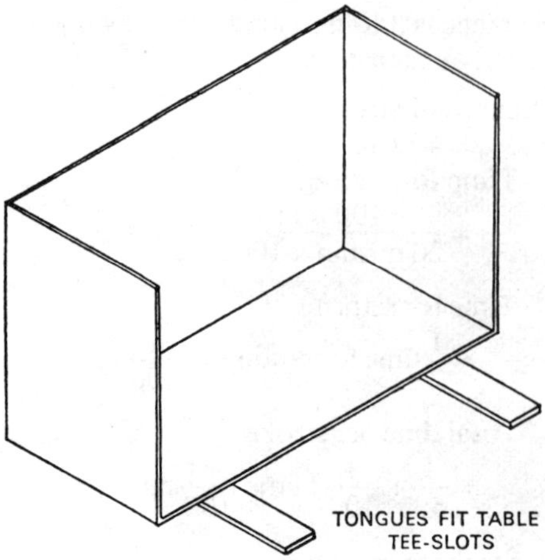

Fig. 10.17. Sheet-metal swarf tray and guard for shaping machine

ram, on either the forward or return stroke, takes a lot of stopping, and will not stop for a delicate object like the human body. If it hits anybody who walks too near they usually get dented with painful results. The machine should be positioned so that nobody walks behind it in the path of the ram. Care should also be taken that clothing is not caught by the handle or other part of the feed mechanism.

It must again be emphasised that any machine tool is safe as long as it is respected and mastered, but stupidity and lack of respect will rapidly meet with painful retribution.

QUESTIONS

1. Fig. 10.2 is a block diagram of a shaping machine on which the controls are not shown. Add to this diagram simple sketches of the following controls in the positions in which they are found on a machine with which you are familiar.
   (a) Handle for adjusting the position of the stroke.
   (b) Handle for locking the position of the stroke.
   (c) Handle for adjusting the length of stroke.
   (d) Handle for locking the length of stroke.

Fig. 10.5 may be of help in answering this question.

2. Explain with the aid of diagrams how the work would be held on a shaping machine to achieve the following condition in the work-piece.
   (a) One face of a rough block is machined – the work is to be held so that the next face to be machined is square to that already cleaned up.
   (b) Two adjacent faces of a workpiece are cleaned up and square to each other – the work is to be held so that opposite faces are parallel.
   (c) A step is to be machined in the top surface of a previously machined block not parallel but at 28° to the ends.
3. Fig. 10.18 is the end view of a block to be machined on a shaping machine. For each face

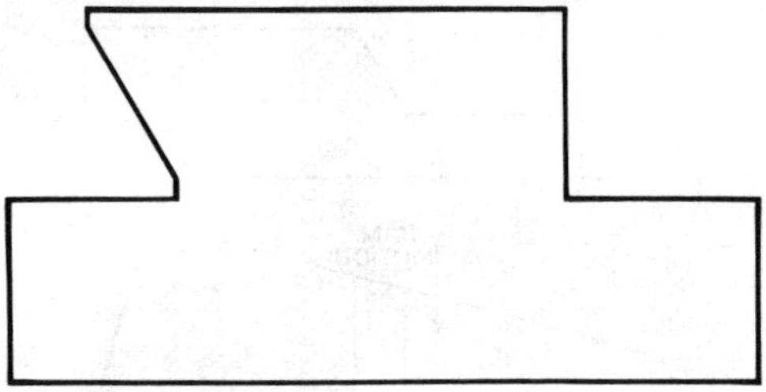

Fig. 10.18. Shaped block

to be machined with the head slide, draw on the diagram two lines showing:
   (a) line of motion of head slide.
   (b) direction of inclination of clapper box.
4. (a) By means of a simple diagram explain the main motions and adjustments available on a standard shaping machine. What factors determine the maximum size of work which can be accommodated?
   (b) Explain the action and purpose of the clapper box. What is the disadvantage of this device for large cutting tools and how is it overcome?
5. (a) The 'swan-neck' or equivalent type of tool for a shaping machine is designed so

as to bring the cutting edge approximately into line with the support face of the tool. Using sketches, explain the reason for this, and give a typical application for such a tool.

(b) Many shaping machines are fitted with a knee or table support bracket. Using simple line diagrams show:
   (i) that this bracket is provided to absorb and counteract the forces produced by cutting,
   (ii) what provision is made with the bracket to permit the table cross-feed to be used.

11 Grinding machines

General Objective: *The student should be able to compile block diagrams of the commonly used grinding machines and describe typical grinding operations.*

INTRODUCTION

The functional requirements of many components demand qualities of working surface that cannot be produced by the usual metal-cutting tools, even those made from modern cutting tool materials such as cemented carbides, as discussed in Chapter 5. The outstanding requirements which preclude such finishing processes are:

(a) A high degree of surface finish
(b) A high degree of accuracy
(c) Hardness to resist wear.

Good surface finish and dimensional accuracy are complementary. A high degree of accuracy in dimensions demands a good surface finish and although a good surface finish can be provided even when high accuracy is not essential, such a finish is comparatively expensive and probably unnecessary.

Three processes are used to achieve the above requirements, all using abrasive cutting methods. These are grinding, honing and lapping. The most commonly used of these processes is grinding, and it is the grinding process and some of the machines used for it which will be discussed in this chapter.

GRINDING PROCESSES

With few exceptions, the geometric form of a component consists of one of, or a combination of, three basic shapes, flat, cylindrical, or conical. Thus a grinding machine must provide the relative motions between the work and wheel to enable such shapes to be produced. The machines available fall into four categories, surface grinding machines, cylindrical grinding machines, universal grinding machines and cutter grinding machines. Other specialised types which will not be discussed here are thread grinding machines and centreless grinding machines.

SURFACE GRINDING MACHINES

This type of grinding machine is used for producing flat, and usually parallel, surfaces and can be considered in many ways as being equivalent to the milling machine. It may have a horizontal wheel spindle, grinding with the periphery of the wheel, or a vertical spindle, in which case the grinding is done by the wheel face. In either case there are certain fundamentals common to both machines:

1. The table motion must be straight both longitudinally and transversely, in order to produce flat surfaces.
2. The table surface must be parallel to its own motion, in order to produce parallel surfaces.
3. The adjustment to set the depth of cut must be sensitive enough to enable increments of 0.001 mm to be achieved either directly or by estimation.

In order to maintain accuracy, the machine must be extremely rigid, which is achieved by having the table on a fixed bed, the vertical adjustment being made by raising and lowering the wheelhead. Again, the similarity to the fixed-bed milling machine may be noted, where rigidity is a major requirement, although for a different reason.

An essential accessory for the surface grinding machine is the magnetic chuck, which enables flat workpieces to be held for grinding without mechanical clamping methods which might damage a finished work surface. Since the chuck is clamped to the machine table and the work is held magnetically on the upper surface of the chuck, the surface and the base of the chuck must be parallel and this parallelism must be maintained. The two surfaces must therefore be treated with care so that their parallelism is not destroyed by small burrs raised by mishandling. Non-ferrous parts cannot be held on a magnetic chuck and for these and for ferrous parts which are not suitable for magnetic holding, a vice can generally be used.

Magnetic chucks may be of the permanent magnet type shown in fig. 11.1, or of the electro-magnetic type, containing coils to energise the magnets, which is operated by simply switching the electric current on or off. This type has the disadvantage that a d.c. supply is necessary, requiring a rectifier. It must also be carefully sealed so that coolant does not get into the wiring. The permanent-magnet type illustrated is the one most commonly used.

The working face of a permanent-magnet chuck is made of a non-magnetic alloy in which

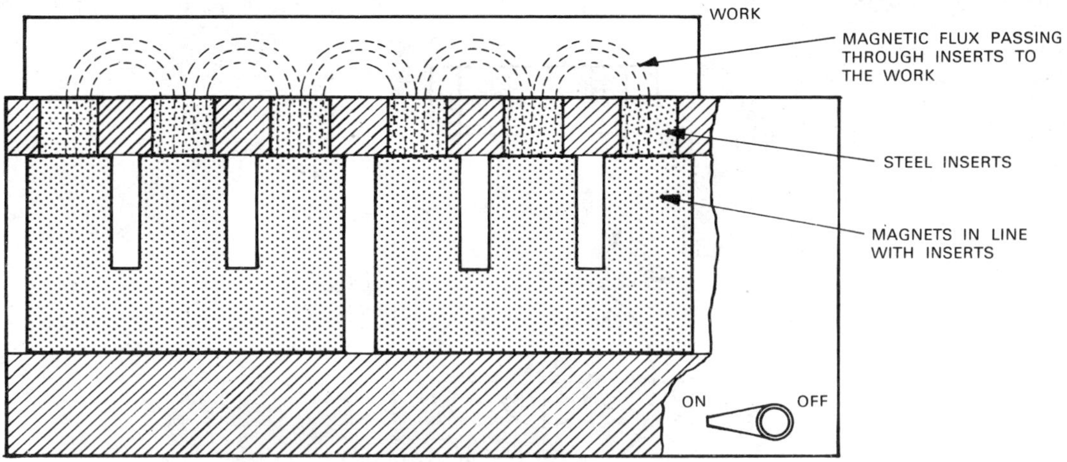

Fig. 11.1(a). Permanent magnet type chuck in on position

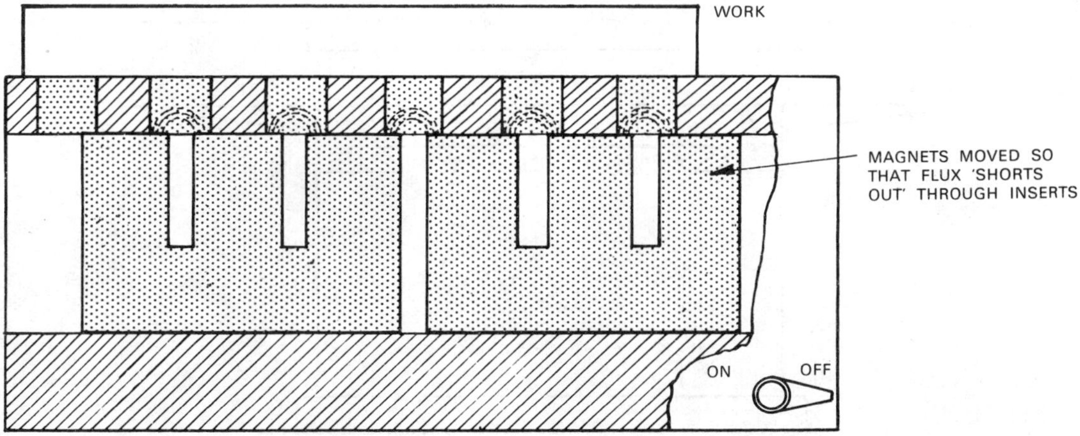

Fig. 11.1(b). Permanent magnet type chuck in off position

steel inserts are fitted to line up with the pole faces of the magnets. Movement of the 'switch' causes the magnets to move sideways so that in the 'on' position the pole pieces are in line with the inserts and the magnetic flux passes through the work as shown in fig. 11.1(a). In the 'off' position the inserts are each in contact with two poles of the magnets and the flux is effectively 'shorted out', as shown in fig. 11.1(b), so that none passes through the work, which can then be removed.

It is often a serious disadvantage for a workpiece to become permanently magnetised and unfortunately hardened steel, which often has to be ground, retains a considerable amount of magnetism after being held on a magnetic chuck. This residual magnetism can be removed by placing the workpiece in an alternating magnetic field and demagnetisers operating on this principle are available.

THE HORIZONTAL-SPINDLE SURFACE GRINDING MACHINE

This machine is illustrated in fig. 11.2, its alignments being similar to those of the plain horizontal milling machine. It can be used for

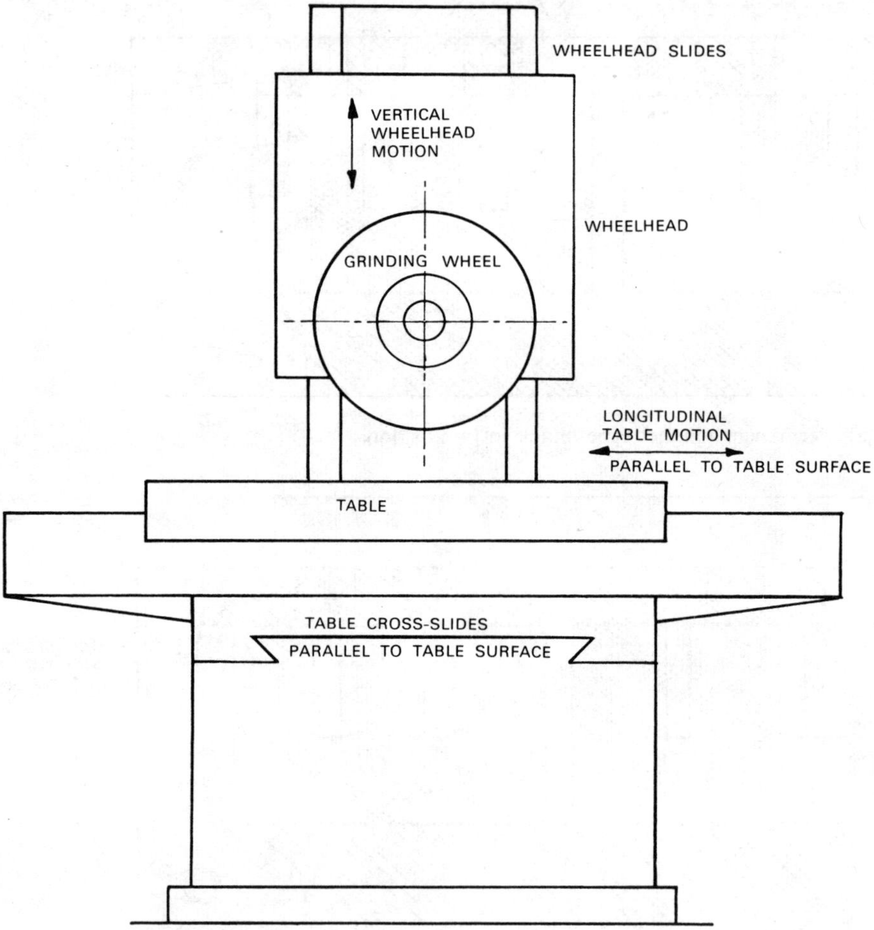

Fig. 11.2. Layout of peripheral wheel-type surface grinding machine

straightforward surface grinding, in which case the table is traversed longitudinally and the wheel is gradually fed across the work to cover the complete surface. The cross-feed may be hand-controlled or automatically controlled, in which case it can be varied from 0.2 mm to 2 mm per stroke of the table.

Alternatively, the machine can be used to grind the sides and bases of slots and grooves. For this type of operation the workpiece must be aligned with the table traverse motion and the wheel must also be positioned accurately across the work. Thus on this type of grinding machine the cross-feed adjustment must be equally as sensitive as the vertical adjustment of the wheel head. It should be noted that for grinding the vertical sides of a workpiece, or a groove, the sides of the wheel must be relieved by dishing as shown in fig. 11.3. On small workpieces, it may be better to perform this type of operation on a small tool-and-cutter grinder, but for large workpieces the peripheral machine is quite suitable.

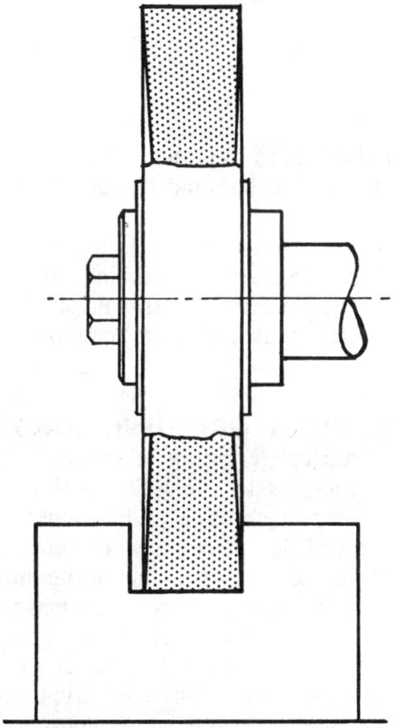

Fig. 11.3. Sides of wheel relieved for grinding groove faces

Wheel dressing is normally carried out with the traverse motion locked, a diamond in a holder being gripped on the magnetic chuck as shown in fig. 11.4 and passed back and forth across the wheel. Note the position of the diamond relative to the direction of rotation of the wheel. If for any reason the diamond becomes loose, the rotation of the wheel will throw it clear rather than drag it under the wheel and create a hazard.

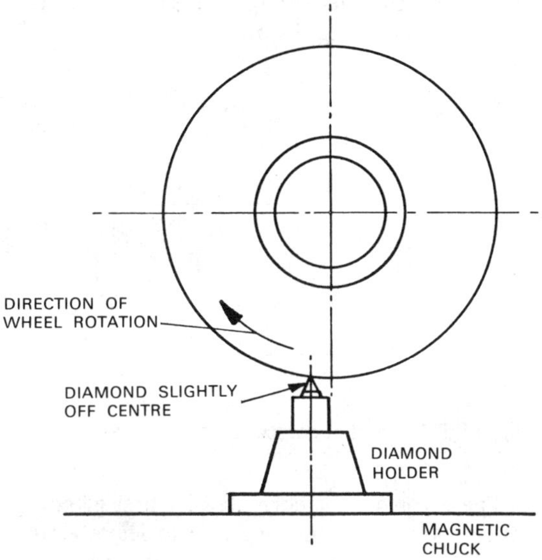

Fig. 11.4. Wheel dressing on surface grinder

An examination of the configuration of the peripheral-wheel machine shows that the wheel spindle overhangs the work table. This is not a very rigid form of construction and this type of machine is more suited to work of high accuracy, requiring sensitive control than it is to work requiring high rates of metal removal. Furthermore, a number of passes are necessary to cover a complete work surface and the machine is therefore slower than the vertical-spindle machine.

It is generally accepted that the purpose of a surface grinding machine is to grind surfaces flat and parallel with their opposite surfaces. However, if a block is to be ground all over it is usually necessary for it to be square as well as parallel. In Chapter 12 a simple device for measuring squareness is described. Assume that, using such a device, a block ABCD has been checked and found to be out of square by 0.075 mm over its length, the excess material being at end 'C', fig. 11.5.

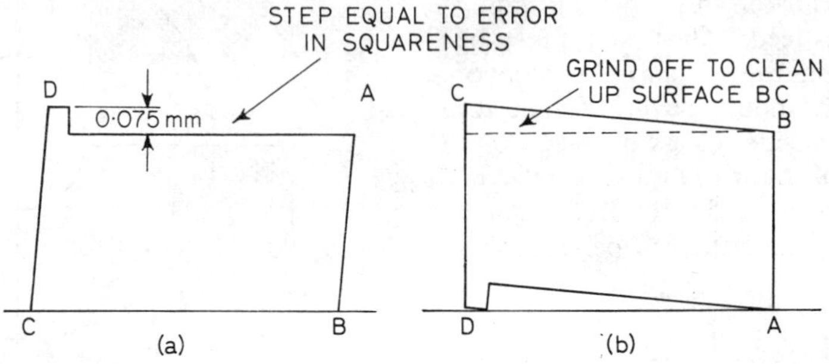

Fig. 11.5. Method of correcting squareness

The error can be corrected by first grinding off this amount from face AD to leave a step equal to the error at D as shown. If the block is now turned over and face BC cleaned up the error will be corrected on that face to give a datum for squareness. It can now be turned over again and the step cleaned off by grinding face AD and continuing grinding until the correct thickness is achieved.

In many shops squareness is checked on a qualitative basis using an engineer's square and the error – hopefully – corrected by packing the work on pieces of cigarette paper. The method described here is based on a measured error which is then corrected by a measured amount. Such methods are much better than hit or miss with 'fag papers'.

THE VERTICAL-SPINDLE SURFACE GRINDING MACHINE

The configuration of this machine is shown in fig. 11.6 and it will be noted that the machine cuts with the face of a cup wheel and not with the periphery. This lends itself to greater rigidity of the wheel spindle and higher metal-removal rates, the width of the work being less than the wheel diameter so that the work can be ground all over at one pass. Frequently the motor is made integral with the wheel spindle.

Many such machines do not have a table cross-feed, so the normal traverse, set to run slowly, is used for wheel dressing, the diamond holder again being held on the magnetic chuck or bolted direct to the table. The wheel is then lowered and the diamond traversed back and forth across the wheel face.

Unlike the horizontal-spindle machine, the vertical-spindle surface grinding machine is suitable only for grinding large flat surfaces but at much higher rates of metal removal.

CYLINDRICAL GRINDING MACHINES

The cylindrical grinding machine performs similar functions to the lathe in that it may be used for generating cylindrical, conical (tapered) and flat work-faces, the cutting tool being the grinding wheel. Unlike the lathe, the cutting tool (grinding wheel) is not traversed to produce cylindrical work, the generating motion being confined to the work table. Thus, in order to produce cylindrical and flat work surfaces, the following alignments are necessary.

1. Table motions must be truly straight.

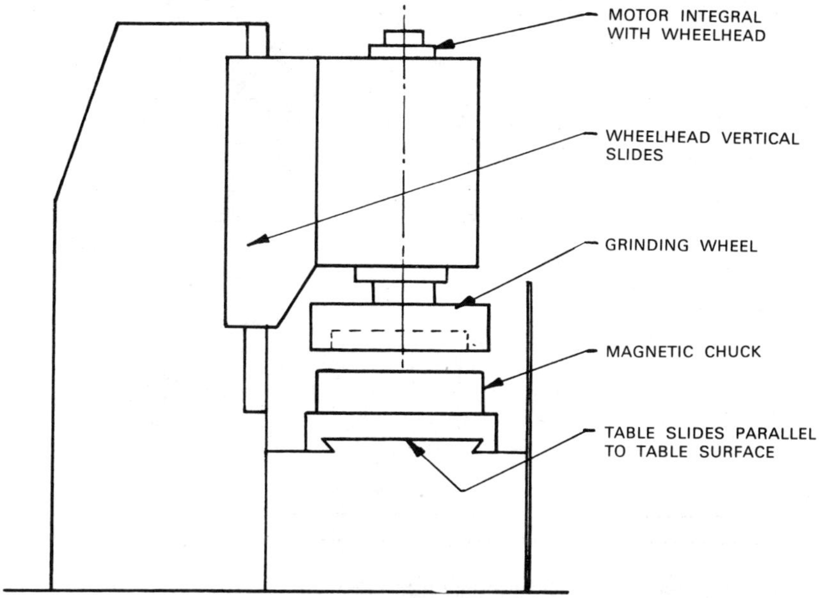

Fig. 11.6. Layout of vertical spindle surface grinder (end view)

2. Longitudinal table feed must be parallel to the work spindle axis or line of centres.
3. Wheel in-feed motion must be truly normal to the line of centres.

These are the only motions available on a plain cylindrical grinding machine. To produce tapered workpieces, a universal machine is used, in which the workhead can be set at an angle to the table motion.

THE PLAIN CYLINDICAL GRINDING MACHINE

The configuration of this machine is shown in fig. 11.7, the machine consisting basically of a bed on which are mounted the table and table slideways. At the back of the machine is the wheelhead mounted on slideways at right angles to the table slides to provide the in-feed of the wheel to set the depth of cut.

Mounted on the table are the workhead and the tailstock, positioned so that the work-spindle axis and the line of centres are both accurately parallel to the table feed motion. Thus any work generated by the work spindle rotation and the table feed will be truly cylindrical.

Facing operations can be carried out by using the side of the wheel as long as two important points are noted.

1. The side of the wheel must be relieved so that it only contacts on an edge of small area.
2. If the facing is done on a shouldered workpiece the corner should be undercut.

These conditions are shown in fig. 11.8.

Cylindrical grinding can be done by using the normal longitudinal feed provided the wheel width is less than the length of the work. In this case the longitudinal rate of feed is important. If the feed per revolution of the work is one-third of the wheel width or less, the outer edges of the wheel will do all the cutting and wear will be concentrated at the edges, causing the wheel to wear to a convex profile. The centre portion of the wheel periphery, which does little cutting, rapidly becomes glazed. If, on the other hand, the work feed per revolution is about two-thirds of the wheel width, then on the rightward motion of the work the cutting will be done by

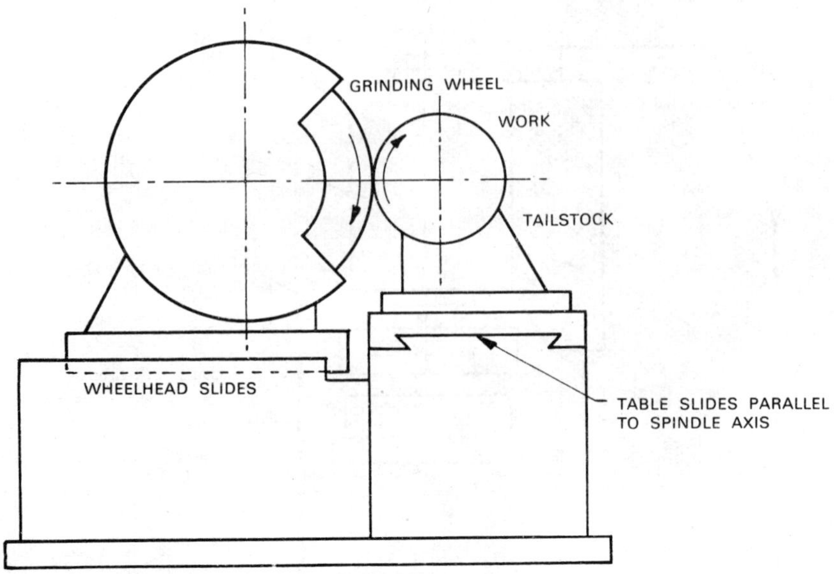

Fig. 11.7. Configuration of plain cylindrical grinding machine

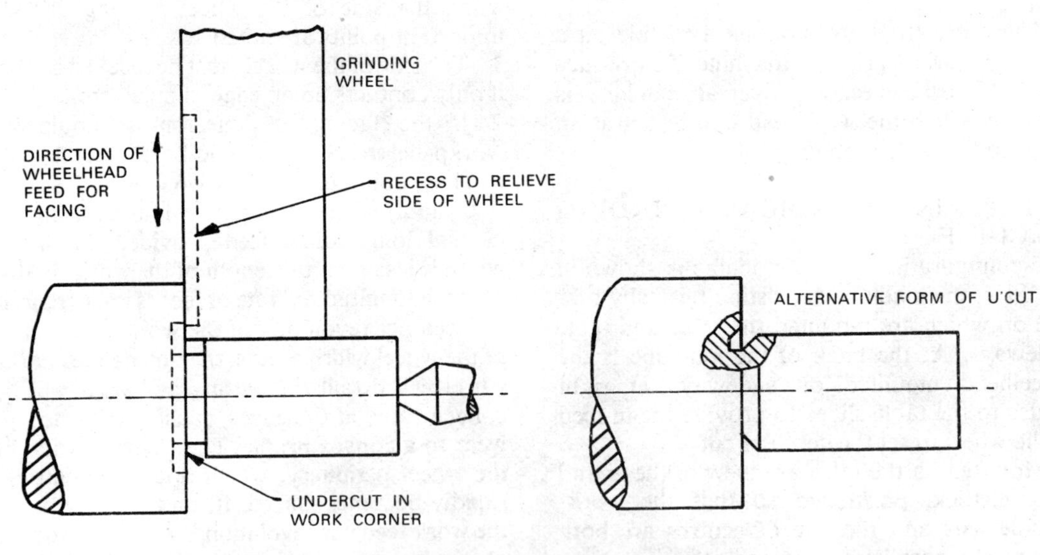

Fig. 11.8. Methods of relieving work and wheel for facing operations

the left-hand two-thirds of the wheel, and on the reverse stroke by the right-hand two-thirds. A little thought will show that these areas overlap and therefore the middle third of the wheel does twice as much cutting as the portions on either side. The wheel tends to wear to a concave profile, which is to be preferred. At the end of the stroke the wheel should not clear the work, but about one-third of its width should be allowed to run off.

If the work width is less than the wheel width, the work can be ground by the plunge-feed method. The wheel is carefully dressed and fed slowly and radially into the work, with no longitudinal feed in the usual sense. To reduce localised wheel-wear, the table should be reciprocated back and forth about 5 mm during cutting. This method, without the table oscillation, lends itself to form grinding.

WORK-HOLDING METHODS

Work to be ground can be held in a conventional chuck or between centres. It is interesting to note that a major cause of errors of concentricity (work running out of truth) on a lathe is the headstock or live centre running out of truth, so that the work does not rotate on a fixed axis. This is overcome on the grinding machine by having both centres stationary. A method of driving the catchplate about a dead centre is shown in fig. 11.9.

The centre spindle passes right through the work head and has two flanges. In each flange is a hole through which a screw can be fitted. With the screw in position 'A', as shown, the centre spindle is locked and cannot rotate. This is the condition normally used, the work-holding device being screwed on the spindle nose extension which is integral with the pulley unit.

If it is necessary to have the centre rotating (normally only for grinding centres) the screw is removed from position 'A' to position 'B'. The second flange is then locked to the driving pulleys and rotates with them, thus driving the centre spindle.

It must be realised that fig. 11.9 is a much simplified diagram and, as drawn, the work-head cannot be assembled nor is provision made for adjusting end play. A useful design problem might be to redraw this workhead with the necessary modifications.

It should also be noted that the tailstock spindle is spring-loaded to take up any expansion of the work as its temperature rises during the grinding operation.

GRINDING STEADIES

Just as in turning, so in grinding, a long slender work-piece tends to be deflected away from the grinding wheel by the forces set up during grinding. The deflection tends to increase as the forces move away from the reactions of the headstock and tailstock centres and reaches a maximum at the midpoint of the work. As the cut is reduced by the work deflection, the work produced becomes slightly barrel-shaped. It is of interest to note that a workpiece supported between centres and resisting applied cutting loads is one of the few genuine cases of a simply-supported beam occurring in practice.

The forces applied by the wheel to the work are shown in fig. 11.10 and they are seen to be, as in other cutting methods, a horizontal force F_h which deflects the work away from the wheel, and a vertical force F_v which, by deflecting the work downwards, also reduces the depth of cut. It would appear, therefore, that the shoes of the work steady should be positioned on the horizontal and vertical centre lines but this is not so. Consider a non-round workpiece with a protrusion at 'A', as shown. When the high point contacts the steady, the work is pressed into the wheel and metal is removed from a point opposite 'A', creating a low point. When this low point contacts the steady block it relieves pressure on the high point 'A' which is not therefore removed. Thus, if the steady blocks are on the centre lines, non-roundness may be exaggerated.

By positioning the steady blocks off-centre at P_1 and P_2 as shown, this condition is avoided and the action becomes self-rounding.

The construction of a grinding steady is shown in fig. 11.11. The steady blocks are made of hardwood and are adjusted independently, pressure being applied first by the lower block

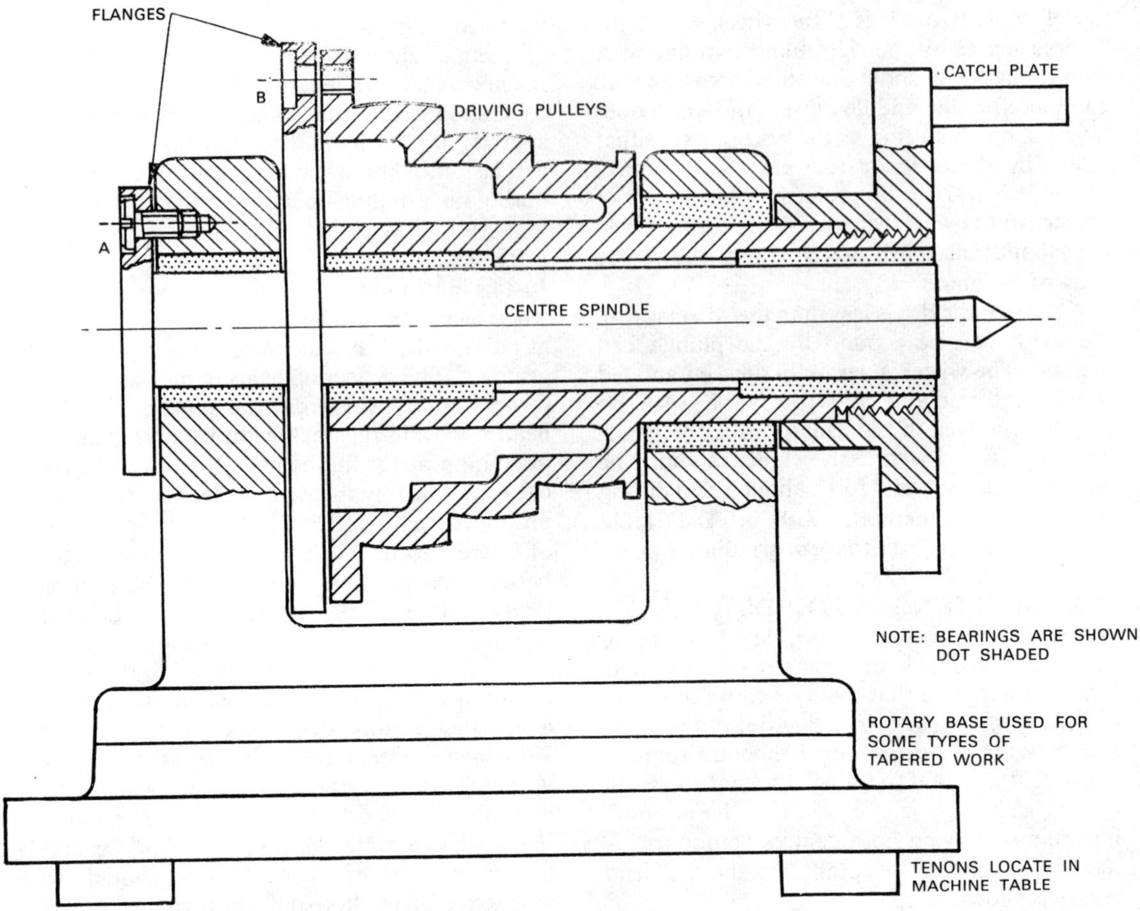

Fig. 11.9. Simplified diagram of workhead for cylindrical grinding machine

which is then locked. The upper block, which may be spring-loaded, is then fed forward until sufficient pressure is applied to the work. Either the spring loading compensates for the reduction in diameter as grinding proceeds, or adjustments are made to the blocks.

THE UNIVERSAL GRINDING MACHINE

The plain cylindrical grinding machine is limited to the external grinding of parallel cylindrical work. The universal machine can be adapted for internal grinding and the alignments can be adjusted to enable tapers to be ground.

Where a great deal of internal grinding is to be done, a machine built solely for the purpose is used, but if a universal machine is to be used it can be adapted in two ways. The internal grinding spindle can be mounted at the back of the wheelhead which is then swung through 180° to bring the internal grinding spindle into use. Alternatively, the internal spindle may be mounted on the front of the wheelhead on a special bracket. In either case, to attain normal cutting speeds of the order of 800–1000 m/min

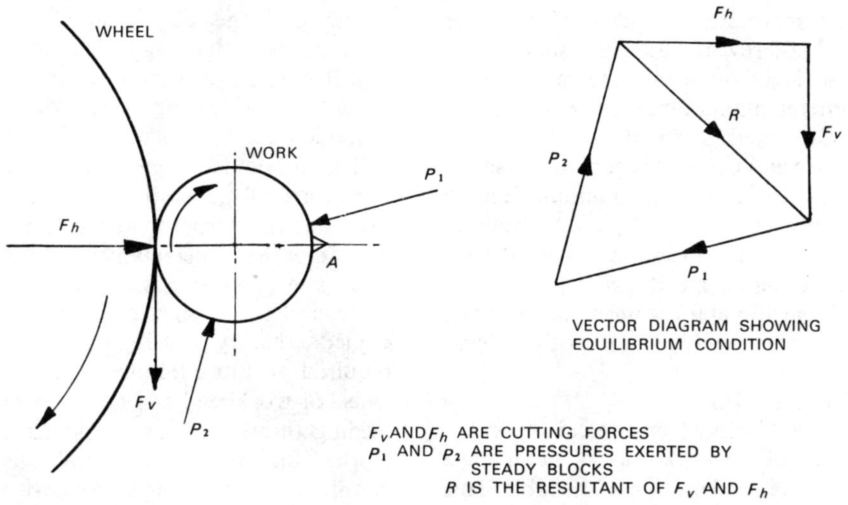

Fig. 11.10. Forces acting in cylindrical grinding when a steady is used

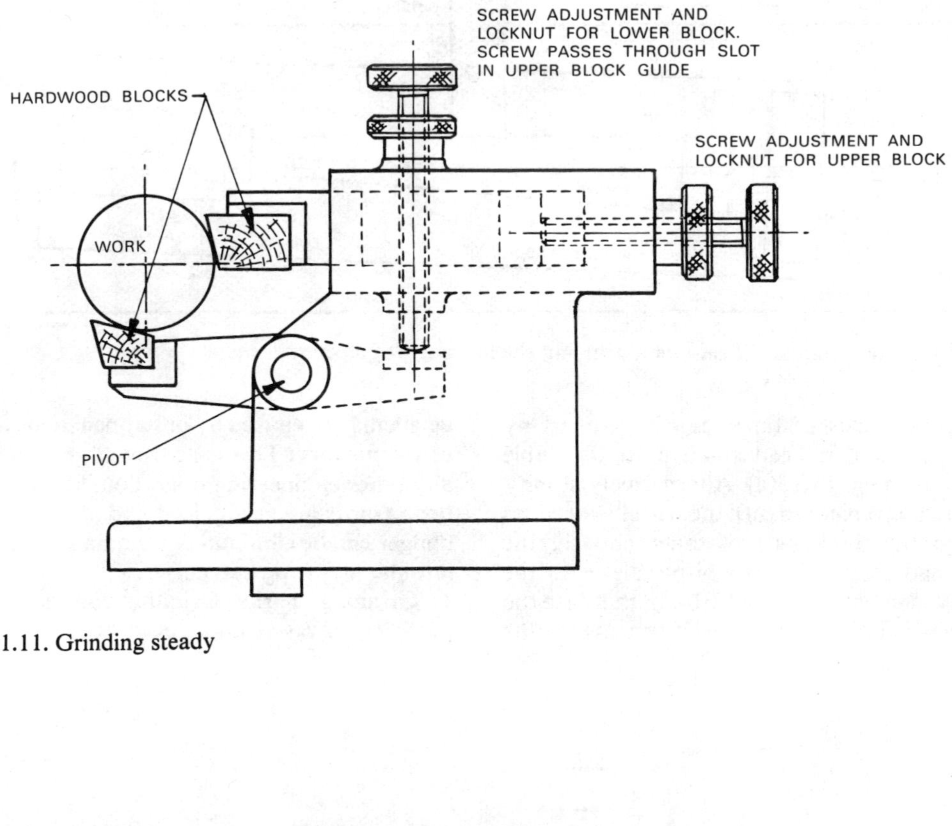

Fig. 11.11. Grinding steady

at the wheel surface, with wheels of 25 mm diameter or less, rotational speeds of the order of 10 000–12 000 rev/min are necessary. These speeds are achieved by driving the spindle from a large-diameter aluminium pulley to a small pulley on the internal spindle extension. A lightweight woven-linen belt is generally used to keep centrifugal effects to a minimum and to ensure that the belt and small pulley remain in contact. Even so, the author has seen such a belt, run at too high a speed, expand under the influence of centrifugal force until, just before it flew off, it was completely clear of both pulleys.

TAPER GRINDING

In order to grind tapered or conical work, the axis of rotation of the work must be at an angle to the longitudinal feed motion of the table. This can be achieved in three ways. For tapers up to about 20° included angle, the table surface can be swung relative to the table feed motion, as shown in fig. 11.12.

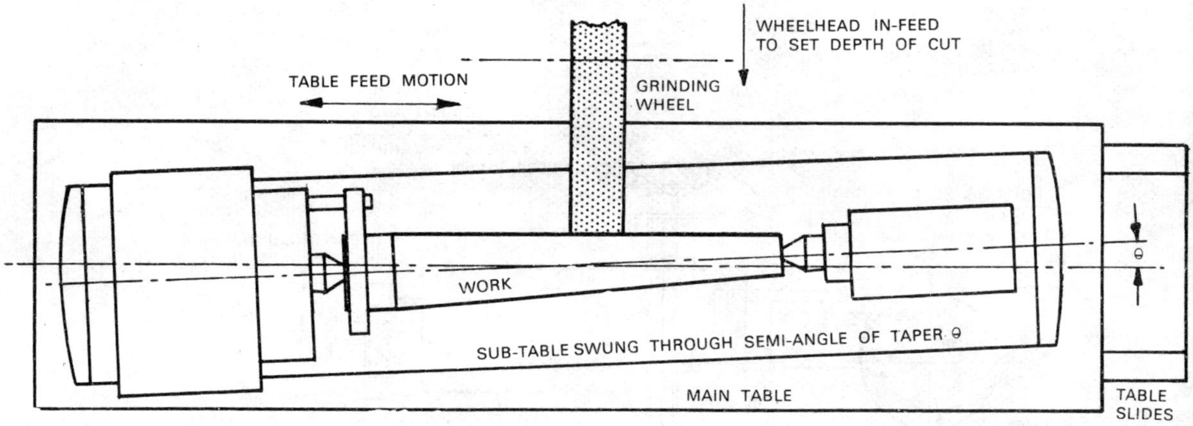

Fig. 11.12. Block diagram of set-up for grinding shallow angle of taper

Steeper angles of taper can be ground by inclining the workhead relative to the table motion as in fig. 11.13(a). Alternatively, it may be more convenient to turn the wheelhead slides through the semi-angle of taper and set the wheelhead so that its axis is parallel with the slides, as shown in fig. 11.13(b). In this case the wheelhead feed motion is used to traverse the wheel across the work in the direction of the arrow and the cut is applied by a slight adjustment of the longitudinal table stop. Note that it may be necessary to dress the wheel in order to avoid fouling of the cylindrical portion of the work.

The methods of grinding tapers described above are all generating methods and, as has been stated, in order to generate a taper the feed motion of the wheel relative to the work must be at an angle to the axis of rotation of the work. Instead of being generated, short tapers can be copied either by dressing the wheel to the angle required or, more frequently, by swivelling the wheel or workhead and *plunge grinding*, that is feeding the wheel slowly and directly into the work. This method is normally used only for chamfers and relieving sharp corners.

SAFETY NOTE

In all his books the author has emphasised that, although there are hazards in engineering, the accidents are caused by or happen to the foolish or the unwary. This is as true of grinding as of any other engineering operation, but as long as the hazards are recognised and allowed for the danger can be eliminated. Grinding hazards fall into the following categories.

1. Grinding sparks. Grinding sparks and dust particles travel at high speed but are of small

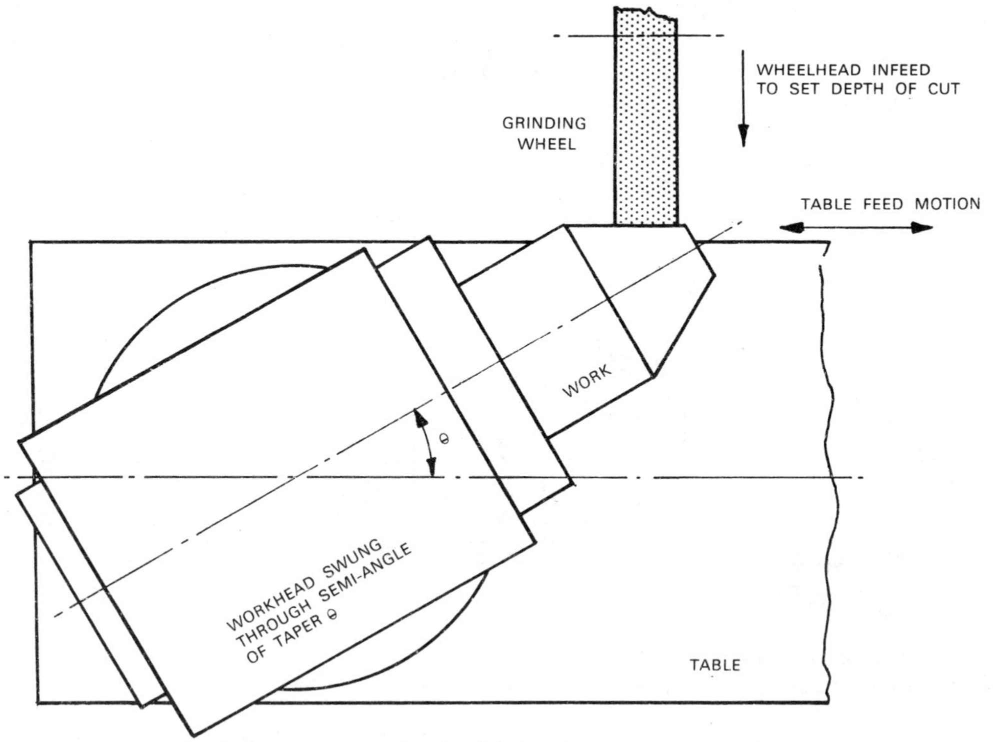

Fig. 11.13(a). Grinding a steep taper by inclining the work head

mass. They will not hurt the skin but can badly damage the eyes. Always wear safety goggles when in a grinding area.

2. Cuts. Never touch a moving grinding wheel. The cut caused by a wheel is usually cauterised by the frictional heat and, apart from being extremely painful, takes a long time to heal.

3. Burst wheels. Grinding wheels should never be run at speeds higher than the stated safe speed for that wheel. Overspeeding is one cause of wheels bursting. The other is incorrect mounting or mounting damaged wheels. Wheels should always be checked for cracks or faults before mounting and the mounting must only be carried out by a trained person. The author only saw one grinding wheel burst and this was mounted by an untrained person who never checked it for faults. It burst while being dressed and fortunately, the only casualty was a man who, realising how close one of the pieces had passed to his head, fainted and cracked his head as he fell. This could have been a tragedy – instead it was slightly amusing but the jokes carried an overtone of relief.

QUESTIONS

1. Explain with the aid of diagrams the difference in layout between horizontal and vertical spindle grinding machines. Discuss the advantages and limitations of each and state under which circumstances the surface produced is generated.

2. Name the most common method of workholding on surface grinders, explain how it functions and why it is so common. What precautions are necessary to ensure that the

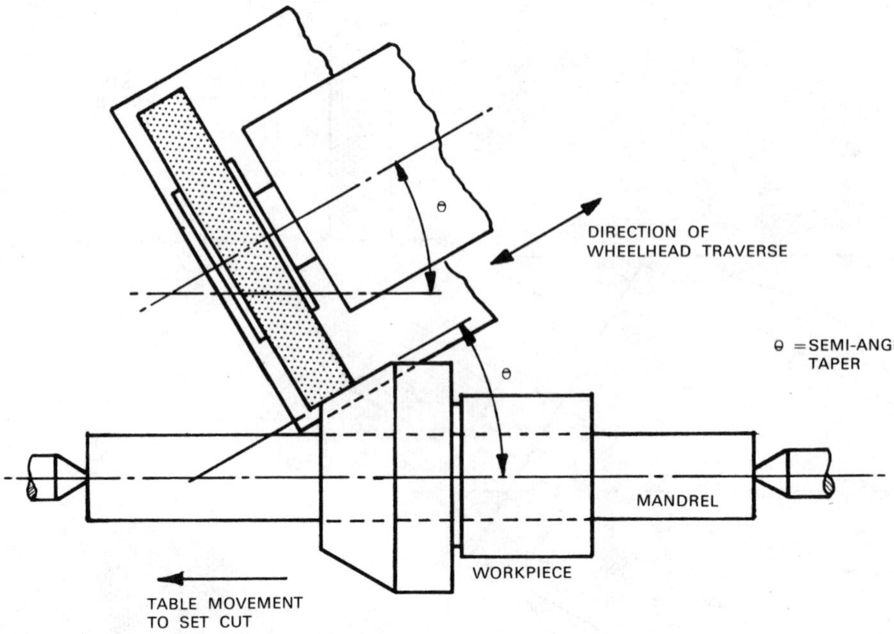

Fig. 11.13(b). Grinding a steep taper by inclining the wheel head. Note that the wheel head slides can be swung relative to the table and the wheel head can be swung relative to its slides

accuracy obtainable with this piece of equipment is maintained.

3. Show with the aid of diagrams how a grinding wheel should be dressed to enable the sides of a slot to be ground on a surface grinder.

A piece of steel ground all over to a section 40 mm square is to fit into a slot. How must the slot be modified from square corners in order that the steel will fit. Why is this modification necessary.

4. Explain how the grinding wheel and the work must be modified to enable the steps on a multi-diameter workpiece to be faced on a cylindrical grinding machine.

5. Why, when cylindrical grinding, is it better to have the feed/rev of the work between $\frac{1}{2}$ and $\frac{1}{3}$ of the wheel width rather than less than $\frac{1}{2}$ wheel width? What is 'plunge feed grinding'.

6. When performing a cylindrical grinding operation between centres it is preferred that both centres are stationary. Explain why this is so but name one operation where a rotating centre is necessary.

7. When grinding long slender work a grinding steady is used. Explain why the steady blocks are positioned off centre relative to the wheel and work.

8. Describe three methods of generating tapers on cylindrical grinding machines.

12 Measurement

General Objective: *The student should appreciate the procedures for the measurement of dimension and form, and the importance of Workshop Standards.*

INTRODUCTION

As man's industrial techniques have progressed and the equipment we accept as part of everyday life has become more complex, there have been two parallel developments in engineering equipment. One has been in the machinery used to produce articles and components, and the other has been in the measuring equipment used to measure them. These are extremely closely related, since it is of little value to use a machine capable of working to ± 0.005 mm if our only measuring equipment is a rule and a pair of calipers. On the other hand, a wood turner making a chair leg is not likely to use a dial gauge or a micrometer. Thus it is important for the student to study measuring techniques and their applications as well as production processes.

LENGTH STANDARDS

Specific Objective: *The student should know the principles of use and care of workshop standards of length.*

How accurately must we be able to measure in modern engineering? Consider a part whose diameter is required to be 25 ± 0.05 mm. It is no good measuring it with a rule, for this will only be accurate to ± 0.15 mm in the hands of a skilled man. We need an instrument which can measure to ± 0.005 mm, i.e. one-tenth of the allowable variation in the part size.

In order to maintain this measuring instrument we need to check it against a standard of length whose accuracy is again within one-tenth of the instrument accuracy, i.e. ± 0.0005 mm, which, put in another way, is ± 0.5 *thousandths of a mm.* The first-year apprentice checking his micrometer with *gauge blocks*, or *slip gauges*, is therefore handling equipment whose order of accuracy is measured in fractions of a micrometer. To check these standards, still higher orders of accuracy are required and the highest standards, *grade* 00 blocks, are checked against the legal standard of length which is nowadays based on the wavelength of a particular colour of light.

These grade 00 gauge blocks are made to an accuracy of $\pm 5 \times 0.00001$ mm and are used only in a standards room, for checking inspection-grade blocks, which are then used for testing equipment and gauges.

SLIP GAUGES OR GAUGE BLOCKS

These are hardened, ground and lapped rectangular blocks of high-carbon steel (sometimes of cemented carbide) made so that:

1. opposite faces are flat;
2. opposite faces are parallel;
3. opposite faces are accurately the stated distance apart;
4. opposite faces are of such a high degree of surface finish that when two blocks are pressed together with a slight twist by hand they will wring together, i.e. remain firmly attached to each other.

These properties of gauge blocks enable them to be built up, by wringing, into combinations which give sizes varying by steps of 0.01 mm and whose overall accuracy is of the order of 0.00025 mm, even with workshop-grade blocks. They are supplied in sets of 50, 78, or 105 pieces, and protective blocks are provided for use with inspection and workshop grades. These protective blocks should, where possible, be used as the end blocks of all combinations.

In the case of metric sizes B.S. 4311 recommends four sets of gauge blocks, two of which are set out below. The smallest increment possible with these sets is 0.0025 mm. Manufacturers do make sets which enable steps of 0.001 mm to be achieved.

Set no. M78

1.01 mm to 1.49 mm by 0.01 mm steps	—	49 blocks
0.50 mm to 9.50 mm by 0.50 mm steps	—	19 blocks
10, 20, 30, 40, 50, 75 and 100 mm	—	7 blocks
1.0025 mm	—	1 block
1.005 mm	—	1 block
1.0075 mm	—	1 block
		78 blocks

Set no. M50

1.01 mm to 1.09 mm by 0.01 mm steps	—	9 blocks
1.10 mm to 1.90 mm by 0.10 mm steps	—	9 blocks
1 mm to 25 mm by 1 mm steps	—	25 blocks
50, 75 and 100 mm	—	3 blocks
1.0025, 1.0050 and 1.0075 mm	—	3 blocks
0.050 mm	—	1 block
		50 blocks

Inch gauges are made in 41- or 81-piece sets, the 81-piece set being as follows:

0.1001 in to 0.1009 in by 0.0001 in steps	—	9 blocks
0.101 in to 0.149 in by 0.001 in steps	—	49 blocks
0.050 in to 1.000 in by 0.050 in steps	—	20 blocks
2 in, 3 in and 4 in blocks	—	3 blocks
		81 blocks

The 41-piece set is made up in a similar manner but not quite as many combinations can be achieved, the blocks provided being:

0.1001 in to 0.1009 in by 0.0001 in steps	—	9 blocks
0.101 in to 0.109 in by 0.001 in steps	—	9 blocks
0.110 in to 0.190 in by 0.010 in steps	—	9 blocks
0.100 in to 0.900 in by 0.100 in steps	—	9 blocks
0.050 in, 1 in, 2 in, 3 in and 4 in blocks	—	5 blocks
		41 blocks

COMBINATIONS OF GAUGE BLOCKS

In wringing combinations of gauge blocks the smallest possible number of blocks should be used. This can be achieved by working in a systematic manner by subtraction as follows.

Example.

Select suitable gauge blocks for making up a pile whose height is 81.83 mm.

Note that six blocks are required with the 50-piece set and only five with the 78-piece set.

As examples the student should work out the combinations required to make up lengths of 3.4159 in and 5.281 in using the inch sets.

1. *Using a 78-piece set*

		Blocks used
	81.83	
2 protective slips	2.00 (subtract)	2 × 1.00 protectives
	79.83 remainder	
Select 2nd decimal	1.33 (subtract)	1.33
	78.50 remainder	
Select 1st decimal	3.50 (subtract)	3.50
	75.00 remainder	75.00
		Total 81.83 (check)

2. *Using a 50-piece set*

	81.83	
2 protective slips	2.00 (subtract)	2 × 1.00 protectives
	79.83 remainder	
Select 2nd decimal	1.03 (subtract)	1.03
	78.80 remainder	
Select 1st decimal	1.80 (subtract)	1.80
	77.00 remainder	
Select 1st unit	2.00 (subtract)	2.00
	75.00 remainder	75.00
		Total 81.83 (check)

APPLICATIONS OF GAUGE BLOCKS

The simplest application of gauge blocks is their use as checking pieces for measuring instruments. They are also used as setting blocks for accurate measuring instruments such as comparators, and for setting sine bars to angular measurements. Their use is greatly increased if a set of gauge block accessories is available, since these can then be made into gap gauges and height gauges, as shown in fig. 12.1.

SAFETY NOTES

Gauge blocks must be handled carefully and protected from damage. The slightest burr will destroy their accuracy and wringing ability.

1. Always wring with a twisting movement, not a longitudinal rubbing movement.

2. Always clean the blocks with chamois leather, or lintless cloth.

3. Handle the blocks over a cloth so that if dropped they do not fall on a hard surface, which might raise a burr.

4. Do not put any blocks back into the box until all work with them is finished. All blocks that have been handled should then be cleaned and lightly greased as they are put away.

5. Never leave gauge blocks wrung together when not in use.

STANDARDS OF FLATNESS AND STRAIGHTNESS

Specific Objective: *The student should know the principles of use and care of Workshop Standards for flatness and straightness.*

THE SURFACE PLATE AND STRAIGHT EDGE

If a large workpiece with widely-separated dimensional features is to be marked out, all such dimensions must generally be referred to a flat surface. A typical example of the effects of using as a reference plane a non-flat surface is shown in fig. 12.2, producing error in the marked-out positions of the hole centres.

The reference planes used in engineering are the *surface table* and the *surface plate*, the former being free-standing and the latter mounted on a bench. These plates are usually of good-quality grey cast iron which has been stress-relieved to ensure that they do not distort over a period of time. The better grades are scraped and the cheaper ones are merely planed, the accuracy of a planed table depending upon the accuracy of the machine producing it. B.S. 817 sets out standards of accuracy for surface plates; a grade 'A' plate of

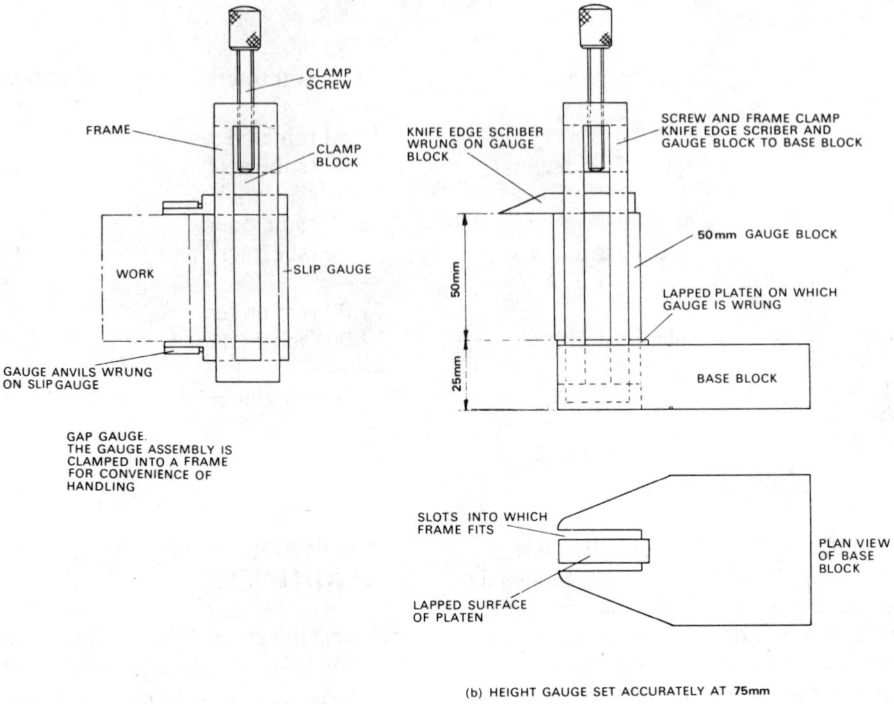

Fig. 12.1. Use of gauge block accessories

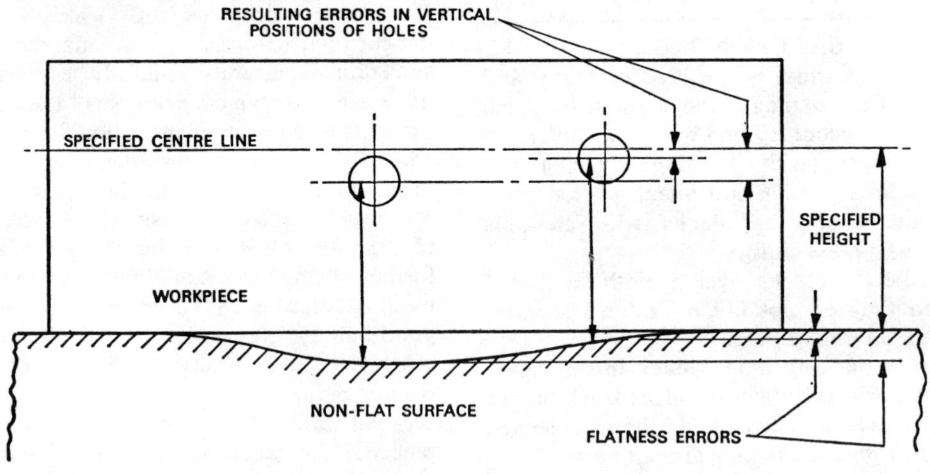

Fig. 12.2. Effect of marking-out from non-flat surface. Note that the errors in the hole positions are equal to the flatness errors

diagonal length 1 metre has to be flat within ± 0.008 mm over its whole surface, and to within ± 0.004 mm over any 300 mm.

Generally, grade 'C' planed plates are suitable for workshop use, grade 'B' scraped plates are used for inspection work and grade 'A' plates for high-precision and standards-room work.

THE WHITWORTH PRINCIPLE

How does one obtain a flat surface? It is easy to say it is scraped but how does one know where to take off the high points? Sir Joseph Whitworth pointed out that if three plates are compared with each other in alternate pairs, they will all only mate perfectly in all positions when they are absolutely flat. Thus, if we have three plates X, Y and Z and scrape X to Z and X to Y, X will mate with both Y and Z. However, if X is concave, Y and Z will both be convex, as shown in fig. 12.3, and will not mate together. When all three mate in all positions, all three are quite flat.

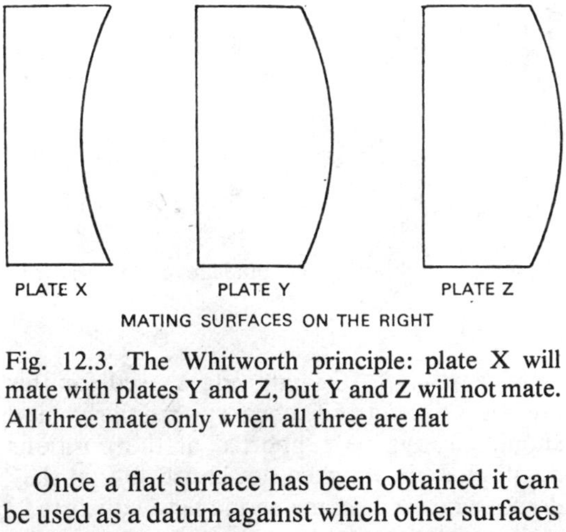

Fig. 12.3. The Whitworth principle: plate X will mate with plates Y and Z, but Y and Z will not mate. All three mate only when all three are flat

Once a flat surface has been obtained it can be used as a datum against which other surfaces are checked.

The same principles apply to straight edges, which are sometimes made of thin-section hardened tool steel and given a knife edge, in the small sizes used by toolmakers. The knife edge gives a reduced bearing area which allows very small clearances to be observed, particularly if viewed against a light box. Larger straight edges are made of cast iron to the shape shown in fig. 12.4. These can be used for checking flat surfaces by lightly smearing the face with engineer's blue and registering it on the surface under test, or vice versa.

When a surface is checked by a straight edge it must be tested in directions not only parallel to its edges but also across its diagonals. Fig. 12.5 shows a surface which appears flat along all the lines shown but is not flat across the diagonals. This is because such a surface consists of an infinite number of straight lines and all that a straight edge tests is straightness. For the surface to be flat, all these lines must lie in one plane and a check over diagonals tests this.

COMPARATIVE MEASUREMENT OF STRAIGHTNESS

The method described of using a straight edge and engineer's blue to indicate the high points on a surface does just that and gives no indication of the amount of error in the surface. A precision steel straight edge and gauge blocks can be used to measure the deviations from straightness to a high level of precision. The technique is called the *wedge method.*

The straight edge is supported against an angle plate on two slip gauges whose height differs by 0.1 mm. The support points are usually marked on steel straight edges but if they are not the supporting gauge blocks should be set apart a distance of 0.554 L, where L is the length of the straight edge. This separation is chosen so that the weight of the overhanging ends balances the weight between the supports and vice versa, to reduce the sag of the straight edge, due to its own weight, to a minimum.[1]

If the length between the supporting gauge blocks is divided into ten equal parts the gauge blocks can be made up in increments of 0.01 mm which should just fit under the straight edge

[1] Note the difference in supporting a straight edge to give minimum deflection and supporting length bars (0.577 L) where the slope at the ends must be zero so that the end faces are parallel.

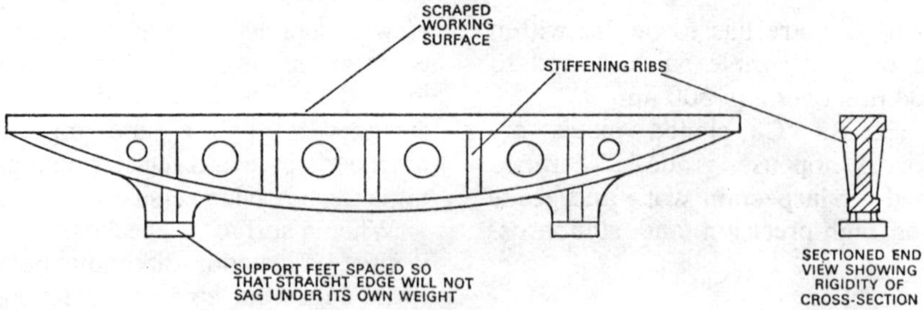

Fig. 12.4. Cast iron bow-shaped or 'camel-back' straight edge

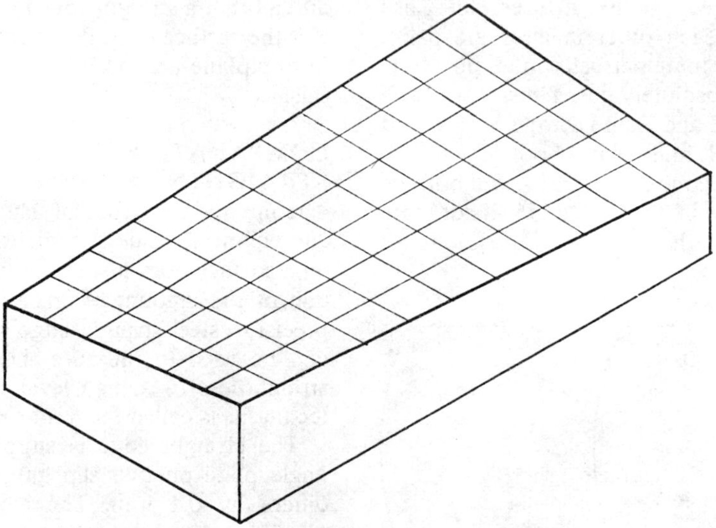

Fig. 12.5. This surface appears flat if checked parallel to the sides. Checking the diagonals reveals the error in flatness

at each mark. If they do then the surface under test is straight. Assume that a gauge block can be slid along under the straight edge 0.25 of the distance between the marks *too far*. It follows that the surface is too low at that point and the error is 0.25 of the height difference between the marks. As this is 0.01 mm then the error at that point is

$$0.25 \times 0.01 \text{ mm} = 0.0025 \text{ mm}.$$

This method can therefore be used to check the straightness of a surface along a given line to a high degree of accuracy *as long as the straight edge is straight*. Straight edges used for this purpose should be checked regularly and should always be supported at the positions mentioned above when not in use so that they do not sag and become permanently distorted.

The method is laborious and depends on the accuracy of the straight edge. For these reasons it has been superseded by optical methods which will be discussed in a later work. It is still useful, however, where specialist optical equipment is not available.

STANDARDS OF SQUARENESS

Specific Objective: *The student should know the principles of use and care of Workshop Standards for Squareness.*

The most commonly used standard for squareness in the workshop is the engineer's square or try-square. The student at this level should be familiar with this instrument from his work in Level 1 or, better still, from actually using it in the workshop as part of his training. As with all other measuring instruments the square must be checked against a standard to ensure that its accuracy is maintained and it is the standards of squareness for checking squares which will be discussed in this section.

TESTING RIGHT ANGLES (90°)

The principle involved in testing a right angle without reference to a standard can be illustrated by considering a draughtsman's set square. If the square is placed against a tee square and a line drawn down the vertical face, the square turned over and another line drawn as shown in fig. 12.6, any angle between the lines is *double* the error in the right angle. This simple principle of reversal is used in almost all cases of testing squareness.

A simple device for checking a square based on this principle without reference to any standard of angle or length can be made up as shown in fig. 12.7. It consists of an angle plate carrying a dial gauge at the top and a stop at the bottom separated by a known length L. It is first set using a parallel block as shown and the dial gauge is set to zero midway between the readings R_1 and R_2. Having set the instrument to zero it can now be used for checking a square by bringing the square up to the stop and the dial gauge reading is a direct measurement of error in squareness over length L. Alternatively, by building the device so that the stop is higher than the stock of the square and having a slot in the base through which the stock can pass readings can be taken on both sides of the blade of the square. In this case the error is half the difference in readings.

i.e. Squareness error $= \frac{1}{2}(R_1 - R_2)$

Note that, used in this way, the device has the effect of indicating double the error due to the use of the reversal principle.

THE SINE BAR

Specific Objective: *The student should know the principle of the sine bar and its applications for checking and setting out angles.*

The simplest instrument for measuring or setting out angles in the workshop is the protractor from a combination set. Unfortunately its accuracy is only of the order of about ± 10 min. This can be improved by using a vernier bevel protractor which will give about ± 5 min and these figures are for setting the instrument. When measuring with these instruments the problem of 'feel' or 'touch' arises, further reducing the accuracy and for more precise work the sine bar should be used.

The sine bar is a hardened and ground steel bar of rectangular section with a hardened steel precision roller at each end, the distance between the centres of the rollers being accurately set. If the sine bar is placed on a surface plate with gauge blocks of height h mm under one roller, as shown in fig. 12.8, the height h is opposite to angle θ and the centre distance l is the hypotenuse of triangle ABC. Therefore:

$$\sin\theta = \frac{h}{l}$$

or

$$h\,\text{mm} = l \sin\theta$$

Thus, to set up angle θ, multiply $\sin\theta$ by l and use a gauge block combination of this height to raise one end of the sine bar.

Metric sine bars are available with lengths of 100 mm, 200 mm and 250 mm. In this case $h = l \sin\theta$ where $l =$ sine bar length.

An error in height h obviously produces an error in angle θ. What is not so obvious is the fact that, as angle θ increases, the same error in h produces a *greater* error in θ. The sine bar should therefore not be used for angles greater

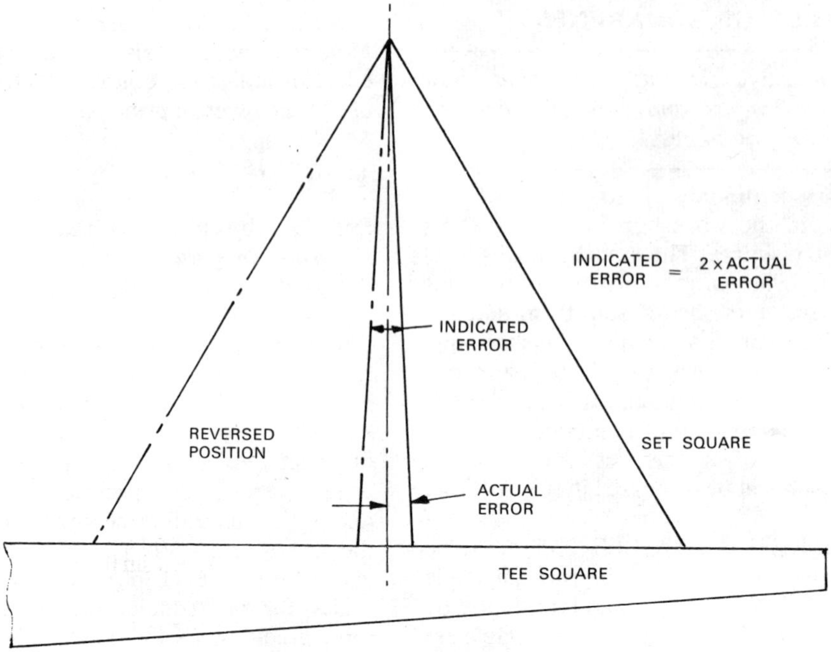

Fig. 12.6. Simple illustration of doubling effect due to reversal

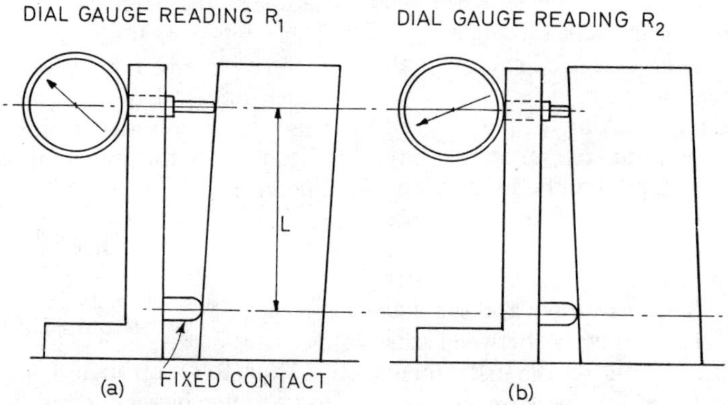

Fig. 12.7. Dial gauge fixture for testing squareness

than 45° if this can be avoided. The work can often be turned through 90° and the complement of the angle used.

With the aid of a sine bar, an angle can be accurately marked-out by a height gauge or checked by a dial gauge and stand, as indicated in fig. 12.8. The accuracy of this simple sine bar depends upon the distance X, since the rollers are pulled into the cutaways by socket-head screws. If this distance is inadvertently ground *undersize*, the sine bar can only be corrected by packing the cutaway at one end with shim,

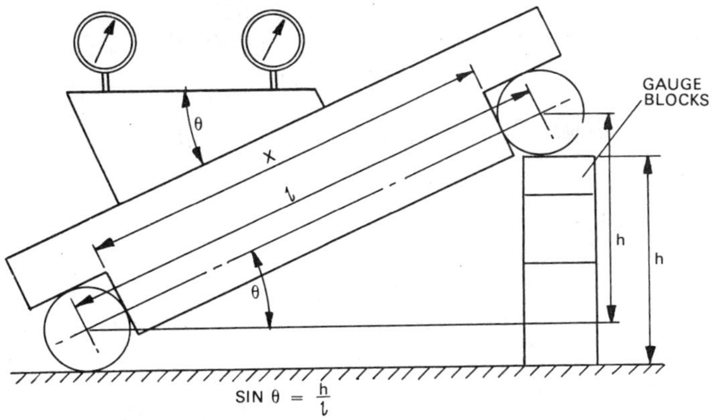

Fig. 12.8. Simple sine bar for checking angles

which is not good practice. An improved type of sine bar is shown in fig. 12.9. In this, if the length is oversize it can be corrected by grinding off one cutaway face, and if undersize by grinding off the other.

setting up the angle 23.5°, with a 100 mm sine bar using four-figure tables.

$$\begin{aligned} h &= 100 \sin \theta \\ &= 100 \text{ mm} \times 0.3987 \\ &= 39.87 \text{ mm} \end{aligned}$$

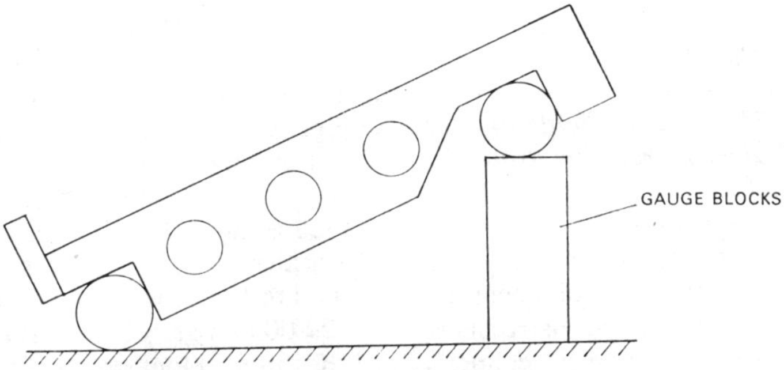

Fig. 12.9. Improved sine bar

SAFETY NOTE

A sine bar and slip gauge set-up tends to be unstable and can easily be knocked over. Always support it against an angle plate and lightly clamp it where necessary. This also helps when the work is aligned with the sine bar to avoid compound-angle errors.

For a high degree of accuracy, five-figure tables should be used. As an example, consider

The fourth figure in four-figure tables is not reliable, and when multiplied by 100 becomes the second decimal figure in our slip gauge pile. If we use five-figure tables we get

$$\begin{aligned} h &= 100 \times 0.39875 \\ &= 39.875 \text{ mm} \end{aligned}$$

This error of 0.005 mm in h can be avoided by using appropriate gauge blocks as follows:

	39.875	*Blocks used*
Select 2 protective slips	2.000 subtract	2 × 1.000 protectives
	37.875	
Select third decimal	1.005 subtract	1.005
	36.870	
Select second decimal	1.470 subtract	1.470
	35.400	
Select first decimal	1.400 subtract	1.400
	34.000	
Select unit	4.000 subtract	4.000
Select 30 mm	30.000	30.000
		Total 39.875 (check)

PRACTICAL MEASUREMENTS

Specific Objective: *The student should be able to carry out practical measurements using slip gauges, balls and rollers.*

Many of the precise measurements in a workshop are beyond the limits of accuracy attainable with standard measuring instruments such as micrometer and vernier calipers. Others, because of the form of the workpiece necessitate the use of precision balls and rollers to give point contacts between the workpiece and instrument rather than the area of anvil normally in contact with the work. Typical measurements carried out using such equipment are the measurements of plain ring gauges and internal and external tapers.

DIRECT MEASUREMENT OF GAUGES

Simple gap and ring gauges can be measured by direct comparison with gauge blocks if the accuracy required is not greater than ±0.01 mm.

1. *Measurement of plain gap gauges*

This type of measurement is made by building up gauge blocks until they 'just fit' the gap, i.e. until one size of gauge block combination will enter the gap and the next one will not. In practice this is done by a process of elimination. Consider a gap gauge measured with a rule and found to be between 24.97 mm and 25.03 mm. A 22.00 mm gauge block is taken and 1.07 and 1.90 mm blocks are wrung to it, the combination being offered to the gauge. It enters but a 25.00 mm gauge does not. Sizes over 25.00 mm are eliminated and the gauge is now known to be between 24.97 mm and 25.00 mm.

Now a gauge combination of 24.98 mm is offered and is accepted, so the 1.08 mm slip is replaced by a 1.09 mm one. If the 24.99 mm combination enters the gap but the 25.00 mm gauge does not, the gap is greater than 24.99 mm but less than 25.00 mm. The whole system should be left to stabilise so that any temperature differences can correct themselves and the gap gauge then rechecked with 24.99 mm and 25.00 mm gauge block combinations. If they check correctly the gap gauge is quoted as 24.995 mm to an accuracy of determination of ± 0.01 mm.

This procedure has been given in some detail since it is the fundamental method to be used when working with gauge blocks. Other points to note are:

(*a*) Clean the equipment carefully and preferably wear chamois leather gloves to avoid corrosion from perspiration; some people are 'rusty fingered' and have corrosive sweat.

(*b*) Perform all such operations on a surface plate. Its mass acts as a heat sink and helps to equalise temperatures.

2. *Measurement of plain ring gauges*

A plain ring gauge can be measured by placing a pair of calibrated precision steel balls in the gauge and building up gauge blocks, by a similar process of elimination to that described for gap gauges, until a 'nice snug fit' is achieved. What constitutes a nice snug fit?

If the gauge blocks and gauge are lightly held down with three fingers of one hand, a steel knitting needle can be used to see whether any free movement of one of the balls is present, as shown in fig. 12.10. The gauge blocks are built up until any free movement is eliminated and the last combination of gauge blocks plus balls is taken as the correct size to an accuracy of determination of ± 0.01 mm. The same procedure can be used with taper ring gauges.

The measurement should be carried out on a surface plate, not on a wooden bench, and excessive force must not be used. The use of the knitting needle enables false results from elastic deformation to be avoided. It must be remembered that the measurement involves four point-contacts, at all of which elastic deformation will occur under very light loads. The largest combination of gauge blocks which

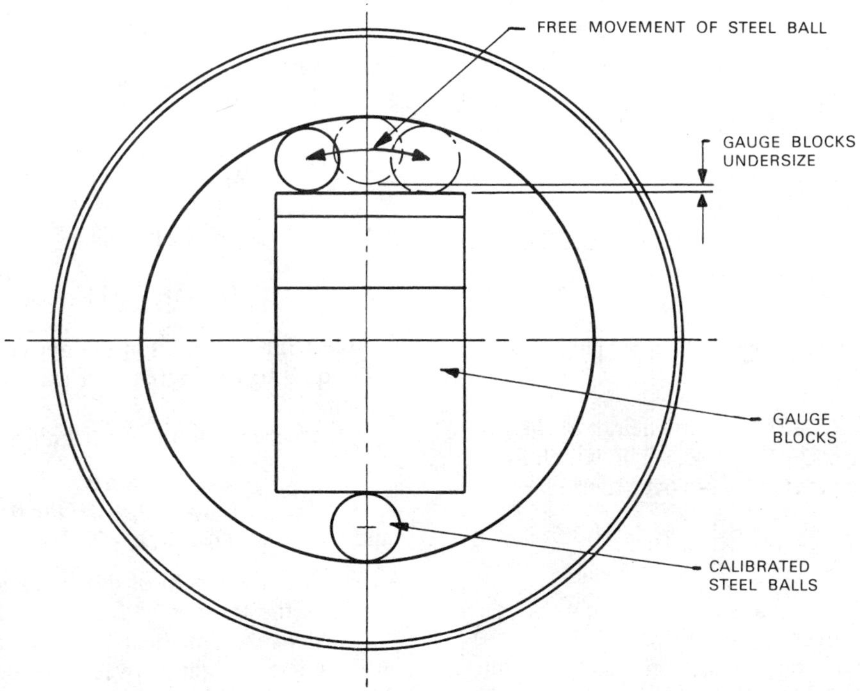

Fig. 12.10. Measurement of plain ring gauge using calibrated steel balls and gauge blocks

just allows movement incurs a minimum of such deformation.

3. *Measurement of taper plug gauges*

The angle of taper of a taper plug gauge can usually be determined by using a sine bar on which are mounted centres for work-holding. If this is not available the angle can be determined by making diametral measurements at known distances along the gauge.

The measurements are normally made over calibrated precision rollers of equal diameter, the separation of the individual measurements being controlled by gauge blocks. This arrangement is shown in fig. 12.11.

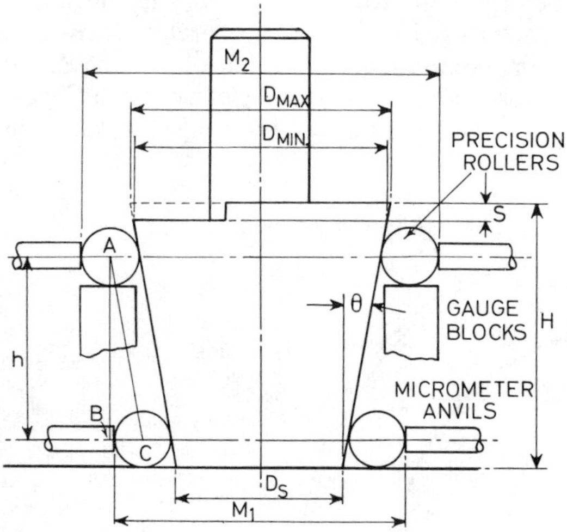

Fig. 12.11. Measurement of a taper plug gauge

In triangle ABC, h is the height of the gauge blocks and $BC=(M_2-M_1)/2$, in which M_1 and M_2 are the measurements over the rollers.

$$\tan\theta=\frac{BC}{AB}=\frac{M_2-M_1}{2h} \qquad \ldots(1)$$

Thus the semi-angle of taper can be determined by direct measurement.

To find the maximum and minimum diameters at the top and bottom of the step it is first necessary to determine D , the diameter at the small end. Fig. 12.12 represents the small end of the gauge during the measurement of M_1.

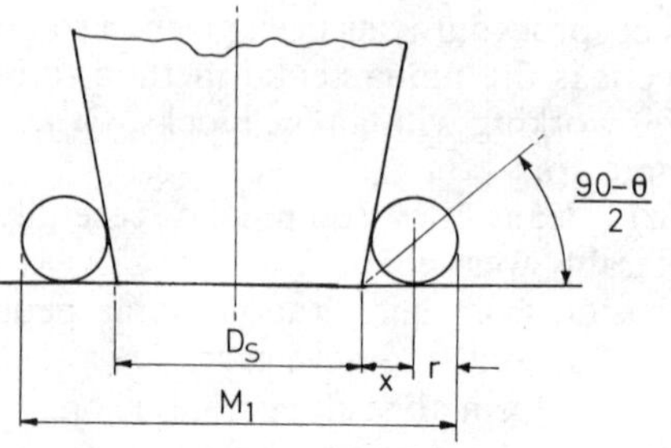

Fig. 12.12. Measurement of the diameter of the small end of a taper plug gauge

From fig. 12.13

$$M_1=D_s+2r+2x$$

$$\tan\frac{(90-\theta)}{2}=\frac{r}{x}$$

$$\therefore \quad x=r\times\text{cotangent}\,\frac{(90-\theta)}{2}$$

$$\therefore \quad M_1=D_s+2r\left[1+\cot\frac{(90-\theta)}{2}\right]$$

or $$M_1=D_s+d\left[1+\cot\frac{(90-\theta)}{2}\right]$$

where d = roller dia.

$$\therefore \quad D_s=M_1-d\left[1+\cot\frac{(90-\theta)}{2}\right]$$

Referring to equation (1) above, $\tan\theta$ is in fact equal to the increase in radius per unit of height.

∴ Increase in diameter per unit height

$$=2\tan\theta$$

$$\therefore \quad D_{max}=D_s+2H\tan\theta$$

and $$D_{min}=D_s+2(H-S)\tan\theta$$

where H is the height of the gauge and S is the height of the step.

If a measurement M_3 is taken at some intermediate height between the positions for M_1 and M_2, two further values of θ can be determined. If all these values of angle are not

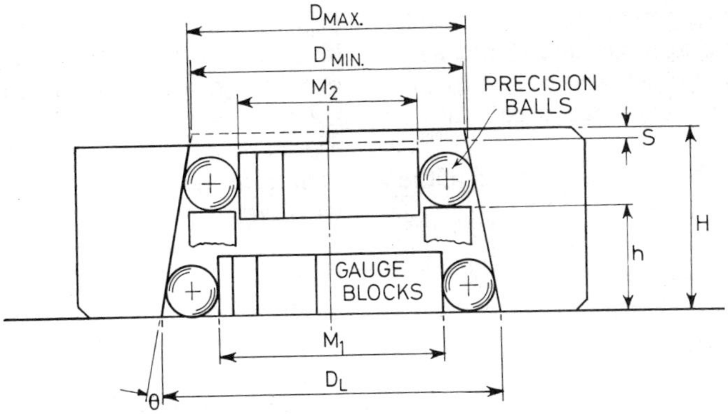

Fig. 12.13. Measurement of a taper ring gauge

the same then the taper is not a true cone, i.e. its sides are not straight. A check on roundness can also be made by carrying out the measurements M_1, M_2, and M_3, at different positions around the gauge.

Measurement of taper ring gauges
The procedure for measuring taper ring gauges is similar to that used for taper plug gauges, the measurements being made with precision calibrated balls and gauge blocks, as in fig. 12.14.

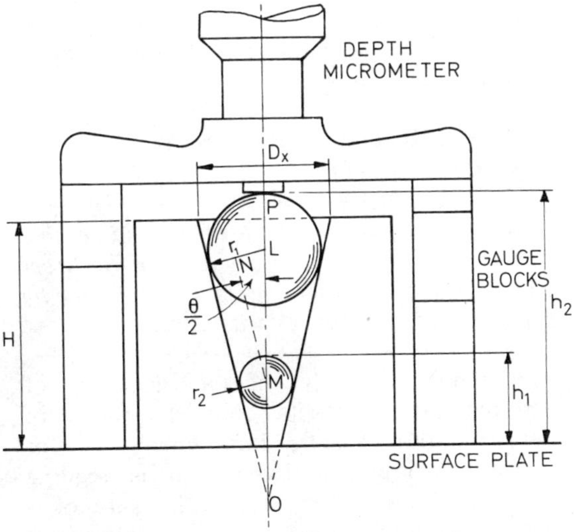

Fig. 12.14. Measurement of a taper bore

Again

$$\tan\theta = \frac{M_1 - M_2}{2h}$$

But in this case

$$D_L = M_1 + d\left[1 + \cot\frac{(90-\theta)}{2}\right]$$

(where d = ball dia.)

and
$$D_{max} = D_L - 2(H-S)\tan\theta$$
$$D_{min} = D_L - 2H\tan\theta$$

Measurement of taper bores
The procedure described above can only be used where the taper bore is of large enough diameter to allow easy access of balls and gauge blocks. For smaller diameter tapers, angular and diametral measurements can be made by the arrangement shown in fig. 12.14, in which the heights at which balls of different diameters seat directly on the taper are measured.

The centre distance LM between the balls will be

$$LM = h_2 - h_1 - r_1 + r_2$$

$$\sin\frac{\theta}{2} = \frac{NL}{LM} = \frac{r_1 - r_2}{h_2 - h_1 - r_1 + r_2}$$

The semi-angle of the taper having been found, its uniformity may be determined by taking a further measurement on a ball of such diameter that it rests approximately mid-way along the length of the taper, and repeating the calculation for the new measured values obtained.

To find the diameter D_x at the large end of the taper:

$$\frac{\mathrm{OL}}{\mathrm{LM}}=\frac{r_1}{r_1-r_2}$$

and $$\mathrm{OP}=\mathrm{OL}+\mathrm{PL}$$

and $$D_x=2\left(\mathrm{OP}\tan\frac{\theta}{2}\right)$$

The practical difficulties of this method of measurement are perhaps even more severe than in the previous case, especially if the taper has a small angle. The wedging effect of the balls and the consequent elastic deformation of both the balls and the gauge can cause appreciable errors in the measured values of h_1 and h_2, to give rise to related errors in the calculated angle. On no account should the balls be dropped into the taper. It is better if they are gently rolled into position with the axis of the taper lying almost horizontal.

COMPARATIVE MEASUREMENT

General Objective: *The student should recognise the principles of comparative measurement.*

At this stage of his education and training the student should be familiar with the principles and use of the standard measuring instruments such as the vernier caliper, micrometer and dial gauge. These are described fully in *Technician Workshop Processes and Material 1* in this series. It is necessary to point out, however, that for precise measurements these instruments have certain shortcomings and are subject to errors of unacceptable magnitude. The direct use of gauge blocks enables measurements to be made in increments of 0.01 mm units, but the accuracy of determination is reduced by the 'feel' or 'touch' of the inspector. If we wish to improve the accuracy of determination an instrument is needed with scale divisions of 0.01 mm which are either subdivided into smaller fractions or can be further subdivided by estimation. Such instruments are called comparators and are made in various forms, all with certain common features.

1. An amplification system is needed to enable small movements of the measuring plunger to be shown by a larger movement of the pointer. The amplification may be by mechanical, optical, electrical or pneumatic means.
2. The range of measurements in such instruments is small and the higher the magnification the smaller is the range of measurement. Thus a comparator with a magnification of 500 × may have a range of ±0.15 mm while a similar instrument with a magnification of 3000 × may have a range of only ±0.025 mm.
3. The measuring pressure should be constant, as far as possible to prevent errors from elastic deformation.
4. The instrument should be robust to avoid deflections in its frame, and of fairly massive construction to act as a heat sink which absorbs small quantities of heat without undue local change of temperature.

If now the problems arising from the use of a typical instrument, the hand micrometer, are analysed the advantages of comparative measurement can be seen.

1. THE MICROMETER AS A COMPARATOR

Specific Objective: *The student should be able to describe how the sources of error in the direct use of measuring instruments are overcome by comparative methods of measurement.*

The hand micrometer has an accuracy of determination over its complete range of approximately ±0.005 mm when used for direct measurement. Even if it is reading correctly at zero, we cannot be sure of a measured dimension to closer than 0.005 mm at other points on the scale because of variations in contact pressure, errors in subdivision of 0.01 mm units and pitch errors in the screw. This can be verified by calibrating a micrometer against gauge blocks over its whole range.

We can reduce these errors by systematically attacking their causes.

(*a*) *Feel and contact pressure* can be made constant by replacing the fixed anvil with an anvil attached to a gauge which gives a constant reading at the same measuring pressure. Such a device is called a *fiducial indicator*.

(*b*) *Subdivision of 0.01 mm units* can be improved by making the thimble of the instrument much larger in diameter and thus making the 0.01 mm divisions large enough to be directly subdivided into 0.001 mm divisions. By then applying the vernier micrometer principle, the micrometer can be made to read to 0.000 2 mm.

(*c*) *Pitch errors in the screw* usually occur over the whole length of the screw. If we use less than one turn (0.5 mm) of the screw they are largely eliminated. This can be achieved by setting the micrometer to a standard setting-cylinder or gauge blocks until the pointer of the fiducial indicator is on its mark and noting the reading. The workpiece is then placed in the instrument and the reading on the workpiece noted. It follows that the difference in readings, added to the size of the setting standard with due regard to sign, is the size of the workpiece.

If a plain plug gauge of nominal size 30 mm is to be tested, a standard setting cylinder of 30 mm diameter is used to set the instrument.

Actual size of setting cylinder	= 29.985 mm
Reading on setting cylinder	= 15.813 mm
Reading on plug gauge	= 15.810 mm
Difference	= 0.003 mm

Note: (1) Because the instrument is used as a comparator, the reading of 15.813 mm bears no relationship to the cylinder diameter of 29.985 mm.

(2) The reading on the plug gauge is *smaller* than the reading on the setting cylinder and the difference must therefore be *subtracted* from the size of the setting cylinder.

Actual size of setting cylinder	= 29.985 mm
Difference in readings	= 0.003 mm
subtract	29.982 mm

If we assume that this type of micrometer can give consistent repetition of readings to within 0.000 5 mm, we can say

Size of plug gauge = 29.982 mm ± 0.000 5 mm

The instrument incorporating these features is called a *bench micrometer* and is illustrated in fig. 12.15. Note that the position of the fiducial indicator can be adjusted so that a micrometer head with a 25 mm range can be used for measurements from 0 mm to 50 mm. An adjustable work stage or stand is provided so that the work need not be held during measurement and temperature effects due to handling are thus minimised .

It is advisable to make the type of contact between the anvil and the workpiece as far as possible the same for the setting master as for the workpiece.

The bench micrometer principle is incorporated in other instruments such as the length-bar comparator, the taper-measuring machine, and the screw-diameter measuring machine.

2. THE LENGTH-BAR COMPARATOR

This is an enlarged version of the bench micrometer, consisting of a rigid bed carrying at one end a large drum-micrometer head and at the other a dial gauge reading to 0.002 mm. Between these units is a pair of vee supports to carry the work, both being adjustable vertically and one horizontally. All these units can be adjusted along the bed and the range of the instrument is from 150 mm to 1 metre, to an accuracy of determination of the order of 0.002 5 mm.

In this case the setting gauge consists of a

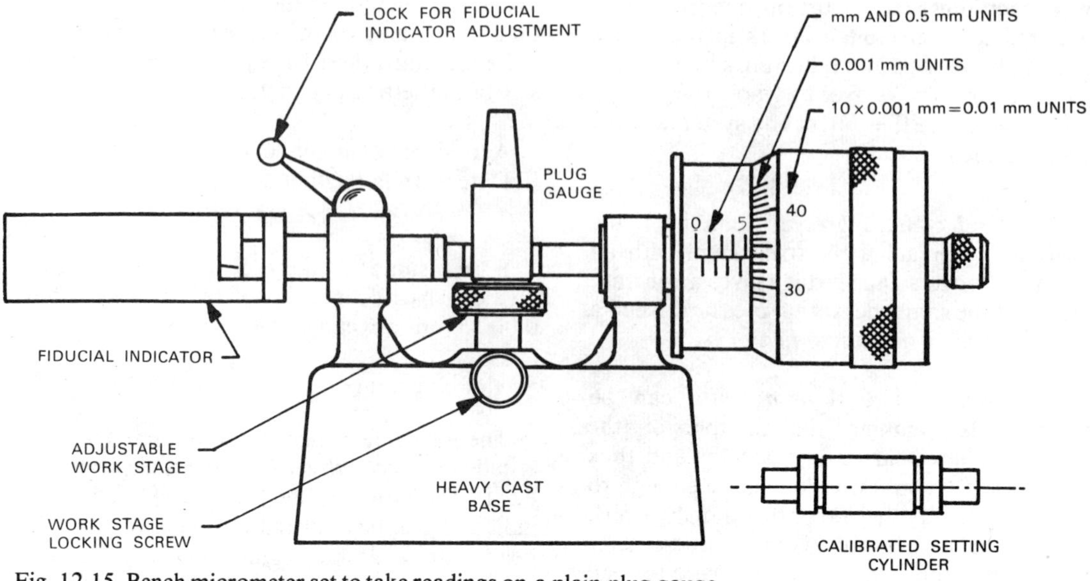

Fig. 12.15. Bench micrometer set to take readings on a plain plug gauge

combination of length bars rather than gauge blocks. The length bars are hardened steel bars of 22 mm diameter whose end faces are lapped flat, parallel and to a wringing finish. The ends are drilled and tapped, loose-fitting studs being provided so that when the bars are assembled the faces take control rather than the screw, and misalignment is eliminated. The comparator and a length bar combination are shown in fig. 12.16.

To determine the length of a gauge a combination of length bars is made up as nearly as possible of the same length as the gauge. Any major difference can be overcome by wringing a gauge block onto one end of the length bar combination. The combination is now placed on the vee supports which are set centrally and at a distance of 0.577 L apart, where L is the overall length of the bars. This method of support ensures that the sag of the bar under its own weight is minimised and the slope at the ends is zero, thus making the end faces parallel.

The bar must be aligned accurately with the line of measurement between the two anvils, which is achieved by adjusting the vee supports. If the bar is initially out of line, the dial gauge reading reduces as it is brought into line. When the bar is taken out of line in the other direction the reading increases and at the reversal of the pointer movement the bar is correctly aligned.

The micrometer is now adjusted until the dial gauge reads zero, after which the length bar is replaced by the bar under test. The micrometer is again adjusted until the dial gauge has the same reading and the micrometer reading is noted. The difference in the two readings is the difference in length of the two bars.

3. BENCH-TYPE MECHANICAL COMPARATORS

These are probably the commonest of the bench-type comparators and are extremely convenient in that they require no services such as electricity supplies or compressed air. They consist of a heavy base carrying the work stage and a column, up and down which the measuring head can be adjusted. Externally, the measuring head consists of the work contact plunger whose movement operates a pointer moving over a dial. Magnifications range from 500 × to 5000 ×, a typical instrument being shown in fig. 12.17.

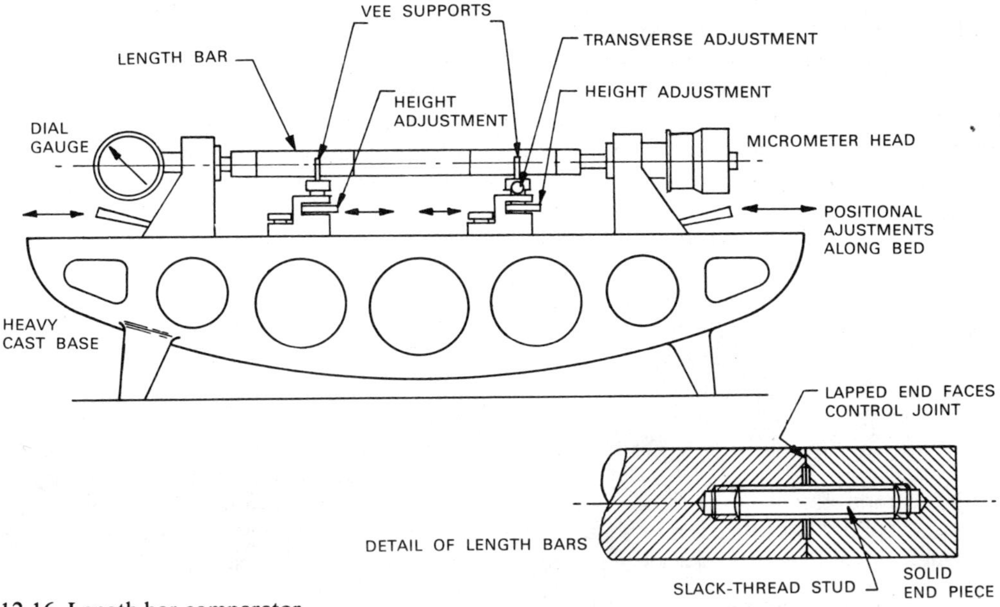

Fig. 12.16. Length bar comparator

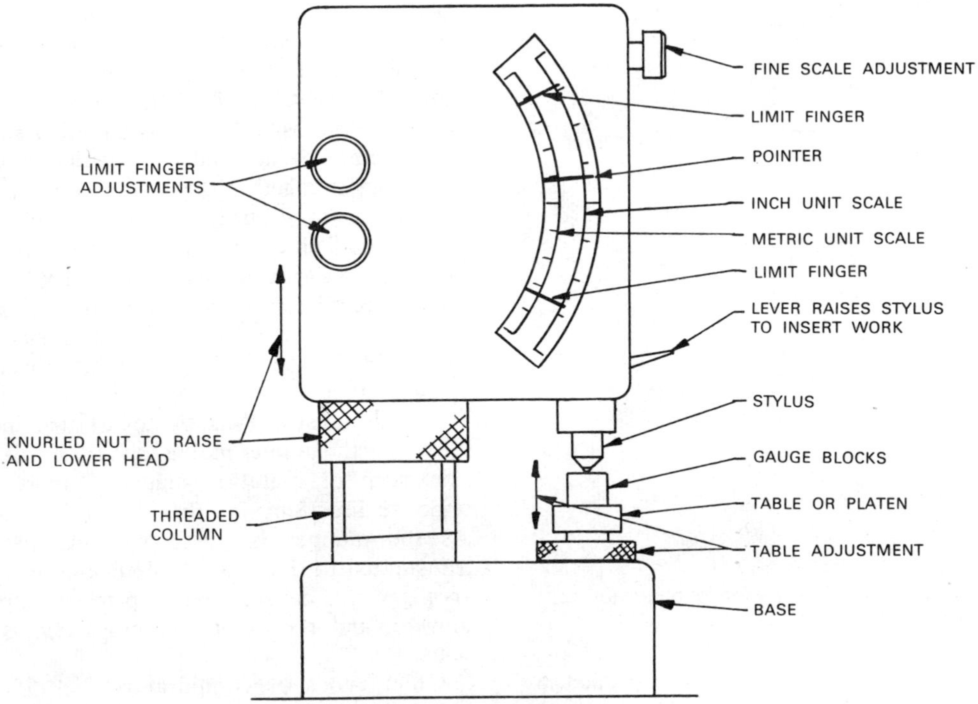

Fig. 12.17 Mechanical comparator showing operating adjustments

In use, a combination of gauge blocks is made up to the nominal size of the work to be measured. If the calibrated values of these blocks are known, the errors should also be totalled. Thus if the nominal size of the work is 51.57 mm we have:

Gauge block (mm)	*Known error* (μm)
1.07	+0.03
1.50	+0.02
9.00	+0.05
40.00	+0.05
Total = 51.57 mm	+0.15 μm

If the comparator is calibrated with the smallest units of 0.000 1 mm it is now set at +1.5 of these units, as shown in fig. 12.18. The errors

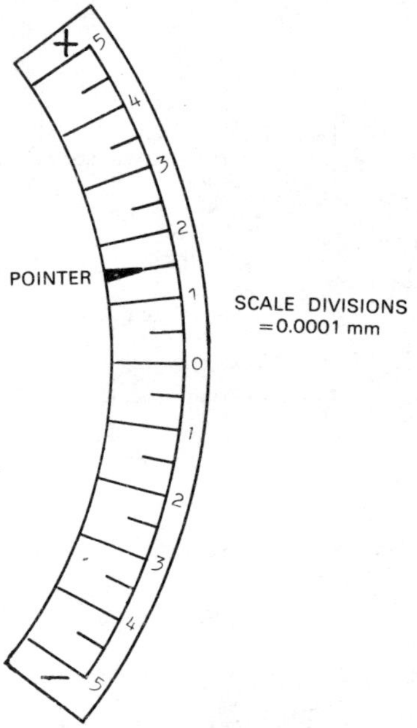

Fig. 12.18. High magnification comparator set to known error in gauge blocks

in the gauge blocks are thus cancelled by presetting the comparator and when the workpiece is inserted its error is read directly.

The Sigma comparator is an example of a mechanical comparator whose amplification is obtained by means of a compound lever arrangement, a simplified form of which is shown in fig. 12.9. The overall magnification is the product of the two lever ratios, i.e.

$$\text{Magnification} = \frac{L}{x} \times \frac{R}{r}$$

The actual movement is shown in diagrammatic form in fig. 12.20. As the plunger is raised, the moving block follows the knife edge and the Y-shaped arm moves about the cross-strip pivot. A phosphor-bronze driving band is carried from the ends of the Y-arm about a driving drum of radius r to which is attached a pointer of length R. Thus as the Y-arm moves it causes the drum to rotate, carrying the pointer with it.

The Johansson comparator is a simple device based upon the same principle as that which operates the child's toy, in which a button is placed at the centre of a loop of twine, one end of which is hooked over each thumb. The button is now 'wound up' so that the twine is twisted in opposite directions on either side. If the thumbs are now alternately drawn apart and brought together the button spins rapidly with a satisfying buzz. A little thought shows that a small linear movement of the thumbs produces a large angular movement of the button. Similarly, in a comparator, a small linear movement of the measuring stylus is required to produce a large angular movement of the pointer.

The Johansson Mikrokator shown in fig. 12.21 has the pointer mounted at the centre of a thin strip of metal permanently twisted in opposite directions on either side of the pointer. As the plunger is raised its movement is transmitted to the strip by the bell-crank lever arrangement, causing the strip to stretch and unwind, and the pointer to move across the scale.

Other types of comparators use different principles of magnification, optical, pneumatic and electrical, but they are all used in the same

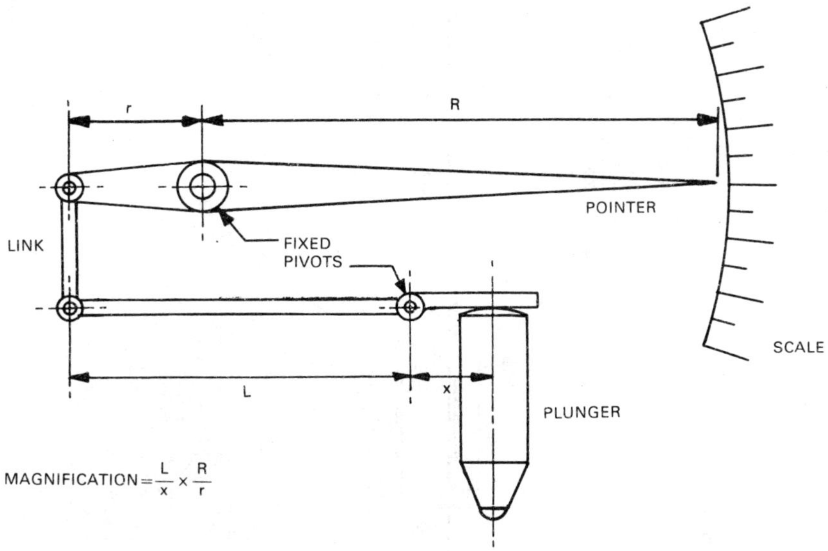

Fig. 12.19. Essentials of compound lever type of comparator movement

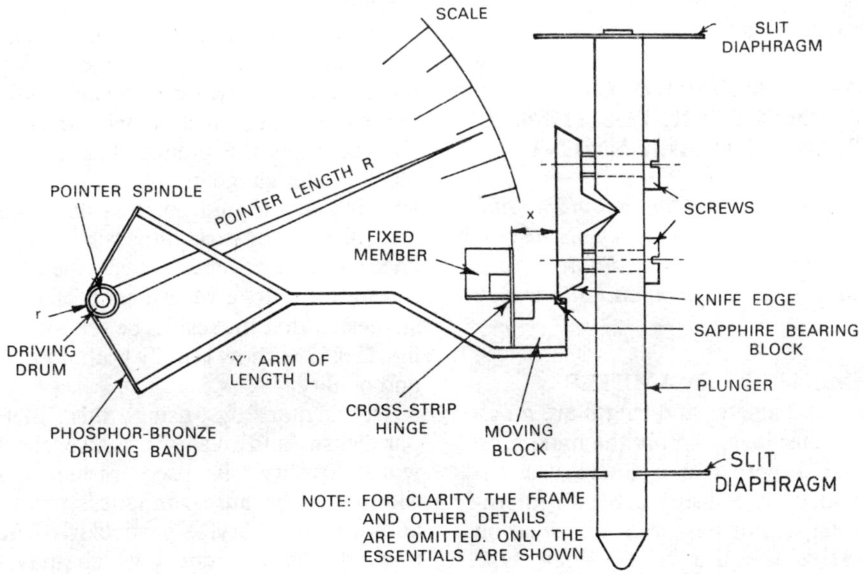

Fig. 12.20. Diagram of movement of Sigma mechanical comparator

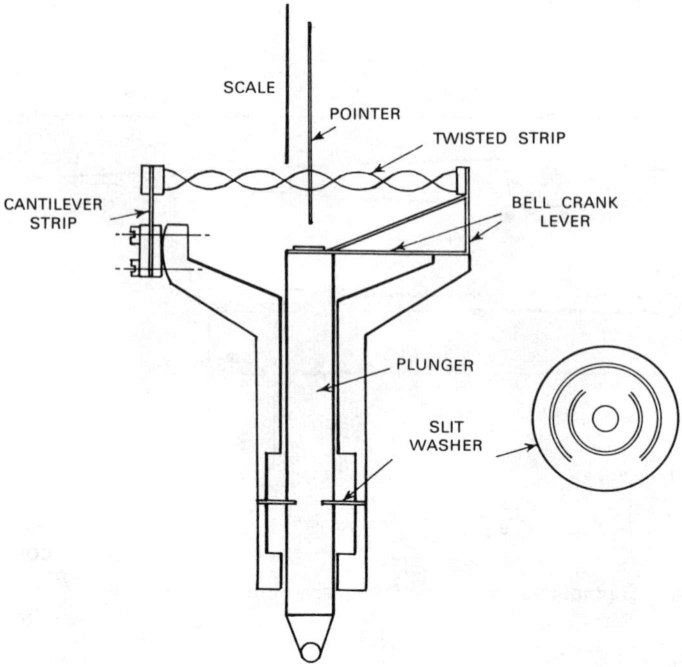

Fig. 12.21. Movement of Johansson Mikrokator

way. They are set to a master gauge, and the gauge under test measured by comparison.

COMPARATIVE METHODS OF CHECKING DIAMETER, POSITION, PARALLELISM AND ALIGNMENT

Specific Objective: *The student should be able to describe with the aid of sketches methods of comparative measurement for diameter and length, roundness, parallelism, and alignment.*

MEASUREMENT OF DIAMETER

Measurement of diameter and length are much the same, diameter being simply the maximum length measured across a cylindrical component. The method of using a bench micrometer as a comparator has been described on page 263. Here the use of a bench type mechanical comparator is considered. A combination of gauge blocks is made up to the nominal size of the plug gauge and the instrument is adjusted to read zero on this combination. The plug gauge is then passed under the comparator and the reading noted, any movement from zero being the difference in size between the plug gauge and the gauge blocks. The gauge should be checked at each end and at the mid-point of its length to give evidence of non-straightness or taper, and this check should be repeated with the gauge turned through 90° to give evidence of ovality. It is suggested that the results be set out as shown in fig. 12.22, to show readily both non-parallelism and ovality.

Note that by using this method the parallelism and ovality are also checked. The word 'ovality' is used rather than 'non-roundness' because roundness errors are not confined to ovality. A particularly troublesome form of non-roundness which may occur in parts produced by certain machining processes is called lobing.

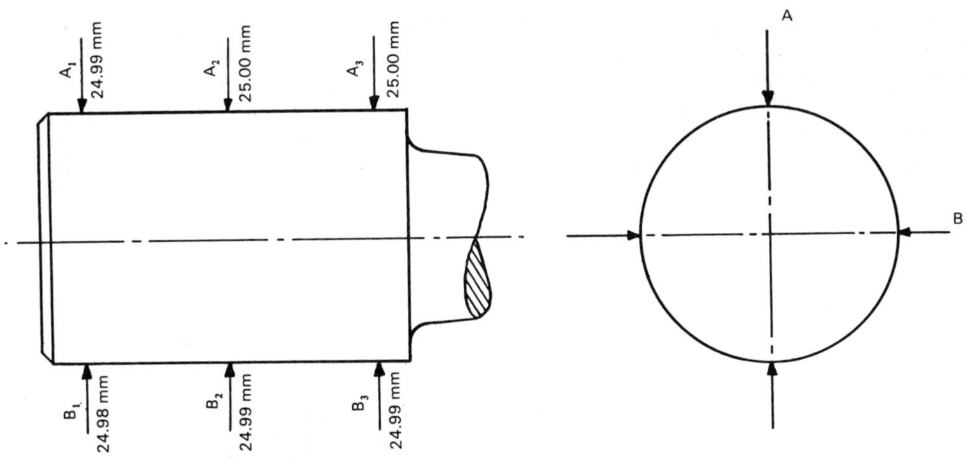

Fig. 12.22. A method of setting out results of measurement of a plain plug gauge

LOBING

Specific Objective: *The student should be able to define lobing, draw and label a lobed part and describe, with the aid of sketches, a test for lobing.*

The simplest lobed figure has three lobes and is based on an equilateral triangle, as shown in fig. 12.23. From each corner of the triangle is struck a large radius 'R' and a small radius 'r' so that the large radius blends with the small radius at the other two corners. This figure will always enter a pair of gauging anvils set at $(R + r)$ apart, but the smallest hole it will enter is much larger than $(R + r)$. This condition will not

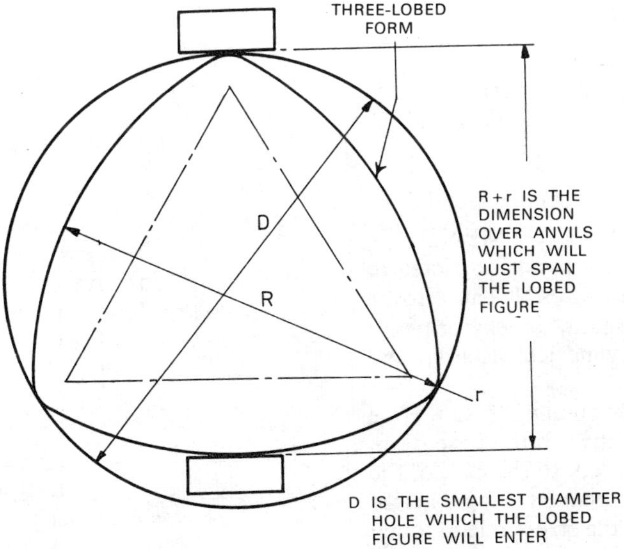

Fig. 12.23. Effect of lobing on cylindrical work

be detected by any measurement involving the use of parallel anvils such as those of a micrometer or comparator. It can be detected by rotating the lobed work under a dial gauge or comparator when the work is resting on a vee block as shown in fig. 12.24. Lobing is not confined to parts having three lobes. Any odd number of lobes gives the same effect and the defect can be detected in the same way.

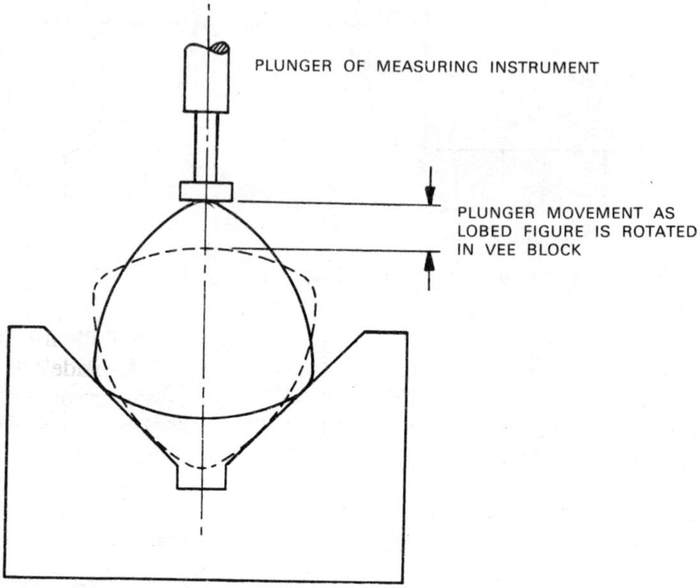

Fig. 12.24. Detection of lobing

MEASUREMENT OF POSITION, PARALLELISM AND ALIGNMENT

Comparative measurement is not restricted to the measurement of part sizes. It can also be used for checking position, other forms of parallelism than on cylindrical shafts, and alignment.

Consider the work shown in fig. 12.25. Here the position of holes relative to the base of an angle plate is being checked. In the case of the lower hole a direct measurement using a height gauge is being made, all the errors due to touch being introduced. A much greater degree of accuracy can be achieved if a comparative measurement is made by wringing up a pile of slip gauges to the nominal height of the top of the plug and then comparing their heights using a dial gauge, as shown in the case of the top hole.

A similar method can be employed to check

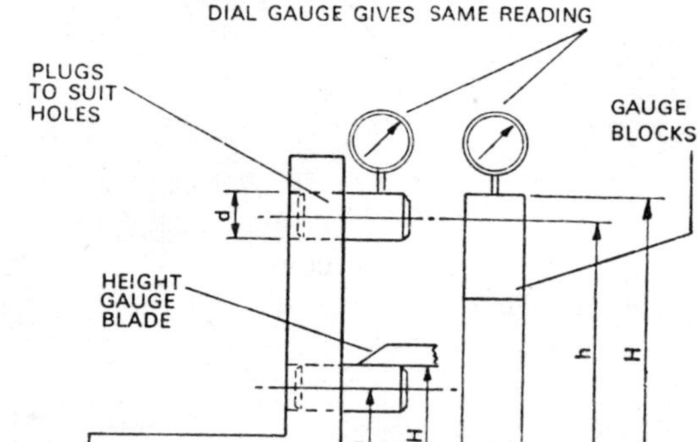

Fig. 12.25. Two methods of checking hole positions relative to base

that the axis of the hole is parallel with the base as shown in fig. 12.26 and, if it is necessary to check that the axis of the hole is square to the face in which it is bored the test mandrel can be set between centres and the work rotated as shown in fig. 12.27. Any regular fluctuations of the dial gauge reading will indicate a squareness error.

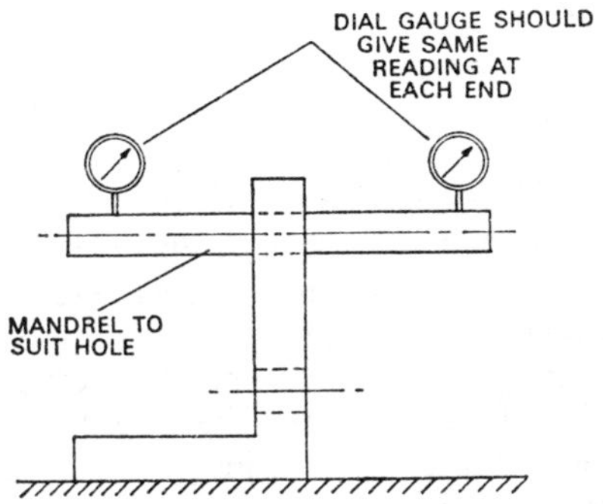

Fig. 12.26. Checking axis of hole for parallelism with base

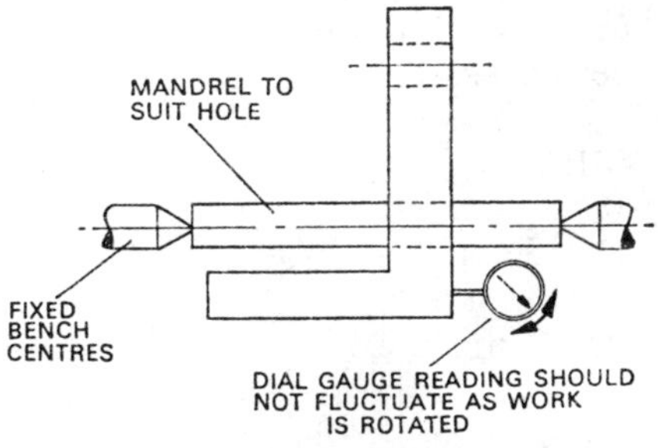

Fig. 12.27. Checking axis of hole for squareness with face

QUESTIONS

1. Name the most common workshop standard of length and list four characteristics of these standards which must be correct if their usefulness as standards is to continue. List the precautions which must be taken to avoid damage to these standards.

2. List the gauge blocks to be used in making up combinations of:
   (a) 62.81 mm from an M78 set
   (b) 86.325 mm from an M50 set.

3. A surface plate is being checked for flatness by the wedge method using a steel straight edge of known accuracy. Explain why the test should be carried out across corners as well as along lines parallel to the edges. How would a small planed surface plate be checked against a large scraped surface plate of a higher grade? In the absence of a master surface how can it be ensured that a number of small surfaces are flat?

4. A square is checked using a device of the type shown in fig. 12.7, the dial gauge readings being +3.8 and −1.7 units of 0.01 mm respectively. The distance between the dial gauge and stop is 100 mm.
   (a) what is the error in squareness over this length?
   (b) if the blade of the square is 125 mm long what is the error in squareness over the length of the blade?
   (c) Express the error in squareness as an angle.

   Hint: For small angles work initially in radians.

$$\text{Error in radians} = \frac{\text{measured error (mm)}}{\text{gauge length (mm)}}$$

$$\text{As } 2\pi \text{ radians} = 360^\circ$$

$$1 \text{ radian} = \frac{360^\circ}{2\pi}$$

$$x \text{ radians } \alpha\ \frac{360^\circ}{2\pi} \times x$$

5. The angle of a wedge-shaped block is being measured using a 200 mm sine bar. With gauge blocks of 28.87 mm height under one end of the sine bar the dial gauge readings at each end of the workpiece vary by 0.08 mm, the gauge block end being low.
   (a) if the workpiece is 50 mm long what should be the next height of gauge blocks tried?
   (b) if this is correct calculate the angle of the workpiece.

6. A plain ring gauge is measured by standing it on a surface plate and placing three steel balls, each of 15 mm diameter in the bore. A fourth ball, also of 15 mm diameter is placed on those in the bore, its weight holding them against the

sides of the bore. A measurement M is made from the top of the fourth ball to the surface plate.

(a) How could the measurement M be suitably made?
(b) If M is found to be 26.20 mm, what is the diameter of the bore?
(c) Compare this method with that described on page 259 and list the advantages and disadvantages of each method.

7. The angle of the external taper of a morse taper sleeve is measured using the method described on page 260 using rollers of 10 mm diameter. With the rollers resting on the table the reading over the rollers is 53.51 mm. With the rollers raised 75 mm the reading is 49.75 mm.

(a) Calculate the included angle of taper of the gauge
(b) Sketch the set-up and explain a difficulty in making this measurement compared with the measurement of a taper plug gauge as shown in fig. 12.11

8. A plain plug gauge is measured using a bench micrometer and setting cylinder, the following data being obtained

Nominal diameter of setting cylinder = 30.000 mm
Known error in setting cylinder = −0.0003 mm
Reading on setting cylinder = 15.128 mm
Reading on plug gauge = 15.134 mm

(a) Calculate the diameter of the gauge at this point
(b) How many readings should be taken and where should they be positioned on the gauge?
(c) Suggest an accuracy of measurement by this method.

9. Construct a 5-lobed figure by drawing a circle of 80 mm diameter and drawing a pentagon around it. From each corner of the pentagon strike an arc of 10 mm radius and blend these arcs with arcs of larger radius also struck from the corners of the pentagon.

From the lobed figure thus drawn determine:

(a) The dimension across a 'diameter' of the figure
(b) The diameter of the smallest circle which will just contain the figure.

10. A shaft which has been checked and accepted by a plain gap gauge is offered up to a hole which has been checked and accepted by a plain plug gauge. A clearance fit should result but the shaft will not enter the hole. The gauges are checked and found to be correct. The shaft and hole are measured individually and found to be correct.

(a) Suggest what might be wrong
(b) If a particular gauge had been available for the shaft the problem could have been detected earlier; name the gauge.

ANSWERS

4. (a) 0.028 mm
   (b) 0.035 mm
   (c) 0.016° or 0.96′
5. (a) 29.19 mm
   (b) 8.39°
6. (b) 34.956 mm
7. (a) 2.87°
8. (a) 30.0057 mm
9. (a) 114 mm
   (b) 119 mm